Ubiquitous Computing Fundamentals

Ubiquitous Computing Fundamentals

Edited by
John Krumm

Microsoft Corporation
Redmond, Washington, U.S.A.

CRC Press is an imprint of the
Taylor & Francis Group an **informa** business

A CHAPMAN & HALL BOOK

Chapman & Hall/CRC
Taylor & Francis Group
6000 Broken Sound Parkway NW, Suite 300
Boca Raton, FL 33487-2742

© 2010 by Taylor and Francis Group, LLC
Chapman & Hall/CRC is an imprint of Taylor & Francis Group, an Informa business

No claim to original U.S. Government works

Printed in the United States of America on acid-free paper
10 9 8 7 6 5 4 3 2 1

International Standard Book Number: 978-1-4200-9360-5 (Hardback)

This book contains information obtained from authentic and highly regarded sources. Reasonable efforts have been made to publish reliable data and information, but the author and publisher cannot assume responsibility for the validity of all materials or the consequences of their use. The authors and publishers have attempted to trace the copyright holders of all material reproduced in this publication and apologize to copyright holders if permission to publish in this form has not been obtained. If any copyright material has not been acknowledged please write and let us know so we may rectify in any future reprint.

Except as permitted under U.S. Copyright Law, no part of this book may be reprinted, reproduced, transmitted, or utilized in any form by any electronic, mechanical, or other means, now known or hereafter invented, including photocopying, microfilming, and recording, or in any information storage or retrieval system, without written permission from the publishers.

For permission to photocopy or use material electronically from this work, please access www.copyright.com (http://www.copyright.com/) or contact the Copyright Clearance Center, Inc. (CCC), 222 Rosewood Drive, Danvers, MA 01923, 978-750-8400. CCC is a not-for-profit organization that provides licenses and registration for a variety of users. For organizations that have been granted a photocopy license by the CCC, a separate system of payment has been arranged.

Trademark Notice: Product or corporate names may be trademarks or registered trademarks, and are used only for identification and explanation without intent to infringe.

Library of Congress Cataloging-in-Publication Data

Ubiquitous Computing Fundamentals / edited by John Krumm.
   p. cm.
   Includes bibliographical references and index.
   ISBN 978-1-4200-9360-5 (hardcover : alk. paper)
   1. Ubiquitous computing. I. Krumm, John.

QA76.5915.U258 2010
004--dc22   2009026632

Visit the Taylor & Francis Web site at
http://www.taylorandfrancis.com

and the CRC Press Web site at
http://www.crcpress.com

Table of Contents

Foreword, vii

Introduction, ix

Contributors, xiii

CHAPTER 1 ▪ An Introduction to Ubiquitous Computing   1
       ROY WANT

CHAPTER 2 ▪ Ubiquitous Computing Systems   37
       JAKOB BARDRAM AND ADRIAN FRIDAY

CHAPTER 3 ▪ Privacy in Ubiquitous Computing   95
       MARC LANGHEINRICH

CHAPTER 4 ▪ Ubiquitous Computing Field Studies   161
       A. J. BERNHEIM BRUSH

CHAPTER 5 ▪ Ethnography in Ubiquitous Computing   203
       ALEX S. TAYLOR

CHAPTER 6 ▪ From GUI to UUI: Interfaces for Ubiquitous Computing   237
       AARON QUIGLEY

CHAPTER 7 ▪ Location in Ubiquitous Computing   285
       ALEXANDER VARSHAVSKY AND SHWETAK PATEL

CHAPTER 8 ▪ Context-Aware Computing 321
ANIND K. DEY

CHAPTER 9 ▪ Processing Sequential Sensor Data 353
JOHN KRUMM

INDEX, 381

Foreword

It has been nearly two decades since the term *ubiquitous computing* burst into our research vernacular. One of the strengths, and one of the challenges, of "ubicomp" is that it is hard to pin down exactly what the intellectual core is. From the very beginning, ubicomp researchers have investigated both bleeding edge technology challenges as well as human-centered opportunities. There are other intellectual mergers of interest as well, including the bridge between the physical and the digital worlds and the (re-) merging of the academic communities of hardware and software.

But this very diversity of intellectual themes presents two challenges to our community, both of which motivate the need for a book like this one. First of all, for established researchers, we have to educate ourselves on the language and methods of disciplines different from the ones we have practiced for many years. Why? Because if we are to advance as an intellectual community, then we all need to embrace the inherent diversity in our thoughts and skills. While it is not strictly necessary that we become expert in all of the relevant subdisciplines of ubicomp represented in this book, it *is* necessary that we appreciate all the perspectives and that we strive to make our own work more relevant and accessible to those many perspectives.

Second, and more importantly, we have to provide a foundation for future generations. I deeply believe that any interesting problem to explore in our everyday lives requires expertise from many disciplines and perspectives. Consequently, we have to train new researchers so that they will be able to stand on the results of the past and direct us as a community to go beyond where we are today. In short, our students must be empowered to be better than we are, or we face extinction as a relevant intellectual community.

Under the skillful guidance of John Krumm, the authors of these chapters have assembled a collection of well-written, tutorial style chapters on topics that have become core to research advances in ubiquitous computing over the past two decades. The result is a must-read text that provides an historical lens to see how ubicomp has matured into a multidisciplinary endeavor. It will be an essential reference to researchers and those who want to learn more about this evolving field.

Professor Gregory D. Abowd, PhD
College of Computing
Georgia Institute of Technology
Atlanta, Georgia, U.S.A.

Introduction

This book is an overview of the fascinating field of ubiquitous computing. Since this field is rapidly progressing, the book is aimed at people who want to explore it as researchers or track its evolution. Intended for advanced undergraduates, graduate students, and professionals interested in ubiquitous computing research, the book covers the major fundamentals and research in the key areas that shape the field. Each chapter is a tutorial that provides readers with an introduction to an important subset of ubiquitous computing and also contains many valuable references to relevant research papers.

The field of ubiquitous computing is simultaneously young and broad. Research papers in the field commonly reference Mark Weiser, who famously coined the term *ubiquitous computing* in his *Scientific American* article in 1991. This is considered the start of the research area, and it has grown to encompass a broad array of technologies since then. Although the field is broad, there are well-established conferences and researchers devoted to it.

We chose 11 of the most prominent ubiquitous computing research devotees to contribute chapters to this book in their area of expertise. Given the field's breadth, it would be difficult to find one person who can expertly cover it all. Some of the chapter authors teach ubiquitous computing at universities. All of them are intimately involved in research in their specialty. Working in the area means they have the experience to not only describe the fundamental research issues, but to also explain practical ways to accomplish research and publish papers in the field.

Ubiquitous computing research can be categorized into three distinct areas where the research is focused: systems, experience, and sensors. The chapters of this book are similarly organized and categorized. The three categories and their supporting chapters are

Systems—These chapters focus on how to build the software support for deploying ubiquitous computing applications.

"Ubiquitous Computing Systems" (Chapter 2) discusses the important issues to consider when building the infrastructure to support ubiquitous computing applications.

"Privacy in Ubiquitous Computing" (Chapter 3) explains how to maintain privacy in systems that inherently need to connect with personal devices and information.

Experience—These chapters highlight the critical points where ubiquitous computing technologies touch people.

"Ubiquitous Computing Field Studies" (Chapter 4) shows how to evaluate ubiquitous computing applications in the field.

"Ethnography in Ubiquitous Computing" (Chapter 5) details how to observe people and consider how they might use ubiquitous computing technology.

"From GUI to UUI: Interfaces for Ubiquitous Computing" (Chapter 6) focuses on moving from the graphical to the ubiquitous computing user interface.

Sensors—These chapters show how systems sense location and analyze and determine context.

"Location in Ubiquitous Computing" (Chapter 7) illustrates how to measure a person's location, one of the most important inputs for ubiquitous computing applications.

"Context-Aware Computing" (Chapter 8) explains the use of context to allow ubiquitous computing applications to deliver the right services at the right time.

"Processing Sequential Sensor Data" (Chapter 9) details how to effectively process sensor data for location and context.

In addition to these specific research areas, the book begins with a chapter called "An Introduction to Ubiquitous Computing," which discusses the history of the field in terms of its major research projects.

Although the chapters cover interrelated topics, they can be covered in any order by a teacher or reader.

We hope you will find this book to be a useful overview of and a practical tutorial on the young and evolving field of ubiquitous computing.

John Krumm

Contributors

Jakob E. Bardram, PhD
IT University of Copenhagen
Copenhagen, Denmark

A.J. Bernheim Brush, PhD
Microsoft Research
Redmond, Washington, U.S.A.

Anind K. Dey, PhD
HCI Institute
Carnegie Mellon University
Pittsburgh, Pennsylvania, U.S.A.

Adrian Friday, PhD
Computing Department
Lancaster University
Lancaster, United Kingdom

John Krumm, PhD
Microsoft Research
Redmond, Washington, U.S.A.

Marc Langheinrich, PhD
Faculty of Informatics
University of Lugano (USI)
Lugano, Switzerland

Shwetak Patel, PhD
Computer Science and
    Engineering
University of Washington
Seattle, Washington, U.S.A.

Aaron Quigley, PhD
Human Interface Technology
    Laboratory Australia
University of Tasmania
Tasmania, Australia

Alex S. Taylor, PhD
Microsoft Research
Cambridge, United Kingdom

Alexander Varshavsky, PhD
AT&T Labs
Florham Park, New Jersey, U.S.A.

Roy Want, PhD
Intel Corporation
Santa Clara, California, U.S.A.

CHAPTER 1

An Introduction to Ubiquitous Computing

Roy Want

CONTENTS

1.1 Founding Contributions to Ubiquitous Computing   3
    1.1.1 Xerox PARC   3
    1.1.2 Tabs, Pads, and Liveboards   6
    1.1.3 Context Awareness   10
    1.1.4 IBM Research: Pervasive Computing versus Ubiquitous Computing   11
    1.1.5 University of Tokyo: T-Engine and the ITRON Operating System   12
    1.1.6 Hewlett Packard: Cooltown   13
1.2 Ubiquitous Computing in U.S. Universities   15
    1.2.1 UC Berkeley: InfoPad   15
    1.2.2 MIT Media Laboratory: Wearable Computing   16
    1.2.3 Georgia Tech: Living Laboratories   18
1.3 Ubiquitous Computing in European Laboratories and Universities   20
    1.3.1 Olivetti Research: Active Badges   20
    1.3.2 Karlsruhe: Cups and Smart-Its   23
    1.3.3 Lancaster University: Guide   25
1.4 Modern Directions in Ubiquitous Computing   26
    1.4.1 Microsoft Research   26
    1.4.2 Intel Research   27

1.5　The Research Community Embraces Ubiquitous Computing　28
1.6　The Future of Ubiquitous Computing　31
References　33

Ubiquitous computing, or *ubicomp*, is the term given to the third era of modern computing. The first era was defined by the mainframe computer, a single large time-shared computer owned by an organization and used by many people at the same time. Second, came the era of the PC, a personal computer primarily owned and used by one person, and dedicated to them. The third era, ubiquitous computing, representative of the present time, is characterized by the explosion of small networked portable computer products in the form of *smart phones*, personal digital assistants (PDAs), and embedded computers built into many of the devices we own—resulting in a world in which each person owns and uses many computers. Each era has resulted in progressively larger numbers of computers becoming integrated into everyday life (Figure 1.1).

Although the general trends in computing are clear, the predictions, research, and philosophy behind the technology that make ubiquitous computing a reality have taken many forms, all of which have been shaped by the organizations that cultivated them. The early informative research in this area began in the late 1980s and was pioneered by Xerox Palo Alto Research Center (PARC), IBM Research, Tokyo University, University of California (UC) Berkeley, Olivetti Research, HP Labs, Georgia Institute of Technology (Georgia Tech), and Massachusetts Institute of Technology (MIT) Media Laboratory. Many commercial entities also began forays into ubiquitous computing during the 1990s, exploring the business potential for ubiquitous services, and novel mobile devices such as pen-based computers. At this time, we also saw the introduction of the Apple Newton,

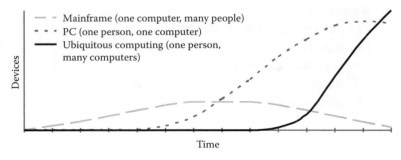

FIGURE 1.1　Graph conceptually portraying three eras of modern computing.

and the term PDA was coined. Other product examples included the EO pad, using GO Pen software, and later the Palm Pilot (with Graffiti) and the Sharp Zaurus; Fujitsu also developed a series of tablet and palm-based devices particularly targeted at vertical markets. Later still, MP3 players from Archos and Apple also played into this market.

Today, demonstrating the most convincing evidence of the value of ubiquitous computing, the cell phone, or more precisely the "smart phone," takes center stage crossing a threshold of processor performance, memory/disk capacity, and connectivity both cellular and local, making it the most widely adopted and ubiquitous computer there has ever been. In the remaining sections, we follow the path of research that has defined ubiquitous computing since its beginning, and discuss the various approaches and some of the philosophies that have grown up around the work.

1.1 FOUNDING CONTRIBUTIONS TO UBIQUITOUS COMPUTING

1.1.1 Xerox PARC

The original term *ubiquitous computing* was coined by Mark Weiser in 1988 at Xerox PARC, while serving as the director of the Computer Science Laboratory (CSL), one of five laboratories at the renowned research center (Figure 1.2). He envisioned a future in which computing technologies

FIGURE 1.2 Xerox PARC—Computer Science Laboratory 1991: Mark Weiser using a Liveboard with a ParcPad visible in the foreground. (Photo courtesy of PARC, Inc., http://www.parc.com)

became embedded in everyday artifacts, were used to support daily activities, and were equally applicable to our work, managing our homes, and for play. A more complete description of this vision is described on a Web site maintained by PARC summarizing Weiser's work and ideas and can be found at www.ubicomp.com/weiser. A concise summary of ubiquitous computing, or ubicomp, as it was originally referred to by researchers at PARC, can also be found in his 1991 *Scientific American* article (Weiser, 1991), which contains his famous quote:

> The most profound technologies are those that disappear. They weave themselves into the fabric of everyday life until they are indistinguishable from it.
>
> —MARK WEISER

The essence of Weiser's vision is that mobile and embedded processors can communicate with each other and the surrounding infrastructure, seamlessly coordinating their operation to provide support for a wide variety of everyday work practices. A consequence of this approach is that each device needs to limit the range of its communication to enable valuable wireless bandwidth reuse. As a result, he introduced the notion of bits-per-second per-cubic-meter to the ubicomp vision (Weiser, 1993a), and inspired many researchers to explore techniques for spatial reuse of the radio spectrum. In the early 1990s, there were no short-range wireless standards that could provide this capability, but today we have Bluetooth, Near Field Communication (NFC), IrDA, Zigbee, and WiFi (soon WiFi PAN), which have enabled wide deployment of devices that take advantage of local ad hoc communication, and can be used to build the ubicomp vision.

Going beyond technology per se, Weiser saw ubicomp as an opportunity to improve on the style of computing that has been imposed on users since the early days of the mainframe also carrying over to PCs—namely, sitting in a chair, staring at a screen, typing on a keyboard, and making selections with a mouse. Through this style of interaction, traditional computers consume much of our attention and divorce us from what is happening all around us, resulting in a somewhat solitary all-consuming experience. Weiser believed that in a ubicomp world, computation could be integrated with common objects that you might already be using for everyday work practices, rather than forcing computation to be a separate activity. If the integration is done well, you may not even notice that any computers were involved in your work. Weiser sometimes also referred to this as *invisible computing* and wrote a number of articles about his philosophy (Weiser, 1993b).

To illustrate the concept of invisible technology more effectively, consider an analogy based on the familiar printed page. The technology behind printing is the deposition of ink on thin sheets of paper. For optimal results the design of the ink and paper must be well thought out. For example, the ink must stain the surface of the paper to provide a high contrast black against the white background, it must be durable in use, and not wick into the paper even if wet. However, when we read a printed page, we rarely notice the underlying ink and paper and ink technologies; we read the pages and comprehend the ideas; but it is not necessary to focus on the technology, the characteristics of the ink, or the manufacturing process of the paper to be able to use it. You might say that printing technology "gets out of the way" of the user, allowing the higher-level goal of reading a story, or acquiring knowledge; on the other hand, traditional PCs rarely do this. Instead, they usually require us to continuously tinker with the system, maintaining it and configuring it to complete a task. In summary, when designing and using ubicomp technologies, we may have the opportunity to more closely parallel the higher-level experience we have when reading the printed word.

Another term Weiser used to describe ubiquitous computing was "The coming age of *calm technology*" (Weiser and Seely-Brown, 1997). Although there is no simple formula to convert a PC application into a calm embedded computing experience, ubiquitous computing takes the opposite philosophy to the PC, which tries to virtualize our world (e.g., the familiar PC desktop and icons representing documents, printers, and trash can). Instead, ubicomp pushes the computerized versions of these technologies back into the physical world (Weiser, 1994). For example, rather than reading documents on a PC screen in a graphic made to look like a printed page, the objective would be to create a dedicated document reader with an embedded processor that you can hold and use just like a book. This is an old idea from PARC originally conceived by Alan Kay with his Dynabook project, but was later updated by Weiser's vision, making it highly connected and coordinating wirelessly with the surrounding systems. From a user's perspective, the experience of using such a device is simplified relative to a PC because it has a dedicated function (a design point sometimes referred to as an *information appliance*); it does not need the complex arrangement of nested menus and control functions required by a generalized computing platform. Although this concept has been tested several times in the marketplace, for example, Rocketbook and

Softbook, which were not commercially successful, the idea is still being revisited in the marketplace today with Sony's e-reader and Amazon's Kindle. Similar to the evolution of the PDA, each generation learns from the failures of the previous generation, and at the same time technology improves, allowing an e-book to more closely match the affordances of the real book it is trying to replace.

1.1.2 Tabs, Pads, and Liveboards

Under Weiser's leadership, CSL set out to design and build a ubiquitous computing environment within the confines of the research center. PARC has long had a philosophy of "Build what you use, and use what you build" and the ubicomp research theme continued that tradition.

However, given the resource constraints of research, it was necessary to limit the scope of the ubicomp exploration to a manageable set of projects. These were selected by asking the question, "What is the minimum set of usable devices that can be built in a laboratory but still provide a sandbox rich enough to explore ubicomp and its defining characteristics?" Toward this goal, a guiding philosophy was inspired by the traditional units of length. The units *inch*, *foot*, and *yard* were born out of everyday needs and had a different origin than the more scientifically rationalized *metric* system with the millimeter, centimeter, and meter.

Consider how the traditional units came about: they most likely represent significantly different uses from a human perspective. Yard-scale measurements are typically used to measure objects around us that are large and immovable. Foot-sized objects can be held in your hands and carried, but are large enough that they are not likely to be carried with us at all times. However, inch-scale objects can fit in a pocket and be forgotten about while carrying out other unrelated daily activities. In other words, these three measurements represent three very different scales of human interaction, and define scale transitions for how we interact with the world around us. If ubiquitous computing systems were built to mimic everyday capabilities that occur at these three scales, any observation of such a system would probably have generic characteristics that would hold true for a much larger set of devices, each falling into one of these categories.

PARC thus embarked on the design of three devices: ParcTab, or Tab, an inch-scale computer that represented a pocket book or wallet (Want et al., 1995); the ParcPad, or Pad, a foot-scale device, serving the role of a pen-based notebook or e-book reader; and Liveboard, a yard-scale device that provides the functionality of a whiteboard.

An Introduction to Ubiquitous Computing ■ 7

(a)

(b)

FIGURE 1.3 (a) Xerox ParcTab a palm (inch-scale) computer communicating using diffuse infrared (IR) signalling; (b) an infrared transceiver basestation installed in the ceiling of each room comprising the ubicomp environment. Note the ring of IR emitters at the edge of the circular board, and four IR detectors at the center pointing in four cardinal compass directions. (Photos courtesy of PARC, Inc., http://www.parc.com)

Tabs communicated wirelessly with a ceiling-mounted basestation using 10 kbps diffuse infrared signaling (Figure 1.3). Each room was typically fitted with one basestation providing an infrared wired *microcellular* communication network. Each basestation also communicated through a wired serial connection to a nearby workstation attached in turn to the building's Ethernet, thus providing a connection to distributed services available on the network. ParcTabs were effectively dumb terminals generating pen/key events in response to user actions, and these were sent to remote applications running on servers attached to the network, resulting in application state changes that sent back screen updates to the Tab displays.

ParcPads employed a similar design approach using a low-bandwidth X-protocol across a radio link, communicating with a basestation through a proprietary short-range near-field radio (Katarjiev et al., 1993) (Figure 1.4). The radio basestation was also mounted in the ceiling of each office or laboratory, and had a 3–4 m range, similar to the infrared system but with 25× more bandwidth at 250 kbps. The reason infrared was used on the Tabs

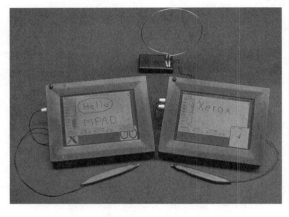

(a)

(b)

FIGURE 1.4   (a) The ParcPad, a notebook-sized (foot scale) tablet computer; (b) the near-field communication basestation mounted on the ceiling of an office at PARC. (Photos courtesy of PARC, Inc., http://www.parc.com)

versus the Pads was that it could be operated at much lower power, and was more suited to the small battery used by the inch-scale ParcTab device.

Liveboards were designed around standard computer workstations, but with much larger pen-based displays, and pen-based input. At PARC, several of these were deployed at fixed locations around the building and linked by a wired network. The display was implemented using a back-projected LCD panel and a 45° mirror to realize the image on a 67-inch frosted display panel. For writing and selection Liveboard employed an infrared pen that was tracked across its screen using a four-quadrant infrared sensor mounted in the optical path behind the screen. The output of the optical sensor was fed through a calibration table resulting in a representative screen coordinate. The primary pen-based interaction software for Liveboard was called Tivoli, also developed at PARC, and allowed many unique pen-centric operations for the drawing and manipulation of graphical freeform objects (Elrod et al., 1992).

Although the objective for designing Tabs, Pads, and Liveboards was to replace equivalent objects in the workplace by offering similar physical affordances, an equally important goal was to enhance their capabilities relative to the original technology, and thus make a compelling value proposition for the user. For example, a conventional whiteboard allows a teacher to write notes about a lesson which can be captured while interacting with the class. However, Liveboard provides this as a baseline capability, but adds the option of indexing the pen-based markup with contextual events to make future search and retrieval easier (Figure 1.5). The annotations and graphics drawn on the board could also be played back using an interactive timeline to support a discussion that revisited earlier topics.

At the foot scale, a book is just a single book with no interactive capability, but ParcPad could be potential thousands of books delivered across the network (or recall them from its local disk). It could also support electronic markup through its pen interface, and thus allow for hyperlinked text, word definition look-up, and cross-referencing with other material (all common today through Web interfaces, but not in the early 1990s).

Last, the ParcTab (replacing the pocket planner), served as a simple Personal Information Manager, but using its infrared network connection ParcTab also supported one of the first wireless pocket email readers. It could also edit documents stored in the network; serves as a remote controller for a room's heating and air-conditioning system; and play games. Because the ParcTab was easily carried, it could also serve as a location beacon, and the system could keep track of the Tabs as they moved around.

FIGURE 1.5 A commercial version of the Liveboard, sold by Liveworks in the mid-1990s, which evolved out of the original wooden laboratory prototype developed at PARC (see Figure 1.2). (Photo courtesy of PARC, Inc., http://www.parc.com)

This led to the notion of context aware applications (Schilit et al., 1994), which has become a central research theme in many other ubiquitous computing programs today.

1.1.3 Context Awareness

Context awareness (see Chapter 8) allows applications to comprehend the environment in which they are being used, and adapt their operation to provide the best possible user experience. A user or device context is difficult to model because it has many dimensions, such as location, the identity of devices close by, who else is present, the time, and environmental factors such as sound, motion, temperature, orientation, and other physical variables, many of which can be measured through on-platform *sensors*. Context awareness can also span multiple levels of system architecture. Operating locally at the device level it can take advantage of on-board sensors. For example, inverting a ParcTab detected by an on-board tilt sensor would invert its screen to maintain the orientation of the display. At a higher-system level, applications could use context to modify their behavior. For example, the ParcTab used an application called a *Proximity Browser*, which provided a user with the option of viewing files that had been accessed at its current location on a previous occasion. The objective was to take advantage of the cache principle: files that had been used at a location in the past were likely to be useful again.

This summary represents the focus of ubicomp at PARC between 1988 and 1996 when these projects were completed. However it has taken ~15 years for the underlying technologies to mature, communication standards to be ratified, and for many of the models to gain traction in the marketplace. This is the nature of ubiquitous computing. It is very sensitive to the affordances of the devices that technologists are trying to replace and, as technologies advance, whether it be processor performance per watt, storage capacity, network bandwidth, display resolution, device size, and weight. Each year more possibilities for the mainstream application of ubiquitous computing open up.

1.1.4 IBM Research: Pervasive Computing versus Ubiquitous Computing

In the mid-1990s, IBM began a research direction it called *pervasive computing* (IBM Mobile and Pervasive Computing), which had many similarities to the goals of ubiquitous computing. In fact, many texts today describe pervasive and ubiquitous as the same thing. Although the notion of being freed from the desktop computer and building on the opportunities opened up by connected mobile and embedded computers is a theme common to both, in 1991 the connection with invisible and calm technologies was a uniquely Xerox PARC perspective. However more than 10 years later, any unique position described by either party has been slowly integrated into the shared vision and by the mid-2000s any publications that set out to describe this topic presented fundamentally the same position.

IBM, to its credit, was one of the first companies to investigate the business opportunity around pervasive systems, and created a business unit dedicated to the task.

One of the first commercial deployments of a pervasive computing system was born from a collaboration between IBM Zurich and Swissair in 1999 (IBM Swissair), enabling passengers to check-in using Web-enabled (WAP) cell phones (Figure 1.6). Once the passengers had accessed the service, the phone also served as a boarding pass, showing gate seat and flight departure information, and identifying the traveler as having valid fight credentials. Although this was one of the most publicized projects, IBM also applied these technologies to other service opportunities in banking and financial services, gaining early experience in this area.

FIGURE 1.6 IBM provided Web-based services for Swissair using WAP on a cell phone to create an electronic boarding pass. (Courtesy of IBM Corp.)

Ubiquitous computing system-level solutions tend to be cross-discipline, and involve the integration of many disparate technologies to meet the original design goals. One of the key enablers for pervasive solutions has been the development of wireless/mobile platforms running standard operating systems that are already widely deployed in the form of smart phones. Today, this has enabled the use of generalized software environments, such as Web Sphere and the J9 Virtual Machine, to name two of their well-known software projects. The strengths of IBM's system integration have played well together to take advantage of a growing commercial service opportunity around pervasive systems.

1.1.5 University of Tokyo: T-Engine and the ITRON Operating System

On the other side of the Pacific Ocean in the late 1980s, researchers in Japan also realized the time had come for a new computing paradigm based around embedded systems. Prof. Ken Sakamura from Tokyo University, a famous computing architect even before his interest in ubiquitous computing, created the TRON research program. He developed a series of embedded computing platforms called T-Engines that were designed to be embedded in devices ranging in size from mobile electronics to home appliances and smart sensors (Figure 1.7). As with the other ubiquitous computing programs described in this chapter, the design of a suitable software framework to support the development of applications was equally as important as the design of novel hardware. Sakamura developed ITRON, an embedded real-time operating system that was portable

FIGURE 1.7 Several examples of T-Engines designed to support embedded computation at various scales of devices from information appliances to sensors. (From Krikke, J., *IEEE Pervasive Computing* 4(2), 2005. With permission.)

across the various scales of T-Engine, and also able to run on a number of commercial platforms (Krikke, 2005).

ITRON became very popular because it was an embedded solution that unified the development of software between various computer vendors in Japan, and what had been a very fragmented approach to building embedded systems began to coalesce around one design solution. The license for ITRON was also attractive to businesses because although it was open source, allowing development, improvement, and software additions, any changes did not need to be integrated back into the source tree, and thus given away to competitors. Instead, each company could keep these changes as a differentiator for their product. Although these advantages served the community, it also led to incompatibilities in some cases, which limited the ability to share and run application code. Nonetheless, ITRON has been extremely successful in this market, and has supported a high degree of innovation in mobile and embedded products, all of which teach us lessons about the development of mainstream ubiquitous computing applications.

1.1.6 Hewlett Packard: Cooltown

The Cooltown project (Kindberg et al., 2002) popularized the notion of linking real-world objects with Web content. The key observation was that every object could have a Web page capable of describing it, such as its name, ownership, origin, and associated network services, etc. This technique has clear value in a corporate setting, providing simple ways to configure and interact with networked systems. For complex devices such as printers and routers, a Web server can be embedded into the device, and accessed through a network connection (wired or wireless). However, there is also an opportunity to provide Web presence for simple, nonelectronic, and unsophisticated objects by attaching electronic tags that encode a unique ID. When used in conjunction with a database that maps these unique IDs with information, the combination becomes a powerful tool; this is similar in concept to the "Bridging Physical and Virtual Worlds" project at Xerox PARC in 1998 (Want et al., 1999). In Cooltown, the tags were implemented with barcodes or IR beacons, and could contain either the full textual URL or a unique number that could be mapped to the URL in order to access the corresponding server.

Although well suited to a corporate environment in which the efficient coordination and tracking of equipment are important, it can also be generalized to bring advantages to almost any everyday situation by tagging people, places, and things. A system that has access to the identity of all the people and objects in a particular place also has access to the defining context, and can make inferences about the activities taking place. Cooltown created a distributed system to represent people, places, and things in the system, and constructed various experimental environments (Figure 1.8) to understand how they could be used to support

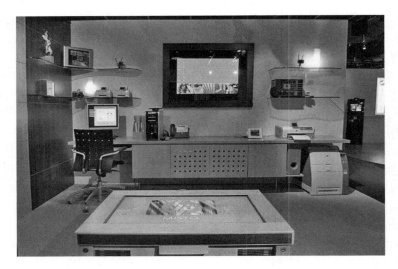

FIGURE 1.8 A Cooltown environment at HP Labs, Palo Alto, CA, in which all objects have a Web presence. (Photo courtesy of Hewlett-Packard Corp.)

work practices. In addition to work and office situations, Cooltown also deployed its Web presence system in museums and other public exhibits to enhance user experience in that setting. The applications of Cooltown provide us with another unique example of how a ubiquitous computing environment can be designed and deployed in the real world.

1.2 UBIQUITOUS COMPUTING IN U.S. UNIVERSITIES

1.2.1 UC Berkeley: InfoPad

The objective of the InfoPad program was to explore architecture and systems level issues for the design of a mobile computer providing ubiquitous access to real-time media in an indoor environment (Truman et al., 1998). The program was created by Prof. Bob Brodersen, involving a large graduate team in the mid-1990s, and served as a vehicle for exploring individual system components and the low-power mechanisms needed to support effective mobility. Toward this goal, several chip designs were fabricated to push conventional design limits, for example, a sub-threshold static RAM operating with exceptional low-power dissipation.

The wireless communication architecture took advantage of a thin client model with a high data-rate downlink for display updates, and a low data-rate uplink for keyboard, mouse, and pen interaction. The asymmetric radio communication system used a Plessey 1–2 Mbps radio modem in the 2.4 GHz band for the downlink, and a 121 kbps uplink based on a Proxim 902 MHz radio thus avoiding in-band interference between the two modes.

The architecture of InfoPad was highly tuned to the pad functionality, without requiring high-performance processors and memories, or mass storage devices (Figure 1.9). As a consequence of using dedicated system functions (and a modestly powered coordinating processor), the processor clock rate could be lowered, and power consumption reduced along with the size and weight of the battery. A primary goal of the project was to enable multimedia applications to be used while mobile; therefore, in addition to buttons and pen input, the device also supported a microphone, enabling speech-based applications. The downlink data stream was also capable of delivering digital audio and video to the screen, in addition to screen graphics. This device was a potent mobile platform that effectively allowed researchers to time travel, exploring a ubiquitous computing environment that is only now commercially available in some parts of the world, such as South Korea's HSAPD peer-to-peer video services, and, in trials, with WiMAX in the United States.

FIGURE 1.9 Infopad: UC Berkeley's exploration of a mobile ubiquitous computing architecture. (Courtesy of Computer Science Department, University of California Berkeley.)

1.2.2 MIT Media Laboratory: Wearable Computing

Taking another approach to ubiquitous computing, *wearable computing* puts the emphasis on a portable computer that can be unobtrusively integrated with a person's clothing, while still being comfortable for the wearer (Siewiorek et al., 2008). The MIT Media Laboratory has been a pioneer in this area, with several researchers embracing the wearable concept and living with the technology on a daily basis. A key characteristic of such a device is the heads-up display integrated into a user's glasses, and an input device that can be operated while out of view, typically in a pocket. The early wearable designs were weakest in their implementation of the display technology, requiring headgear that did not really meet the "unobtrusive" requirement. However, as private companies embraced the opportunity for wearable computing in some vertical markets (e.g., maintenance engineering and military operations), investment in novel solutions has resulted in compact displays that can be integrated with conventional eyeglasses (Figure 1.10a).

A wearable input device used by many of the pioneers is the Twiddler (Figure 1.10b), a one-handed chorded keypad that, with practice, can yield average typing speeds of 47 words per minute (wpm), and as high as 67 wpm in some cases (Lyons et al., 2006), which is fast enough for many tasks. However, this is only half the speed of a QWERTY keyboard, which according to one study has an average speed of 82 wpm and a maximum of about 114 wpm (Lyons et al., 2004).

An Introduction to Ubiquitous Computing ■ 17

(a) (b)

FIGURE 1.10 (a) Prof. Thad Starner, one of the wearable computing pioneers from MIT Media Laboratory, wearing a heads-up display produced by Microptical Inc. integrated with his glasses (Photo copyright Sam Ogden. With permission); (b) Twiddler, a chorded keypad used by many wearable systems as a means of one-handed typing, with the added advantage that it can be operated from inside a pocket. (Courtesy of Intel Corp.)

Although wearable computing is perhaps limited to people who are prepared to adapt and learn how to use specialized equipment that deviates from the Weiser model of ubiquitous computing, it opens up many new possibilities for using computing while mobile, and in an environment that may have limited ubicomp support.

An important related topic is *augmented reality* (Feiner et al., 1997) in which a computer is able to overlay information on top of what a user sees in order to improve ability to carry out a task. For example, a maintenance engineer servicing an engine would benefit from a visual overlay of labels that identify key components augmenting what she would normally see. Such a system could also access related manuals simply by selecting the corresponding item in the field of view with a pointing device. In an expanded view of ubiquitous computing with a suitably unobtrusive heads-up display embedded in our eyewear, augmented reality could become an indispensable tool of the future in much the same way we have come to rely on the cell phone today.

1.2.3 Georgia Tech: Living Laboratories

In the late 1990s, the Future Computing Environments group at Georgia Tech (Atlanta, GA) began a number of research activities under the general category of Living Laboratories. Prof. Gregory Abowd is the lead researcher and cofounder of the group, and best known for two exploratory ubicomp projects: Classroom 2000 (Abowd, 1999) and Aware Home (Kidd et al., 1999).

Classroom 2000 began in July 1995 with a bold vision of how ubiquitous computing could be applied to education and provide added value to standard teaching practices in the future (Figure 1.11). In a typical classroom setting, a teacher stands at the front of a class and writes on a whiteboard, often erasing older points to make room for new annotations. During the class various questions are likely to be asked, answers provided, and in some cases a discussion or debate may ensue. However, after the class is over, the detailed structure, discussion, and delivery of the lesson are usually lost, and all that remains are the student notes, abstract references, and sometimes teacher-supplied handouts. Classroom 2000 investigated the possibility of capturing the entire lesson in a form that would be a useful reference itself. However, a monolithic video is of limited use as a reference, and a key challenge was to create index points that enabled students to skip over a block of video of little interest, and be

(a)   (b)

FIGURE 1.11 (a) Classroom 2000: Gregory Abowd teaches, while (b) the Zen Pad UI captures notes and index points on a timeline of the lesson. (Courtesy of Prof. Gregory Abowd, Georgia Tech.)

able to jump to the exact point in time that might provide the answer to a question. Furthermore, these index points needed to be automatically generated, with clear meaning to anybody who wanted to use them. With an electronic board (the Xerox Liveboard was used for some of this work), it is possible to timestamp all the annotations made by a teacher during the lesson, along with slide transitions during a presentation, and other user-generated input, and these were used to index the audio and visual record of the lesson. Thus, the combined media, timeline, and indices represent a powerful summary that can be immediately made available to the students when the class finishes. One of the justifications for designing and testing the system in a real class setting was to understand the many subtle design points and user interface requirements that needed to be included in the system to be successful. Classroom 2000 has proven its value through popular demand, and was replicated five times on the Georgia Tech campus, and further deployed at three other universities. Classroom 2000 is an excellent example of ubicomp being applied to a real problem, and creating a solution of measurable value.

In 1999, the Aware Home project was founded and set out to explore how computation and embedded technologies could support everyday activities in a home. In the spirit of Living Laboratories, a complete residential building was designed from scratch, providing all of the expected features in a modern home, but with additions to support embedded computation and sensing, wiring conduits, and a control center. The result was a building with integrated sensing and control, and with enough flexibility to support unforeseen applications that might crop up in the future.

The building was completed in 2000 (Figure 1.12), and represents one of the largest dedicated spaces created to explore the vision of ubiquitous computing—a full embodiment of Weiser's ideas of 10 years earlier. As with all new technologies, the reality of such a deployment often differs in many significant ways from the original vision, and learning about the real problems, along with the best proposed solutions, was the primary benefit for undertaking a project on this scale.

Systems introduced into the Aware Home to support the research included cameras and RFID tags to identify and track an occupant's location, and various forms of sensors. For example, the house included a *smart floor* composed of a network of pressure sensors that could identify the characteristic ambulatory gait of individuals as they moved between rooms, thus providing additional means of occupant identification. Among the many

FIGURE 1.12 Georgia Tech's Aware Home. (Courtesy of Prof. Gregory Abowd, Georgia Tech.)

research projects that made use of the Aware Home, a novel method of sensing occupant activity was achieved by monitoring the electricity, gas, water, and waste lines connected to the house. Researchers found that characteristic signals in the consumption of power, gas, and water over time could be used to accurately infer and model activities going on in the house. Models of this type can be used to conserve resources by advising occupants of the consequences of their everyday actions, a simple but very tangible benefit of this approach in a world that needs to pay attention to conservation.

The Aware Home also investigated techniques to allow elderly residents to remain in their homes longer than would normally be possible, sometimes called "aging in place," through remote sensing. By unobtrusively allowing family members to monitor an elderly relative, the family could be made aware how well the relative was managing with everyday tasks, and avoid undue strain on an already overloaded health care system by requesting support only when absolutely necessary.

1.3 UBIQUITOUS COMPUTING IN EUROPEAN LABORATORIES AND UNIVERSITIES

1.3.1 Olivetti Research: Active Badges

About the same time as Xerox PARC embarked on its ubiquitous computing program, Olivetti formed a research laboratory in Cambridge, UK, with a focus on high-speed local area networks (LANs), and the novel applications and systems they enable. Toward this goal, the Pandora project was formed to explore the design of integrated multimedia services based on a Cambridge Fast Ring (a 100 Mbps slotted ring: the slot passing mechanism providing an inherently fair scheme for sharing bandwidth among multiple network clients). To support integrated services, which included audio and video traffic (note that this is 1988, the age of Plain Old Telephone Service, and well before Skype and cell phones), there was interest in using distributed computation to provide sophisticated telephone services such as video telephony and conferencing using a LAN. At that time, commercial telephone features were defined by the telephone company, and computer telephony integration (CTI) standards were only beginning to be addressed. To support effective automatic delivery of digital phone calls within the Pandora system, a dial-by-name approach was adopted. However, the system still lacked the ability to deliver a call to the location where the intended recipient was actually located, instead just connecting to the recipient's office.

To solve this problem, the Active Badge Location System was created (Figure 1.13). It was the first automated indoor location system and used a diffuse infrared beacon embedded into the electronic badges (Active Badges) worn by various members of the laboratory (Want et al., 1992). Each badge emitted an infrared signal encoding a unique ID that was received by a network of sensors, typically one sensor per room (multiple for large conference rooms). Badge beacons had a 1/10 second duration and a period of ~15 seconds, thus resulting in a 1:75 chance of

FIGURE 1.13   (a) Active Badge Sensor basestation; (b) an Active Badge. (Photo courtesy of PARC, Inc., http://www.parc.com)

collision with another badge signal. The oscillator in each badge was deliberately built from low-tolerance components to ensure that two badges that were synchronized by chance would soon drift apart, and after a couple of beacon periods would produce distinct signals. Each room sensor was polled using a low-bandwidth custom network returning badge IDs buffered at each location, thus associating badges, and hence people, with the rooms they occupied. The polling computer aggregated the ID–location pairs, reinterpreting the numbers as name–location pairs, which could then be advertised to other computers through a network-based location service. True name-based dialing was achieved by making use of this location service.

The Active Badge system provided one of the earliest available location systems and opened up the possibility of experimenting more generally with location-based services. As a result, Olivetti received many inquiries about the system, and built a consignment of badges and sensors to supply research laboratories around the world including Xerox PARC, EuroPARC, DEC SRC, DEC WRL, MIT Media Laboratory, and Cambridge University Computer Laboratory. The resulting location-based services inspired many alternate indoor location technologies. See Chapter 7 for more information on location systems for ubicomp.

Olivetti continued to be a leader in developing novel location technologies, and by 1997 designed the Active BAT system (Harter et al., 1999), with the goal of providing accurate three-dimensional (3-D) positioning within a room. This capability was achieved with a new type of badge (BAT) that combined ultrasound and radio to enable range measurements to be made from an array of sensors placed in the ceiling of every room. The radio signal was sent by a single transmitter and used to address a particular badge. When the corresponding badge received a signal encoding its address, it responded by transmitting an ultrasonic pulse traveling at the speed of sound in air to the array of ceiling sensors. Because the speed of light is so much faster than the speed of sound, each sensor could measure the difference in time between the original radio signal and the received ultrasonic signal, thus providing multiple independent estimates for the range of the badge to each sensor. By combining range estimates from the known sensor topology, a relative 3-D coordinate could be calculated for the badge. In practice, a room-based sensor grid used one sensor per ceiling tile (~2 ft^2) to collect multiple measurements, and achieve a positional accuracy of about 3 cm in a typical room. This is a very high-level description of the system, and in practice it is more complicated because ultrasound can reflect from nearby objects, generating ghost pulses that need to be removed from the calculation using signal processing techniques to limit any errors. Using this system, each BAT could be polled and an accurate location determined even if it was moving. Therefore, a key advantage of this approach is the ability to accurately track objects in real time in three dimensions.

MIT produced an alternate indoor positioning system shortly after Olivetti. This system was also based on ultrasound, but used a design approach that was the inverse of the BAT system. Instead of the badges transmitting an ultrasound pulse, a ceiling-mounted device called a *Cricket* (Priyantha et al., 2000) simultaneously transmitted a radio packet and an ultrasound pulse as a combined beacon. Any mobile (or fixed) device receiving the radio packet could also measure the time interval for the accompanying ultrasound packet to arrive, and therefore calculate its range to the Cricket. If several Crickets were placed in a room, and range calculations made by the mobile device to each one, it could then calculate its position relative to the known locations of the Crickets. This approach had some privacy advantage over the BAT system, as each mobile could then decide who else could find out its location (i.e., it would have to actively transmit it to another device). However, because the mobile range calculations for each Cricket are not carried out simultaneously, unlike the BAT, accurate 3-D location while in motion is not possible. In terms of deployment, however, the Cricket system is easier, requiring less ceiling-mounted components than the BAT, and does not require networking. In practice, both systems have their strengths and weaknesses, and depending on the application, a case can be made for either approach.

1.3.2 Karlsruhe: Cups and Smart-Its

One of the strongest university ubiquitous computing programs in continental Europe is found at the University of Karlsruhe, also the origin of the Ubicomp conference. In 1997, researchers began experimenting with ubicomp concepts and gained experience embedding computing into everyday objects, the MediaCup (Beigl et al., 2001) being a well-known example (Figure 1.14a). Inspired by this work and the resulting research questions it inspired, Karlsruhe led the formation of the Smart-Its project, a collaboration with ETH (Zurich) and Lancaster (UK), Viktoria Institute (Sweden), Interaction Institute (Sweden), and VTT (Finland). The goal of the Smart-Its (Holmquist et al., 2004) project was to create a general-purpose tool for experimenting with ubicomp concepts and providing

 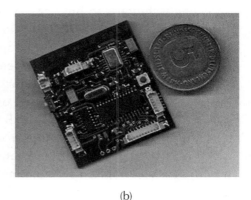

(a) (b)

FIGURE 1.14 (a) The MediaCup warns you if the coffee is too hot to drink or that it is being tipped at a critical angle; (b) a Smart-It embedded computing platform. (Courtesy of Michael Beigl.)

researchers with a hardware and a software development environment well suited to ubicomp research (www.smart-its.org).

Smart-Its are small embedded computers with communication and sensing components that can be further integrated with everyday objects such as furniture, clothing, coffee cups, product packaging, and fresh produce (Figure 1.14b). The initial research collaboration resulted in 16 different projects exploring a wide variety of applications. Some examples include tracking the position of objects on a table by measuring the relative loading of each table leg; and in the event of an avalanche augmenting a skier's jacket to wirelessly send data from a pulse oximeter to a rescue team and alerting them which skier might need critical assistance first; and a smart restaurant that could tag and track waiters, serving dishes, bottles, and food items to ensure timely delivery of a complete order. Smart-Its have been a very successful research platform and widely adopted across the European university network. Much of the success results from standard APIs that have been created to allow low-level abstraction of sensing and communication components, and making use of higher-level operating behaviors called *percepts*. Percepts are supported by the Perception API (PAPI) and enable devices to wirelessly share sensing components situated on nearby smart-its with little programming overhead — an abstraction called *collective perception*.

1.3.3 Lancaster University: Guide

Since the mid-1990s Lancaster University has been a leading university research center in England contributing to the ubicomp vision. From 1997 to 2001, the GUIDE project (Cheverst et al., 2000) was created, obtained a government grant, and captured the imagination of many researchers interested in location-based services in the wild. It was the first mobile electronic guidebook designed and optimized from concept to implementation for use by tourists. The user platform was based around a rugged Fujitsu TeamPad tablet (one of the early pen-based tablet products, 21 × 15 × 1.5 centimeters and weighing about 1/3 pound; Figure 1.15). A fully operational TeamPad had a battery life of about 2 hours, which for nominal use meant it could be given to a tourist in the city of Lancaster on a day trip, and effectively monitor the entire user experience.

The system was initially evaluated by a trial conducted over a 4-week period, and used by 60 people during that time. GUIDE was unique in comparison with earlier reported electronic guidebooks, such as Cyberguide (Long et al., 1996) from Georgia Tech, in that it used NCR Wavelan radios compatible with the 802.11b standard (new at that time), and took advantage of the characteristic poor radio propagation to create servers that covered a region called a cell. Each cell server presented the user with options to learn more about areas of interest. Once the user made a selection, a Web page was retrieved from the GUIDE Web server providing a full description of nearby points of interest. GUIDE was the first electronic tour guide that made use of a microcellular communication network to determine

FIGURE 1.15 The Fujitsu TeamPad used to implement the mobile platform used by the GUIDE system. (Photo courtesy of Prof. Nigel Davis, Lancaster University.)

location in a citywide deployment. The system also used a mobile tablet with a display small enough to carry, but large enough for effective user interaction, and it had the ability to add new information on the fly.

1.4 MODERN DIRECTIONS IN UBIQUITOUS COMPUTING

Since 2000, there has been a wealth of ubicomp projects launched around the world, and although we restrict our description in this introduction to the genesis of ubicomp and some of the activities that constituted its formative years, it is worth noting that two of the mainstream companies in the computer industry today, Microsoft and Intel, have also embraced ubicomp as an exploratory direction in their research laboratories. Here, we briefly describe some of the most significant projects they have undertaken in the past 10 years.

1.4.1 Microsoft Research

Microsoft Research (MSR) was founded in 1991 in Redmond, WA, with expansion to Cambridge, England, in 1997, and Asia in 1998. MSR Bangalore, Silicon Valley, and Cambridge, MA, were later established between 2005 and 2007. Three noteworthy ubicomp projects carried out by various groups in these laboratories include SenseCam & MyLifeBits, RADAR, and EasyLiving.

SenseCam is a small wearable computer that periodically captures images of the world as a user moves around. In collaboration with the MyLifeBits (Gemmell et al., 2006) project, SenseCam provides a wealth of contextual data about the wearer, augmented by a database describing documents and other electronic media that the individual has accessed. The result is a prosthetic memory aid that can be used to answer basic questions about an individual's life, enabling more detailed recall than most of us could achieve by unaided means (Hodges et al., 2006).

RADAR was the first example of a wireless system allowing mobile computers to locate themselves in a building [an indoor global positioning system (GPS)]. It was designed around the first WiFi (802.11b) radios that were commercially available, and made use of RSSI reception maps in a Microsoft building hosting several access points (APs) with the data collected using a wireless survey tool. By comparing the received signal strength indication (RSSI) reading for each AP measured at a point of interest with the RSSI signal maps on record, the system could automatically determine a mobile computer's most likely location to an accuracy of 2–3 meters (Bahl and Padmanabhan, 2000).

EasyLiving, established in 1999, was MSR's closest project to the spirit of the original ubicomp vision. The research was centered on a smart room designed to support both work and recreational activities. A key ingredient was the use of image processing to recognize activities in the space, making use of multiple cameras to track the occupants, and objects situated in the room. Of particular note was the ability to migrate computing sessions from screen to screen as people moved around, and mechanisms to automatically control the lighting and music in the space to best suit all the occupants (Brumitt et al., 2000).

1.4.2 Intel Research

In 1990, Intel established a new organization called Intel Research to begin off-roadmap exploratory research. In support of this mission, it founded a number of laboratories in close proximity to universities with top-tier computer science departments: University of Washington, Seattle; Carnegie Mellon, Pittsburgh; and UC Berkeley along with other groups embedded in corporate locations at Hillsboro, Oregon and Santa Clara, California. Ubiquitous computing became the exploratory direction for Intel Research Seattle (IRS), some of the Berkeley projects, and one of the research groups based in Santa Clara.

Place Lab was the best known of the projects from IRS, exploring and building a system that could determine the location of a mobile device by cataloguing and mapping WiFi access points throughout a city (LaMarca et al., 2005); this was later extended to GSM towers, and demonstrated effective location-based services for cell phone applications. The project was rolled out to the research community as a sandbox to experiment with location-based services on standard mobile platforms. It had a distinct advantage over GPS, in that it worked well indoors and did not require the purchase of additional GPS equipment for operation outdoors. Today, this approach has been adopted by several companies, especially the Apple iPhone.

At Intel Research Berkeley, various research projects were created around sensor networks and Motes (Polastre et al., 2004), a small form factor wireless sensor node that was the heart of these systems. Sensor networks share a common thread with the embedded computation vision of ubicomp. Although they are usually associated with scientific monitoring, they can also be integrated with objects in the environment to provide embedded sensing and control (similar to the Smart-Its work described earlier).

The Personal Server (Want et al., 2002) project in 2001 took a fresh look at the trends in computation and mobility, recognizing that even handheld mobile computers were beginning to make use of high-performance processors along with high-capacity solid-state memories, and with low-power dissipation they could be operated from a small battery. However, the primary limitation preventing them from serving as an effective computer was the user interface (display and keyboard). The Personal Server project reversed the traditional design paradigm, and instead of using the mobile device as a client for a remote network server, it used the mobile device as the server and ran the client on a nearby PC, thus enabling a user to gain access through a large high-resolution display and full-feature keyboard. This was further facilitated by the advent of standardized short-range radio technologies such as Bluetooth and WiFi/802.11b. The personal server concept was initially implemented on a custom XScale embedded system called Stargate, and later ran on a Motorola e680 cell phone (Linux/XScale), demonstrating that one day the ubiquitous cell phones could become our primary computer, which we would interact with wirelessly through nearby infrastructural computers.

1.5 THE RESEARCH COMMUNITY EMBRACES UBIQUITOUS COMPUTING

In the early days of ubicomp research (circa 1990), finding an appropriate venue to publish papers and discuss research was challenging. In fact, none of the conferences at that time were well suited to the topic at all. For application-oriented ubicomp research, the Association of Computing Machinery (ACM) Special Interest Group on Computer–Human Interaction (CHI) (origin: 1982) and ACM Symposium on User Interface Software and Technology (UIST) (origin: 1989) were the closest options, but naturally biased toward the evaluation of user interfaces rather than mobile and embedded systems. However, some journals provided viable options; for example, early work on the Active Badge was published in 1991 in *ACM Transactions on Office Information Systems* (*TOIS*), but although well received, it was not representative of their typical published work. Mark Weiser also had some success publishing his early ubicomp ideas in venues such as *Scientific American* and *Communications of the ACM*.

Part of the problem was that mobile computing was also in the early stages of development—no wireless standards existed for LAN-based solutions at that time. In fact, the first plenary meeting of the Institute of Electrical and Electronics Engineers (IEEE) that included the 802.11 working group was only held in 1991. It took another 3 years for the Workshop on Mobile Computing Systems and Applications (WMCSA) to be established (1994); it was the first academic mobile computing workshop and was held in Santa Cruz, CA. This was followed by the first ACM-supported conference of this type, the Mobile Computing and Networking Communications conference, or ACM MobiCom, in Berkeley, CA (1995), which soon became the premier conference for work on mobile systems. As a result, the organizers of WMCSA decided to let MobiCom take the lead and discontinued their workshop for several years. However, it soon became clear to the rest of the research community that MobiCom did not cover the full spectrum of research that was being explored under the banner of ubiquitous computing. To fill the conference vacuum, Karlsruhe University in Germany created a new conference, Handheld and Ubiquitous Computing (HUC), in 1999, which had a much broader charter than MobiCom. HUC, by comparison, melded applications, user interfaces, systems, wireless networking, and hardware design into its charter. After HUC 2000, the word "Handheld" was dropped, and from 2001 onward the conference was renamed Ubicomp. In 2008, it was also adopted by the ACM and is now supported by ACM Sigmobile and SIGCHI.

With the growing momentum behind mobile, wireless, and embedded technologies, by 2002 considerably more researchers were entering the field of ubicomp and looking for venues to publish. As a result, another conference, called "Pervasive," began independently; the first open event was held in Zurich, Switzerland in 2002. The name "Pervasive" was chosen because it was initially supported by IBM, which had already invested heavily in the name. By 2003 it was clear that many of the program committee members for both conferences overlapped and were serving the same community with a similar charter. The organizers decided to coordinate these two annual conferences so that they would be held 6 months apart each year—with Pervasive being held in the Spring and Ubicomp in the Fall—thus providing two equally spaced opportunities to publish per year, each with similar standards and criteria for acceptance. Meanwhile, MobiCom continued its charter with a focus on communications, which only strengthened from 2000 onward, possibly because many of the application and systems papers were now being submitted to Ubicomp and Pervasive. However, members of Sigmobile felt that there was room for another ACM conference with a focus on mobile systems and applications, with the additional criterion that all research papers should describe systems that had actually been built, deployed, and evaluated, and typically described multiple layers of the system stack. As a result, in 2003, ACM launched MobiSys, with its first conference held in San Francisco,

CA. MobiCom continued in parallel, still focusing on networking and research simulations, thus providing two distinct venues with complementary charters. ACM MobiSys and MobiCom are now generally recognized as the two top-tier conferences in this area.

While these developments were taking place, two other conferences of note were also being forged. In 1997, the wearable computing community began the first International Symposium on Wearable Computing (ISWC) in Boston, MA, and in 1998, Human–Computer Interaction for Mobile Devices (MobileHCI) began in Edinburgh as a workshop, and became a full conference in 2002. Also in 1999, WMCSA was resurrected as an annual workshop to provide researchers with valuable feedback on early mobile systems work. WMCSA was renamed Hotmobile in 2008 and also became an ACM-supported conference. There is now a strong link with the organizers of ACM MobiSys providing a mechanism to progress research from a workshop forum to a full conference and helping ensure that senior members of the research community will attend the workshop by colocating the program committee meeting for MobiSys with the HotMobile event.

In contrast to the early 1990s, from 2009 going forward there is a wealth of ubicomp-related conferences available to researchers. Other notable venues are listed here along with the year they were founded: IEEE Pervasive Computing and Communication in Dallas-Fort Worth, TX, 2003; MobiQuitous in Boston, 2004; Location and Context Aware (LoCA) in Munich, Germany, 2005; Tangible and Embedded Interaction (TEI) in Baton Rouge, LA, 2007; and Mobile Computing, Applications, and Services (Mobi CASe) in San Diego, CA, 2009. However, there are many other conferences that are not mentioned mainly because they have little impact on the mainstream research community.

For papers more suitable to journals and magazines, a number of peer-reviewed publications have also been created over the years. Magazines include *IEEE Personal Communications*, which was first printed in 1994. It combined in-depth wireless research with application, and systems. In 1997, Springer began publishing *Personal & Ubiquitous Computing*, which continues to serve as an important venue for publishing ubicomp articles and research in Europe. In 2001, *Personal Communications* evolved into two new publications, with *IEEE Pervasive Computing* providing a peer-reviewed forum for systems and applications level work, and *IEEE Wireless Communications* mainly soliciting manuscripts on research based on the lower layers of the protocol stack. *IEEE Transactions on Mobile Computing* has also been an archival publication for more mature research since 2002.

1.6 THE FUTURE OF UBIQUITOUS COMPUTING

Since 2000, there has been a dramatic improvement in the capabilities of high-end cell phones, also known as smart phones. These devices no longer just make phone calls; they also integrate computation and communication into a handheld device small enough to be dropped into a pocket or purse. In 2008, there were about 1.2 billion cell phone shipments and 3.3 billion cellular subscribers, about half the population of the Earth. Not surprisingly, the cell phone has become the most ubiquitous computer there has ever been. Admittedly, the vast majority of these devices are relatively unsophisticated, but the percentage that comprises smart phones is growing steadily each year. Another notable data point is that smart phone shipments outnumbered laptop PC shipments for the first time in 2007, at about 116 million units (Figure 1.16) compared to 108 million for laptops. Since 2002, the smart phone has also been absorbing the PDA market, and with many devices running mainstream operating systems, for example, embedded Linux or Windows Mobile 6.0, these phones are beginning to feel more like mainstream computers.

To understand where this market is going, we look at a new family of mobile processors called Atom, which was announced at the Intel Developers Forum in Shanghai, China, in April 2008. The first generation of processors in this family is based on a 45 nm process, is a fully compatible ×86 instruction set, and is designed for low-power operation in the

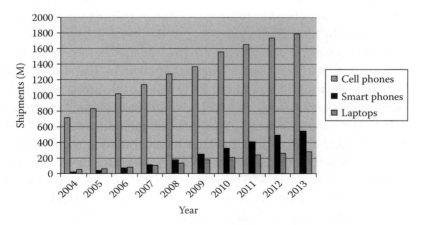

FIGURE 1.16 Mobile device shipments for 2004–2008 and forecasts for 2008–2013. (Data from Intel Library.)

1–2 W range. Initially targeted at MIDs and NetBooks, this is a game-changer for mobile products.

It does not take much imagination to realize that in a couple of generations of this processor family, the cell phone industry will be able to design a device that is fully compatible with a PC, but with the commensurate power consumption of a smart phone. The primary advantages of Atom are performance and software compatibility, things that have plagued the mobile market for years, with a fragmented community of hardware and software developers. Using this processor, legacy PC code can be reused, and through remote wireless display technologies, applications can be run without modification. Developers also can have high confidence that programs designed and tested on a PC can be successfully transferred to a mobile without modification. Furthermore, by using virtualization techniques, application code running on a PC in a Virtual Machine (e.g., KVM) can be migrated across a wireless connection to a mobile device, and then back to a PC to facilitate both mobile and enterprise work practices—again, all without modifying the applications.

Uniformity of processor architecture and the opportunity for a mobile runtime environment that is compatible with the desktop, server, and embedded market will accelerate the development of ubicomp applications and facilitate their use in many new situations. Just as the PC provided a common platform for the successful development of widely available desktop software in the early 1980s, cell phones are poised to take that role in the next decade.

To conclude, consider the original Weiser vision in which Tabs, Pads, and Liveboards were used as informative design points to explore future opportunities for ubiquitous computing. Today, there are already several analogous commercial devices that are in wide use. For example, smart phones are much like Tabs, as they can be kept with us at all times, access a ubiquitous data network, and put us in touch with useful remote services. Modern laptops (some of which also convert into tablets) are very similar to the ParcPad concept. Furthermore, modern electronic books, such as the Sony e-reader and Amazon Kindle based on e-ink technology, are already providing a Pad-based user experience that rivals traditional books. Large-screen LCD displays with 50 to 60 inch diagonals are now also common, primarily driven by the television industry. Unlike the Liveboard, the majority of these devices do not have pen input or a touch surface, but there are a few products that provide this capability. Most of the new models are also equipped with a network interface, enabling them to send and receive remote content. We may not be living the complete Weiser vision of ubiquitous computing, but many of the key ingredients are rapidly moving into place. The remainder of this book describes the underlying technologies and software architectures that are being developed to realize ubicomp more fully. At this time, ubicomp is no longer just a vision; it is rapidly becoming the reality.

REFERENCES

Abowd, G. D., Classroom 2000: An experiment with the instrumentation of a living educational environment, *IBM Systems Journal* 38(4): 508–530. Special issue on Pervasive Computing, 1999.

Bahl, P., and Padmanabhan, V. N., RADAR: An in-building RF-based user location and tracking system, *IEEE InfoCom*, March 2000.

Beigl, M., Gellersen, H., and Schmidt, A., Mediacups: Experience with design and use of computer-augmented everyday artifacts, *Computer Networks* 35(4): 401–409, 2001.

Brumitt, B., Meyers, B., Krumm, J., Kern, A., and Shafer, S., EasyLiving: Technologies for intelligent environments, Second International Symposium on Handheld and Ubiquitous Computing (HUC 2000), September 2000, pp. 12–27.

Cheverst, K., Davies, N., Mitchell, K., and Friday, A., Experiences of developing and deploying a context-aware tourist guide: The GUIDE project, MobiCom 2000, Seattle, pp. 20–31.

Elrod, S., Bruce, R., Gold, R., Goldberg, D., Halasz, F., Janssen, W., Lee, D., McCall, K., Pederson, E., Pier, K., Tang, J., and Welch, B., Liveboard: A large interactive display supporting group meetings, presentations and remote collaboration, ACM SigCHI, November 12, 1992.

Feiner, S., MacIntyre, B., Hollerer, T., and Webster, A., A touring machine: Prototyping 3D mobile augmented reality systems for exploring the urban environment, *Proceedings of the 1st IEEE International Symposium on Wearable Computers (ISWC)*, p. 74, 1997.

Gemmell, J., Bell, G., and Lueder, R., MyLifeBits: A personal database for everything, *Communications of the ACM* 49(1): 88–95, 2006.

Harter, A., Hopper, A., Steggles, P., Ward, A., and Webster P., The anatomy of a context-aware application, *Proceedings of the Fifth Annual ACM International Conference on Mobile Computing and Networking (MOBICOM)*, August 1999.

Hodges, S., Williams, L., Berry, E., Izadi, S., Srinivasan, J., Butler, A., Smyth, G., Kapur, N., and Wood, K., SenseCam: A retrospective memory aid, P. Dourish, and A. Friday (Ed.), *Ubicomp 2006, LNCS 4206*, pp. 177–193, Springer-Verlag, Berlin, 2006.

Holmquist, L. E., Gellersen, H., W., Kortuem, G., Stavros, A., Michahelles, F., Schiele, B., Beigl, M., and Mazé, R., Building intelligent environments with Smart-Its, *IEEE Computer Graphics and Applications*, 24(1): 56–64, 2004.

IBM Mobile and Pervasive Computing, http://domino.watson.ibm.com/comm/research.nsf/pages/r.mobile.spotlight.html.

IBM Swissair, IBM first to enable passengers to check in from mobile phones. http://www.zurich.ibm.com/news/99/wap_sr.html.

Katarjiev, C., Demers, A., Frederick, R., Krivacic, R., Weiser, M. et al. Experiences with X in a wireless environment, Mobile and Location-Independent Computing Symposium on Mobile and Location-Independent Computing Symposium, Cambridge, MA, pp. 11–21, 1993.

Kidd, C. D., Orr, R., Abowd, G. D., Atkeson, C. G., Essa, I. A., MacIntyre, B., Mynatt, E. D., Starner, T. E., and Newstetter, W. The Aware Home: A living laboratory for ubiquitous computing research, *Proceedings of CoBuild '99*, Position paper, October 1999.

Kindberg, T., Barton, J., Morgan, J., Becker, G., Caswell, D., Debaty, P., Gopal, G., Frid, M., Krishnan, V., Morris, H., Schettino, J., Serra, B., and Spasojevic, M., People, places, things: Web presence for the real world, *Mobile Networks and Applications* 7(5): 365–376, 2002.

Krikke, J., T-Engine: Japan's ubiquitous computing architecture is ready for prime time, *IEEE Pervasive Computing* 4(2): 4–9, 2005.

LaMarca, A., Chawathe, Y., Consolvo, S., Hightower, J., Smith, I., Scott, J., Sohn, J., Howard, T., Hughes, J., Potter, F., Tabert, J., Powledge, P., Borriello, G., and Schilit, B., Place Lab: Device positioning using radio beacons in the wild, *Proceedings of Pervasive 2005*, Munich, Germany, 2005.

Long, S., Kooper, R., Abowd, G. D., and Atkeson, C. G., Rapid prototyping of mobile context-aware applications: The Cyberguide case study, in *Proceedings of the 2nd ACM International Conference on Mobile Computing and Networking (MobiCom '96)*, November 1996.

Lyons, K., Starner, T., Plaisted, D., Fusia, J., Lyons, A., Drew, A., and Looney, E. W., Twiddler typing: One-handed chording text entry for mobile phones, *Proceedings of CHI 04*, Vienna, Austria, April 2004.

Lyons, K., Gane, B., and Starner, T., Experimental evaluations of the Twiddler one-handed chording mobile keyboard, *Human-Computer Interaction* 21(4): 343–392, 2006.

Polastre, J., Szewczyk, R., Sharp, C., and Culler, D., The Mote revolution: Low power wireless sensor network devices, Hot Chips 16: A Symposium on High Performance Chips, August 22–24, 2004.

Priyantha, N. B., Chakraborty, A., and Balakrishnan, H., The cricket location-support system, *Proceedings of the Sixth Annual ACM International Conference on Mobile Computing and Networking, MobiCom*, August 2000.

Schilit, B., Adams, N., and Want, R., Context-aware computing applications, *First Annual Workshop on Mobile Computing Systems and Applications (WMCSA)*, Santa Cruz, December 1994.

Siewiorek, D., Smailagic, A., and Starner, T., Application design for wearable computing, *Morgan Claypool Synthesis Lectures on Mobile and Pervasive Computing*, 2008, 66 pages.

Truman, T. E., Pering, T., Doering, R., and Brodersen, R. W., The InfoPad multimedia terminal: A portable device for wireless information access, *IEEE Transactions on Computers* 47(10): 1073–1087, 1998.

Want, R., Hopper, A., Falcao, V., and Gibbons, J., The Active Badge location system, *ACM Transactions on Office Information Systems (TOIS)* 10(1): 91–102, 1992.

Want, R., Schilit, B., Adams, N., Gold, R., Goldberg, D., Petersen, K., Ellis, J., and Weiser, M., An overview of the Parctab ubiquitous computing experiment, *IEEE Personal Communications* 2(6): 28–43, 1995.

Want, R., Fishkin, K., Harrison, B., and Gujar, A., Bridging real and virtual worlds with electronic tags, *Proceedings of ACM SIGCHI*, Pittsburgh, May 1999, pp. 370–377.

Want, R., Pering, T., Danneels, G., Kumar, M., Sundar, M., and Light, J., The personal server: Changing the way we think about ubiquitous computing, *Proceedings of Ubicomp 2002: 4th International Conference on Ubiquitous Computing, Springer LNCS 2498*, Goteborg, Sweden, September 30–October 2, 2002, pp. 194–209.

Weiser, M., The computer for the 21st century, *Scientific American* 265(3): 94–104, 1991.

Weiser, M., Hot topics: Ubiquitous computing, *IEEE Computer* 26(10): 71–72, 1993a.

Weiser, M., Some computer science issues in ubiquitous computing, *Communications of the ACM* 36(7): 75–84, 1993b.

Weiser, M., The world is not a desktop, *Interactions* 1(1): 7–8, 1994.

Weiser, M., and Seely-Brown, J., The coming age of calm technology, in *Beyond Calculation: The Next Fifty Years*, Copernicus, pp. 75–85, 1997.

CHAPTER 2

Ubiquitous Computing Systems

Jakob Bardram and Adrian Friday

CONTENTS

2.1	Introduction	38
2.2	Ubicomp Systems Topics and Challenges	41
	2.2.1 Resource-Constrained Devices	42
	2.2.2 Volatile Execution Environments	43
	2.2.3 Heterogeneous Execution Environments	44
	2.2.4 Fluctuating Usage Environments	45
	2.2.5 Invisible Computing	48
	2.2.6 Security and Privacy	49
	2.2.7 Summary	51
2.3	Creating Ubicomp Systems	51
	2.3.1 Why Build Ubicomp Systems?	51
	2.3.2 Setting Your Objectives	52
	2.3.2.1 Testing Your Ideas	53
	2.3.3 Designing "Good" Systems	54
	2.3.3.1 Computational Knowledge of the Physical World	54
	2.3.3.2 Seamfulness, Sensibility, and Tolerant Ignorance	56
	2.3.3.3 User Mental Model and Responsibility	58
	2.3.3.4 It Is Always Runtime	60
	2.3.3.5 Handling Transient Connections	60

		2.3.3.6 The State of the World	61
		2.3.3.7 Is It Working?	63
	2.3.4	Summary	64
2.4	Implementing Ubicomp Systems		64
	2.4.1	Choosing "Off-the-Shelf" Components	64
	2.4.2	Deploying Ubicomp Systems	67
		2.4.2.1 Expect the Unexpected	73
	2.4.3	Summary	74
2.5	Evaluating and Documenting Ubicomp Systems		74
	2.5.1	Evaluating Ubicomp Systems	74
		2.5.1.1 Simulation	75
		2.5.1.2 Proof-of-Concept	76
		2.5.1.3 Implementing and Evaluating Applications	77
		2.5.1.4 Releasing and Maintaining Ubicomp Systems	79
	2.5.2	Learning from What You Build	79
		2.5.2.1 Communicating Your Findings	80
		2.5.2.2 Rigor and Scientific Communication	81
	2.5.3	Documenting Ubicomp Systems	81
	2.5.4	Corollary	83
2.6	Getting Started		83
	2.6.1	Prototyping Your Ideas	84
	2.6.2	Smart Room in a Box	85
	2.6.3	Public Domain Toolkits	86
		2.6.3.1 Vision and Augmented Reality	86
		2.6.3.2 Sensing	86
		2.6.3.3 Hardware	87
	2.6.4	Datasets	87
	2.6.5	Summary	88
2.7	Conclusion		88
References			89

2.1 INTRODUCTION

The prevalent computing paradigm is designed for personal information management, including personal computers (PCs) such as desktops and laptops with fixed configurations of mouse, keyboard, and monitor; wired local area network; dedicated network services with fixed network addresses and locations, such as printers and file servers; and a user interface consisting of on-screen representation and manipulation of files, documents, and applications through established metaphors such as the mouse pointer, icons, menus, and windows.

Ubiquitous computing (ubicomp) strives at creating a completely new paradigm of computing environment in almost all of these respects. Ubicomp systems aim for a heterogeneous set of devices, including invisible computers embedded in everyday objects such as cars and furniture, mobile devices such as personal digital assistants (PDAs) and smart phones, personal devices such as laptops, and very large devices such as wall-sized displays and tabletop computers situated in the environments and buildings we inhabit. All these devices have different operating systems, networking interfaces, input capabilities, and displays. Some are designed for end user interaction—such as a public display in a cafeteria area—whereas other devices, such as sensors, are not used directly by end users. The interaction mode goes beyond the one-to-one model prevalent for PCs, to a many-to-many model where the same person uses multiple devices, and several persons may use the same device. Interaction may be implicit, invisible, or through sensing natural interactions such as speech, gesture, or presence: a wide range of sensors is required, both sensors built into the devices as well as sensors embedded in the environment. Location tracking devices, cameras, and three-dimensional (3-D) accelerometers can be used to detect who is in a place and deduce what they are doing. This information may be used to provide the user with information relevant in a specific location or help them adapt their device to a local environment or the local environment to them. Networking is often wireless and ad hoc in the sense that many devices come in contact with each other spontaneously and communicate to establish services, when they depart, the network setup changes for both the device and the environment.

Ubicomp environments involving technologies such as the ones described above have been created for a number of application domains, including meeting rooms (also known as *smart rooms*), classrooms, cars, hospitals, the home, traveling, and museums. In order to get a feeling of what ubicomp systems would look like, let us consider some examples from a future hospital (Bardram et al., 2006). Doctors and nurses seamlessly move around inside the hospital using both personal portable displays (e.g., a super lightweight tablet PC) as well the large multitouch displays available on many walls inside the wards, conference rooms, operating rooms, and emergency departments. Indoor location tracking helps in keeping track of clinicians, patients, and equipment, as well as assisting the clinicians and patient with location- and context-dependent information. For example, the patient is constantly guided to the right examination room, and on the doctor's portable devices, relevant information on the nearby patient is fetched from the central servers and presented according to the doctor's preference on this specific type of devices. If he needs more display space, he simply drops the portable display in a recharge station, and moves to a wall display where the information is transferred. In the conference room, the large conference table is one large display surface that allows for colocated collaboration among the participating physicians. The location tracking system as well as biometric sensors keep track of who is accessing medical data, and prevents nonauthorized access. Unique identification tags and medical body sensor networks attached to patients as well as to the patient's bed and other equipment inside, for example, the operating rooms, constantly monitor the patient and provide a high degree of patient safety. Not only are critical medical data such as pulse, electrocardiogram (ECG), and heart rate monitored, but also more mundane safety hazards such as wrong side surgery and lack of relevant instruments are constantly monitored and warnings issued if the "system" detects potential problems.

Ubicomp systems research is concerned with the underlying technologies and infrastructures that enable the creation and deployment of these types of ubicomp applications. Hence, ubicomp systems research addresses a wide range of questions such as: how to design hardware for sensor platforms, operating systems for such sensor platforms; how to allow devices to find each other and use the services on each other; how to design systems support for resource impoverished devices that run on batteries and need to save energy; how to run large distributed infrastructures for seamless mobility and collaboration in creating applications for such settings as smart rooms and hospitals; and a wide range of other systems aspects. To some degree, ubicomp systems questions and challenges overlap and coexist with other systems research questions, but as outlined in this chapter, the ubicomp vision and the nature of ubicomp applications, present a unique set of challenges to ubicomp systems research.

The chapter commences with a discussion of the key topics and challenges facing ubicomp systems, highlighting assumptions that are often made in traditional systems thinking that are unreliable in this problem domain. Then, design rationale and process for creating "good" ubicomp systems is explored, leading to advice on how to choose hardware and software components well and consolidated tips on what to look for when deploying ubicomp systems "in the wild." The chapter goes on to discuss the process of evaluating and documenting ubicomp systems—essential if the system is to be of any importance in moving the field forward. Finally, the chapter concludes with pointers to available software and hardware components and datasets that can help you bootstrap your experimental systems development.

Building ubicomp systems is essential to the progress of the field as a whole. Experimentally prototyping ubicomp systems enables us to experience them, discover what they are like to use, and reason about core precepts such as the boundaries of the system, its invisibility, the role of its users, and the degree of artificial intelligence endemic to it. By implementing systems, we discover what comprises ubicomp systems, what is and is not computationally tractable, form hypotheses to be tested, and uncover the research challenges that underpin and inform the evolving vision of ubicomp itself.

Based on the notion that "forewarned is forearmed," the aim of this chapter is to offer advice to those planning to create ubicomp systems to sensitize them to the issues that may face them in the design, implementation, deployment, and evaluation stages of their projects. We ground this advice both in the literature and with reference to direct experience of researchers who have created and deployed influential ubicomp systems over the past decade or so. Armed with this knowledge, it is our profound hope that you will be able to more quickly design, build, deploy, and evaluate your ubicomp system, and that you will able to communicate your findings concisely and effectively to the community, and thereby contributing to moving the science of ubicomp systems development forward.

2.2 UBICOMP SYSTEMS TOPICS AND CHALLENGES

Creating ubicomp systems entails a wide range of technical research topics and challenges. Compared to existing systems research, some of these topics and challenges are new and arise because of the intention to build ubicomp applications. For example, ubicomp applications often involve scenarios where devices, network, and software components change frequently. The challenges associated with such extremely volatile executing environments are new to systems research (Coulouris et al., 2005). These kinds of challenges are introduced because we intend to build new computing technology that is deployed and runs in completely new types of physical and computational environments. Other topics and challenges existed before ubicomp but are significantly aggravated in a ubicomp setting. For example, new challenges to security arise because trust is lowered in volatile systems; spontaneous interaction between devices often imply that they have little, if any, prior knowledge of each other, and may not have a trusted third party in common.

This section will take a closer look at some of the more significant topics and challenges to ubicomp systems research.

2.2.1 Resource-Constrained Devices

The first—and perhaps most visible—challenge to most ubicomp applications and systems is that they involve devices with limited resources. Due to Moore's law, we have been used to ever-increasing CPU speed, memory, and network bandwidth in servers and desktop computers. With ubicomp, however, a wide range of new devices are built and introduced, which are much more resource-constrained. Devices such as PDAs, mobile phones, and music players have limited CPU, memory, and network connectivity compared to a standard PC, and embedded platforms such as sensor networks and smart cards are very limited compared to a PC or even a smart phone. Hence, when creating systems support in a ubicomp setting, it is important to recognize the constraints of the target devices, and to recognize that hardware platforms are highly heterogeneous and incompatible with respect to hardware specifications, operating system, input/output capabilities, network, etc.

Resource-aware computing is an approach to develop technologies where the application is constantly notified about the consumption of vital resources, and can help the application (or the user) to take a decision based on available resources now and in the future. For example, video streaming will be adjusted to available bandwidth and battery level (Garlan et al., 2002), or the user may be asked to go to an area with better wireless local area network coverage.

Generally speaking, the most limiting factor of most ubicomp devices is energy. A device that is portable or embedded into the physical world typically runs on batteries, and the smaller and lighter the device needs to be, the lower its battery capacity. For this reason, one of the main hardware constraints to consider when building ubicomp systems and applications is power consumption and/or opportunities for energy harvesting— including recharging. A central research theme within ubicomp is *power foraging*, that is, technologies for harvesting power in the environment based on, for example, kinetic energy from a walking person. *Cyber foraging* is a similar research theme where devices look for places to offload resource-intensive tasks (Balan et al., 2002). For example, a portable device may offload computations to server-based services, or if the user tries to print a document located on a file server from a PDA, the document is not first send to the PDA and then to the printer, but instead sent directly from the file server to the printer (Kindberg et al., 2002).

Computation, accessing memory, and input/output all consume energy. The major drain on the battery is, however, wireless communication, which is also typical for mobile or embedded ubicomp devices. Power consumption in wireless communication is hence another major topic in ubicomp systems research, investigating resource-efficient networking protocols that limit power consumption due to transmitting data, while maintaining a high degree of throughput and reliability. For example, since processing consumes much less power than communication, mobile ad hoc sensor networks (MANETs) seek to do as much in-network processing as possible, that is, ensuring that nodes in a sensor network perform tasks such as aggregating or averaging values from nearby nodes, and filtering before transmitting values.

2.2.2 Volatile Execution Environments

A core research topic in ubicomp systems research is *service discovery*, that is, technologies and standards that enable devices to discover each other, set up communication links, and start using each others' services. For example, when a portable device enters a smart room, it may want to discover and use nearby resources such as public displays and printers. Several service discovery technologies have now matured and are in daily use in thousands of devices. These include Jini, UPnP, Bonjour/multicast DNS (mDNS), and the Bluetooth discovery protocol. Nevertheless, several challenges to existing approaches still exist, including the lack of support for multicast discovery beyond local area networks, the lack of support beyond one-to-one device/service pairing, and rather cumbersome methods for pairing devices, often involving typing in personal identification numbers or passwords. Research is ongoing to improve upon service discovery technologies.

Ubicomp systems and applications are often distributed; they entail interaction between different devices—mobile, embedded, or server-based—and use different networking capabilities. Looking at ubicomp from this distributed computing perspective, a fundamental challenge to ubicomp systems is their volatile nature (Coulouris et al., 2005). The set of users, devices, hardware, software, and operating systems in ubicomp systems is highly dynamic and changes frequently.

One type of volatility arises because of the spontaneous nature of many ubicomp systems; devices continuously connect and disconnect, and create and destroy communications links. But because—from a communication perspective—these devices may leave the room (or run out of battery) at any time, communication between the mobile devices and the services in the smart room needs to gracefully handle such disconnection.

Another type of volatility arises due to changes in the underlying communication structure, such as topology, bandwidth, routing, and host naming. For example, in an ad hoc sensor network, the network topology and routing scheme is often determined by nodes available at a given time, the physical proximity of the nodes in the network, their current workload, and battery status; in addition, this network routing scheme should be able to handle nodes entering and leaving the network. A simpler example arises in smart room applications where devices entering the room do not know the network name or addresses of the local services, and in this case services discovery would entail obtaining some network route to the service.

Volatility arguably also exists in more traditional distributed systems; client software running on laptops are being disconnected from their servers in a client-sever setup, and PDAs and cell phones whose battery is flat are able to reconnect once recharged. The main difference, however, is that unlike most traditional distributed systems, the connectivity changes are common rather than exceptional, and often of a more basic nature. For example, in a client-server setup the server remains stable, and both the client and server maintain their network name and address. For these reasons, existing distributed computing mechanisms such as the Web (HTTP), remote procedure calls, and remote method invocation (Java RMI, .NET Remoting, or CORBA) all rely on stable network connections (sockets) and fixed network naming schemes. In a ubicomp environment, these assumptions break down.

2.2.3 Heterogeneous Execution Environments

Most ubicomp applications and systems inherently live in a heterogeneous environment. Ubicomp applications often involve a wide range of hardware, network technology, operating systems, input/output capabilities, resources, sensors, etc., and in contrast to the traditional use of the term *application*, which typically refers to software that resides on one—at most, two—physical nodes, a ubicomp application typically spans several devices, which need to interact closely and in concert in order to make up the application. For instance, the Smart Room is an application that relies on several devices, services, communication links, software components, and end user application, which needs to work in complete concert to fulfill the overall functionality of a smart room. Hence, handling heterogeneity is not only a matter of being able to compile, build, and deploy an application on different target platforms—such as building a desktop application to run on different versions of Windows, Mac OS, and Linux. It is to a much larger degree a matter of continuously— that is, at *runtime*—being able to handle heterogeneous execution environments, and that different parts of the ubicomp application run on devices with highly varying specifications. For example, when a user enters the smart room and wants to access the public display and print a document, this may involve a wide range of heterogeneous devices, each with their specific hardware, operating systems, networks interfaces, etc.; the user may be carrying a smart phone running Symbian; he may be detected by a location tracking system based on infrared sensors on a Berkley Mote running the TinyOS; his laptop may use mDNS for device discovery, whereas the public display may be running Linux using the X protocol for sharing its display with nearby devices.

The challenge of heterogeneity partly arises because ubicomp is a new research field and a new standard technology stack including hardware, operating system, etc., has yet to mature. In the above scenario, one could argue that the patchwork of technologies involved is overly complex and unnecessary. This is partly true, and existing or new technology platforms may gradually be able to handle the ubicomp requirements in a more homogeneous and consistent manner. On the other hand, ubicomp applications will always need to use different kinds of technologies ranging from small, embedded sensors, to large public display and mobile handheld devices. As such, heterogeneous hardware devices are a fundamental part of ubicomp applications, and the corresponding operating systems and software stacks need to be specifically optimized to this hardware; the small sensor nodes need a software stack optimized for their limited resources and the large display similarly needs a software stack suited for sophisticated graphics and advanced input technologies.

Therefore, a core systems topic to ubicomp is to create base technologies that are able to handle such heterogeneity by balancing the need for optimizing for special-purpose hardware while trying to encapsulate some of the complexities in common standards and technologies.

2.2.4 Fluctuating Usage Environments

The challenges discussed above are all concerned with issues relating to the execution environment of ubicomp applications. However, there is also a set of challenges that, to a larger degree, are associated with the nature of ubicomp applications and how they are designed to be used.

Contemporary computing is primarily targeted at information management in the broadest sense. Users use PCs for information management locally or on servers; they engage in a one-to-one relationship with the PC; the physical use context is fairly stable and is often tied to a horizontal surface such as an office desk or the dining table at home; the number and complexity of peripherals are limited and include well-known devices such as printers, external hard drives, cameras, and servers.

Compared to this usage model, ubicomp applications and hence systems live in a far more complicated and fluctuating usage environment. Users have not one but several personal devices, such as laptops, mobile phones, watches, etc. The same device may be used by several users, such as the public display in the smart room or a smart blood pressure monitor in the patient ward of a hospital. Ubicomp systems need to support this many-to-many configuration between users and devices. Furthermore, compared to the desktop, the physical work setting in ubicomp exhibits a larger degree of alternation between many different places and physical surroundings. Mobile devices mean that work can be carried around and done in different places, and computers embedded into, for example, furniture that is constantly used by different people. Finally, doing a task is no longer tied to one device such as the PC, but is now distributed across several heterogeneous devices as explained above. This means that users need technology that helps them stay focused on a task without having to deal with the complexity of setting up devices, pairing them, moving data, ensuring connectivity, etc.

A core research challenge to ubicomp is to create systems, technologies, and platforms allowing the creation of applications that are able to handle such fluctuating usage environments. Special focus has, so far, been targeted at handling three types of fluctuation in usage environment: (1) changing location of users, (2) changing context of the computer, and (3) multiple activities (or tasks) of the users.

Fluctuations related to different location of the users arise once mobile devices are introduced. *Location-based computing* aims to create systems and applications that provide relevant information and services based on knowledge about the location of the user. For example, in the GUIDE (Cheverst et al., 2000) project, tourists were guided around historic sites by a location-aware tour guide, which automatically would present relevant descriptions based on the tourist's location. Central to location-based systems research is the challenge of sensing the location of the user or device (often treated synonymously although studies have shown that this can be a false assumption). A wide range of technologies already exist, such as global positioning systems (GPS). But because they all have their advantages and disadvantages, new location technologies are still emerging. Hightower and Borriello (2001) provide an older, but still relevant, overview of available location technologies and their underlying sensing techniques. The chapter on location technologies in this book provides further background on this topic.

Context-aware computing aims at adapting the application or the computer in accordance with its changing context. "Context of the application" includes information about who is using the computer; who else is nearby; ambient information about the room, including light, sound, and temperature; physical materials and tools used; and other devices in a room. For example, a context-aware hospital bed having embedded computers, displays, and sensors can be built to react and adapt to what is happening in its proximity; it may recognize the patient in the bed; it may recognize clinical personnel approaching; it may bring up relevant medical information on the display for the clinicians; and it may issue a warning if the nurse is mistakenly trying to give the patient another patient's medication (Bardram, 2004). The context of a computer/device may change for two reasons: either the device moved to a new context (mobile device) or the physical context of an embedded computer changed because, for example, new people and devices entered a room. A core research challenge to ubicomp systems research is to investigate proper technical architectures, designs, and mechanisms for context sensing, modeling, aggregation, filtering, inferring, and reasoning; for context adaptation; and for distribution and notification of context events. The chapter on context-aware computing in this book provides more background and references on context-aware computing.

Activity-based computing (ABC) (Bardram and Christensen, 2007) aims at handling fluctuations based on users' need for handling many concurrent and collaborative activities or tasks. For example, in a hospital each clinician (doctor or nurse) is engaged in the treatment and care of several patients, each of whom may have a significant amount of clinical data associated. For the clinician, it is associated with a substantial mental and practical overhead to switch between different patients, because it involves using several devices, displays, and medical software applications. The ABC approach helps users manage the complexity of performing multiple activities in a complex, volatile, heterogeneous ubicomp systems setup involving numerous devices in different locations. Hence, focus is on systems support for aggregating resources and services that are relevant to an activity; supporting those activities, and their associated resources, moving seamlessly between multiple devices; supporting multiple users working together on the same activity—potentially using different devices; supporting intelligent and semiautomatic generation and adaptation of activities according to changes in the work environment; and supporting the orchestration of multiple services, devices, and network setup to work optimally according to the users' changing activities.

2.2.5 Invisible Computing

Invisible computing is central to the vision and usage scenarios of ubicomp, but handling and/or achieving invisibility is also a core challenge—for example, having embedded sensor technology that monitors human behavior at home and provides intelligent control of heating, ventilation, air conditioning, and cooling (HVAC), or pervasive computing systems in hospitals that automatically ensure that patient monitoring equipment is matched with correct patient ID, and that sensor data are routed to the correct medical record. In many of these cases, the computers are invisible to the users in a double sense. First, the computers are embedded into buildings, furniture, medical devices, etc., and are as such physically invisible to the human eye. Second, the computers operate in the periphery of the users' attention and are hence mentally invisible (imperceptible).

From a systems perspective, obtaining and handling invisible computing is a fundamental change from traditional computing, because traditional systems rely heavily on having the users' attention; users either use a computer (e.g., a PC or a server through a terminal or browser) or they do not use a computer. This means, for example, that the system software can rely on sending notifications and error messages to users, and expect them to react; it can ask for input in the contingency where the system needs feedback in order to decide on further actions; it can ask the user to install hardware and/or software components; and it can ask the user to restart the device. Moving toward invisible computing, these assumptions completely break down.

A wide range of systems research is addressing the challenges associated with building and running invisible computers. *Autonomic computing* (Kephart and Chess, 2003), for example, aims to develop computer systems capable of self-management, in order to overcome the rapidly growing complexity of computing systems management. Autonomic computing refers to the self-managing characteristics of distributed computing resources, adapting to unpredictable changes while hiding intrinsic complexity for the users. An autonomic system makes decisions on its own, using high-level policies; it will constantly check and optimize its status and automatically adapt itself to changing conditions. Whereas autonomic computing is, to a large degree, conceived with centralized or cluster-based server architectures in mind, *multiagent systems* research (Zambonelli et al., 2003) seeks to create software agents that work on behalf of users while migrating around in the network onto different devices. Agents are designed to ensure that lower-level systems issues are shielded from the user, thereby maintaining invisibility of the technology.

Research on *contingency management* (Bardram and Schultz, 2004) seeks to prevent users from being involved in attending to errors and failures. In contrast to traditional exception handling, which assumes that failures are exceptional, contingency management views failures as a natural contingent part of running a ubicomp system. Hence, techniques for proactive management of failures and resource limitations need to be put into place. For example, off-loading an agent before a mobile device runs out of battery, and ensure proactive download of resources before leaving network coverage.

A similar research topic is *graceful degradation*, which addresses how the system responds to changes and, in particular, failures in the environment (Friday et al., 2005). Most existing technology assume the availability of certain resources such as Internet connectivity and specific servers to be present permanently. However, in situations where these resources are not available, the entire system may stop working. Real life demands systems that can cope with the lack of resources, or better still, systems should be able to adapt gracefully to these changes, preserving as much functionality as possible using the resources that are available.

2.2.6 Security and Privacy

Security and privacy is challenging to all computing. With ubicomp, however, the security and privacy challenges are increased due to the volatile, spontaneous, heterogeneous, and invisible nature of ubicomp systems (particularly imperceptible monitoring).

First, trust—the basis for all security—is often lowered in volatile systems because the principals whose components interact spontaneously may have no a priori knowledge of each other and may not have a trusted third party. For example, a new device that enters a hospital cannot be trusted to be used for displaying or storing sensitive medical data, and making the necessary configuration may be an administrative overhead that would prevent any sort of spontaneous use. Hence, using the patient's mobile phone may be difficult to set up.

Second, conventional security protocols tend to make assumptions about devices and connectivity that often do not hold in ubicomp systems. For example, portable devices can be more easily stolen and tampered with, and resource-constrained mobile or embedded devices do not have sufficient computing resources for asymmetric public key cryptography. Moreover, security protocols cannot rely on continuous online access to a server, which makes it hard to issue and revoke certificates.

Third, the nature of ubicomp systems creates the need for a new type of security based on location and context; service authentication and authorization may be based on location and not the user. For example, people entering a cafe may be allowed to use the cafe's printer. In this case, if a device wants to use the cafe's printer, it needs to verify that this device indeed is inside the cafe. In other words, it does not matter who uses the printer, the cafe cares only about where the user is (Kindberg and Fox, 2002).

Fourth, new privacy challenges emerge in ubicomp systems (Langheinrich, 2001). By introducing sensor technology, ubicomp systems may gather extensive data on users including information on location, activity, who people are with, speech, video, and biological data. And if these systems are invisible in the environment, people may not even notice that data are being collected about them. Hence, designing appropriate privacy protection mechanisms is central to ubicomp research. A key challenge is to manage that users—wittingly or unwittingly—provide numerous identifiers to the environment while moving around and using services. These identifiers include networking IDs such as MAC, Bluetooth, and IP addresses; usernames; IDs of tags such as RFID tags; and payment IDs such as credit card numbers. Chapter 3 examines some of these challenges in more detail.

Fifth, the usage scenarios of ubicomp also set up new challenges for security. The fluctuating usage environment means that numerous devices and users continuously create new associations, and if all or some of these associations need to be secured, this means that device and user authentication happens very often. Existing user authentication mechanisms are, to a large degree, designed for few (1–2) and long-lived (hours) associations between a user and a device or service. For example, a user typically logs into a PC and uses it for the whole workday. In a ubicomp scenario, where a user may enter a smart room and use tens of devices and services in a relatively short period (minutes), traditional user authentication using, for example, usernames and passwords is simply not feasible. Moreover, if the devices are embedded or invisible, it may be difficult and awkward to authenticate yourself—should we, for example, log into our refrigerator, a shared public display, and the HVAC controller in our homes?

All in all, a wide range of fundamental challenges exists in creating security and privacy mechanisms that adequately takes into concern the technical as well as the usage challenges of ubicomp systems.

2.2.7 Summary

This section discusses many of the core challenges that are endemic to ubicomp and hence, in a supporting role, ubicomp systems research. These include coping with impoverished and resource-constrained devices, energy harvesting and usage optimization, environmental and situational volatility, heterogeneity and asymmetry of device capabilities, adapting to dynamic and fluctuating execution environments, invisibility and its implications for ubicomp systems, and privacy/security challenges. The next section will look at the process of designing systems to meet some of these challenges.

2.3 CREATING UBICOMP SYSTEMS

Building, deploying, and maintaining ubicomp systems require considerable and sustained effort. You should think carefully before you start building, about why you are building it, what you hope to learn, and what is going to happen to it in the future. Making good design decisions and being pragmatic about your objectives early on can save you enormous amounts of potentially unrewarding effort later on. Understanding why you are building a system can help you think more strategically about how to achieve the impact you desire or answer your research question more expediently. This section highlights key reasons to build ubicomp systems, best practices for developing systems, and common issues and pitfalls facing ubicomp systems developers.

2.3.1 Why Build Ubicomp Systems?

There are many reasons to build ubicomp systems. What you hope to do with the system, who the intended users are (both technically as developers and in terms of user experience), and the planned longevity should shape the design and implementation decisions for the project. Possible targets for ubicomp systems research include

- Prototyping future systems to explore ubiquity in practice
- Empirical exploration of user reactions to ubicomp
- Gathering datasets to tackle computational problems relating to ubicomp
- Creating ubicomp experiences for public engagement or performance
- Creating research test beds to agglomerate activity and stimulate further research
- To explore a hypothesis concerning ubicomp more naturalistically
- To test the limits of computational technologies in a ubicomp setting
- Addressing the perceived needs of a problem domain or pressing societal issue

2.3.2 Setting Your Objectives

How one goes about achieving these objectives effectively should naturally impact how you undertake the research. It is important to consider where you will place your engineering effort and, importantly, whether parts of the problem need to be fully implemented and indeed are reasonable computationally to achieve the outcome you desire. For example, a small-scale study to test how users react to context-aware systems would require considerable effort to achieve sufficiently reliable context determination automatically, whereas emulating the context determination using "Wizard of Oz" techniques may be adequate to gain the results and far easier to engineer. Conversely, a system that is intended to run for a long time unattended (research test bed, smart room infrastructure) will need a much stronger focus on robust design and defensive programming if it is to survive without a high degree of attention and support. It is important to realize that some parts of your system will require considerable effort to achieve, but may not in themselves lead directly to novel results. The trick is to keep one eye on your objective to ensure that your focus never entirely shifts from the goals of the project, ensuring that your efforts are rewarded.

Naturally, some flexibility is required because the goals of your system may shift over time. An initial prototypical exploration may uncover an

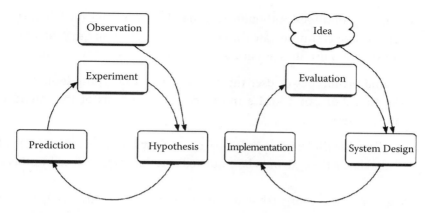

FIGURE 2.1 Compare scientific method (on the left) with the role of experimentation in ubicomp system design (right) (paraphrased from Feitelson, 2005). As one experiment may enable a hypothesis to be refined leading to further experiments cyclically, so a system design may lead to another and be iteratively refined. Importantly, developing and evaluating a system may uncover a hypothesis that can be experimentally tested, and vice versa.

interesting ubicomp problem to solve or hypothesis to test. A project whose initial focus is public engagement may uncover a rationale for wider ubicomp systems design that may lead to further exploration and more focused empirical studies. This is a natural and intended consequence of the scientific exploration of the ubicomp problem space (Figure 2.1), which has two key consequences: (1) that you remain sufficiently aware and agile to recognize such changes and plan for them consciously and (2) that the important lessons from your exploration are communicated effectively to the community. How to achieve this last point is discussed further in Section 2.5.

2.3.2.1 Testing Your Ideas

Having established the context of your system and its objectives, it is common sense not to rush into the design phase without first testing and refining your ideas. There are many possible approaches with various time and effort implications, for example, one might create the following:

- Low-fidelity prototypes, which can be simple scenarios that can be discussed, paper prototypes, or even models of devices or graphical storyboards of proposed interactions—anything that can add richness to the discussion of the system with potential users.

- Video prototypes, although considerably requiring more effort to create, can communicate the concepts in the system quite effectively and act as a useful reference for explaining the system later on.

- Rapid prototypes of user interfaces using prototyping toolkits (see Section 2.6) can afford a more realistic synthesis of the intended user experience.

- "Wizard of Oz" prototypes of parts of the system may allow the final behavior of the system to be emulated and thus experienced by others.

In general, the more labor-intensive options are only really worth investing in if the project itself is a significantly larger undertaking or there is additional value to having the prototypes or associated media. One of the cheapest and lightest weight mechanisms is simply to present the proposed system to someone else to gain informal feedback. If they find the idea entirely preposterous or can see obvious significant flaws, it is certainly worth revisiting your scenario. This works best if the person is not a member of the project team!

2.3.3 Designing "Good" Systems

Once you have decided on your system's objectives and are happy with your ideas, the next challenge is to design a system that is fit for its purpose. There are many important concerns unique to ubicomp that you should consider in your design.

2.3.3.1 Computational Knowledge of the Physical World

From a systems design perspective, it is far from clear what the interfaces and internals of a ubicomp system should be, necessitating an experimental approach. In his much-venerated article in *Scientific American* (Weiser, 1991), Weiser espoused embedded virtuality and calm computing: the notion that computational devices were effectively invisible to its users and that interfaces to such systems were through entirely natural, sensor-driven, and tactile interactions. Arguably, such systems almost empathically support the user in their daily tasks, requiring a high degree of knowledge about the user's desires and intents—in some cases, Artificial Intelligence. Only recently have we begun to see researchers challenge some of these precepts and explore other possible visions of ubicomp (e.g., Rogers, 2006). As a discipline, it is important that we continue to explore the boundary between "the system" and "the user" to find the balance points for computational tractability and effective user support.

A key challenge for ubicomp systems designers then, is to consider the "barrier" between the physical world and virtual (computational) world. Unlike conventional software applications, interfaces to ubicomp systems are often distributed, may have many forms of input and output involving several devices, and often incorporate subtle, oblique forms of interaction involving hidden or ambient sensors and displays. In their influential article, Fox and Kindberg (2002) encapsulate the divide between the responsibilities of the system and the user as the "semantic rubicon": "[that] demarcates responsibility for decision making between the system and the user." More specifically, in terms of system design, crossing the semantic rubicon implies defining the knowledge the system can have of the physical world and of user(s) behavior, that is, through sensing and user interaction; the counterpart, that is, the knowledge the user has of the system and how they might influence it; and the mechanisms and permissible interactions for one to influence the other.

As a system designer, you must decide what knowledge your system will need about the real world to function, how it will get into the system, how to represent it, how this state will be maintained, and what to do if it is incorrect. Unless this knowledge is easy to sense, or trivial to reason with, then you must also decide what the implications are if the knowledge is imperfect or conclusions are erroneously reached. There is clearly a significant difference in implication if the outcome of misconstruing the user's situation while laying still is to call the emergency services rather than dim the lighting! Designing when to involve the user with decisions, or in the context of the semantic rubicon, when the decision of the system becomes the decision of the user, may well be crucial to the acceptability of the system or its fit to its task, especially in sensitive or deployed settings.

Key questions you should ask yourself are

1. What can be reliably sensed
2. What can be reliably known?
3. What can be reliably inferred?

The degree to which you can answer these questions for the intended function of your system will help determine the feasible scope, or set some of the research challenges.

2.3.3.2 Seamfulness, Sensibility, and Tolerant Ignorance

There are clearly limits to what your system can know about the physical world and the people who inhabit it. Sensors have innate properties due to their construction and the underlying physics that governs how well they sense. They may not be optimally placed or sufficiently densely deployed to cover the area or activity that you wish to detect. High-level sensors, such as location systems, have complex behaviors governed by properties of the built environment and its associated "radio visibility"—these are time varying properties that also significantly depend on where they are used. The activity you wish to observe may simply be challenging to detect due to its subtlety, or difficult to isolate from other activities, noise, or the concurrent activities of other people. There is also the question of the reliability of what is sensed in the presence of partial or total sensor failure (e.g., erroneous sensor readings may be misinterpreted as activity).

This was once articulated as the challenge of designing systems that exhibit "tolerance for ignorance" (Friday et al., 2005). Although it is certainly challenging to consider how to build systems that continue to function well in the face of ongoing indeterminacy and uncertainty, a first step is to consider the scope and boundaries to your system. In their paper, Chalmers et al. (2003) present "seamful design," the notion that the seams or the boundaries and inaccuracies of the system can be exploited as a resource for system designers. They consider the example of a location-based mixed reality game called "Can you see me now," where runners physically on the streets attempt to catch online players virtually overlaid on the same space. The runners (Figure 2.2) were tracked using GPS, and it quickly became apparent that the seams of the system—in this case, the ability to track the runners—was having an impact on the game play: "analysis of system logs shows estimated GPS errors ranged from 4 meters to 106 meters, with a mean of 12.4 meters. Error varied according to position in the game area, with some of the more open spaces exhibiting typically only a few meters error while the more narrow built–up streets suffered considerably more" (Chalmers et al., 2003).

Over the 2 days, the runners had time to talk with each other and develop tactics, as exemplified in this quote:

> Crew: What defines a good place to catch them?
> Runner: A big open space, with good GPS coverage, where you can get quick update because then every move you make is updated when you're heading toward them; because one of the problems is,

FIGURE 2.2   A runner tracked in physical space is speeding to catch an online player whose location in the virtual world is overlaid on the physical environment. ("Can You See Me Now." Image copyright Blast Theory, 2001. Used with permission.)

> if you're running toward them and you're in a place where it slowly updates, you jump past them, and that's really frustrating. So you've got to worry about the GPS as much as catching them.
>
> (FLINTHAM ET AL., 2003)

In this case, awareness of the limitations and characteristics of the system allowed the longer-term players (the runners) to improve their ability to play (use the system). This raises the design issue of how far to go toward exposing the seams of the system. As reported by Chalmers et al. (2003), Benford is quoted as proposing four strategies for presenting information to the user:

- Pessimistic: Only show information that is known to be correct
- Optimistic: Show everything as if it were correct
- Cautious: Explicitly present uncertainty
- Opportunistic: Exploit uncertainty (cf. Gaver et al., 2003)

One might regard the pessimistic and optimistic approaches as being a "more traditional," perhaps engineering-led approaches. It is very common to present a location of a user on a map as a dot, for example, although this does not typically communicate any underlying uncertainty or imprecision in the location estimate or may not even reflect whether the system believes this to be the true location of the user (e.g., if no GPS satellites are in view, this may simply be a historic artifact). Adjusting the size or representation of such a dot to reflect the confidence in location would enable the user to develop a greater trust and understanding of the system. The cautious approach is widely adopted on a typical mobile phone: the "bars of signal strength" indicator provides an intuitive iconographic representation of underlying features of the system architecture, which is a resource for the phone user both to reason about the success of making a phone call but also a plausible social device for claiming they got cut off due to "low signal strength." Cautious or even opportunistic ubicomp designs may offer systems that are amenable to user comprehension or even appropriation.

What can and cannot be sensed or its underlying seams may even be an opportunity for design. Benford et al. (2004) present some examples of physical ubicomp interfaces in the context of what is sensible, sensable, and desirable (Boucher et al., 2003). This taxonomy helped the authors categorize uses of their devices, but also spot opportunities for other types of interaction with their devices that they had not originally foreseen.

2.3.3.3 User Mental Model and Responsibility

The corollary of considering the semantic rubicon and seamfulness of your system is to carefully plan the role the user will play in the system's operation. Ubicomp systems often differ significantly in the degree of understanding and "intelligence" they are intended to show toward the users' goals and desires. There are a spectrum of design choices as to when to involve the user in sensing or understanding the physical world and in decision making or instigating actions. For example, at one end of the spectrum, we might consider a scenario where the system fully "understands" the user's wants, and takes actions preemptively in anticipation of these (one can argue the degree and scope over which this is achievable). At the other extreme, perhaps more cautiously, we might design assuming no action is taken by the system without user assent, or where the user provides sensory input (e.g., confirming the activity they are currently engaged in, although clearly this could quickly become tiresome). A compromise position might involve partially automating to support the user's perceived needs, but offering the ability for the user to intervene to cancel or override the actions proposed by the system. A further approach might be adaptive: for example, using machine learning that starts by involving the user in decisions but learns from this, moving toward automation of common or consistently detected tasks (but, crucially, can move back to learning mode again if unreliable or undesirable!).

To help consider where on this scale parts of your system might lie, consider:

- The frequency or inconvenience of potential user involvement
- The severity or undesirability of the consequences if the system gets it wrong
- The reliability of detecting the appropriate moment and appropriate action
- The acceptability to the user of automating the behavior

As an example of how people can cooperate with ubicomp systems and supplement sensing capabilities, consider these two examples. In the GUIDE (Cheverst et al., 2000) context-aware tour guide system, city visitors could enter a dialogue with the system (involving selecting a series of photos of landmarks they could see from their position) to reorient the system when it was outside the scope of the wireless beacons used to determine location. This extended the effective range of the system without requiring the logistical and financial expense of adding additional microcells. Self-reported position was used very effectively in "Uncle Roy All Around You" (Benford et al., 2004), where players with mobile devices built up trust in unknown online players by choosing when and where to declare their position to the system to test and reaffirm the advice being offered to them. It would have clearly been possible to use GPS or cell fingerprinting to automate locating the players, but instead this became a key feature of the cooperation between online and mobile players. It is worth thinking about how the seams and possible limitations of your system can be used as a resource for design.

A key question is, "What do you intend for the user to understand or perceive of the system in operation?" To grow comfortable with it, adopt it, and potentially appropriate it, the user must be able to form a mental model of cause and effect or a plausible rationale for its behavior. In more playful or artistic ubicomp systems this question may be deliberately provocative or challenging, but this should still benefit from being a conscious and designed behavior.

2.3.3.4 It Is Always Runtime

Ubicomp systems are composed of distributed, potentially disjoint, and partially connected elements (sensors, mobile devices, people, etc.). The term "partially connected" here reflects that these elements will often not be reliably or continuously connected to each other; instead, the system is the product of spontaneous exchanges of information when elements come together. Clearly, interaction patterns and duration will vary with the design and ambition of any given system, but it is important to consider a key precept: once deployed, all changes happen at runtime. In a system of any scale, you will typically not have simultaneous access to all the elements to (for example) upgrade them or restart them. This has a number of implications:

1. Systems requiring a carefully contrived startup order are likely to fail.

2. If the availability of elements may be sporadic, your system should be able to gracefully handle disconnection and reconnection or rebinding to alternate services.

3. Assume that individual components may fail or be temporarily isolated (which is especially true of software elements on mobile devices) and design your system accordingly so that state can be recovered.

4. Decide proactively how to handle data when an element is disconnected: are the data kept (e.g., buffered) until reconnection, and if so, how much will you buffer before discarding. What strategy will you choose to decide which data to keep or discard (oldest, freshest, resample, etc.)?

5. Consider including version information in protocols used in systems designed to run longitudinally to at least identify version mismatches.

2.3.3.5 Handling Transient Connections

Network connections (or the lack or failure thereof) can have profound effects on the performance of ubicomp systems and, crucially, the end user experience. The effects on unsuspecting software throughout a device's software stack can be serious: network names stop being resolved, closed connections can lead to software exceptions that stop portions of the code from executing, input/output system calls can block leading to stuck or frozen user interfaces. Considering what will happen if elements in the system that you are assuming to be always available—especially if they are on the critical path in terms of system responsiveness—fail, will help you identify and ideally mitigate for these potential problems.

When networks fail, it is common for data to be buffered and dropped at many levels in the protocol stack. In ubicomp, where data are often sensor traces informing the system of important events relating to interactions in the world, this buffering can introduce an array of associated problems. For example, old (buffered) data can be misleading if not timestamped and handled accordingly. Consider buffered GPS traces logged on a mobile device while the connection to the backend system is down: old locations can appear like current inaccuracy; the fast replay of buffered locations that normally occur 1 per second might look like you stopped then started moving very quickly and like lag if the replay is rate-paced or there is a perceptible latency in the system. Finally, your fresh data over a multiplexed interface will be behind the buffered data—a potentially serious delay can arise if the connection speed is low and the buffer large. If fresh data are due to user interaction, then the system will appear very unresponsive until the buffer is drained. It is very common for a frustrated user to try to interact multiple times or in many ways in the face of inexplicable delays in unresponsive interfaces, thereby exacerbating the problem. This is another good reason for revealing the connection status to the user using an appropriate metaphor (Cheverst et al., 2000; Satyanarayanan, 2001).

2.3.3.6 The State of the World

Transient connections and component failures have an impact on how consistent the state of your overall system will be. It is important to design in strategies for recovering from both of these cases. Parts of your system may be replicable or sufficiently available to use well-known techniques to mask such failures and achieve some degree of fault tolerance. However, in many ubicomp systems, software components are often intimately linked to specialist or personal hardware, or may be placed in unique locations, which makes traditional techniques involving redundant replication or fail-over inappropriate. In such systems, we need an alternative. Techniques that have been reported in the literature include

- Optimistic replication of state, which allows partitioned elements to continue to function while disconnected and then reconcile the journal of changes made offline upon reconnection (e.g., the CODA mobile file system [Kistler and Satyanarayanan, 1992] allowed optimistic writes to cached files while disconnected, which were then replayed upon reconnection).

- Converging on eventually consistent state. Bayou (Terry et al., 1995) used gossip-style "anti-entropy" sessions during user encounters to propagate updated state via social networking, converging on a final state (e.g., scheduling group meetings by exchanging possible times and availabilities and iterating toward an agreed option—tentative and committed state, in this case appointments, was reflected to users in the user interface).

- Use of persistent stores or journals to allow recovery of state (locally or remotely). A central database or state repository is often used.

- Externalizing state (e.g., to a middleware platform, such as a Tuple Space), so that most components are lightweight and can recover state from the middleware (Borchers et al., 2002; Friday et al., 1999).

- Use of peer caches to replicate state for later repair. An on-demand state "repair" scheme (Floyd et al., 1997) was used in L_2imbo (Friday et al., 1999), where peer replicas detect missing state by snooping for sequence numbers and asking neighbors to repair any missing data. Recursively, the system converges.

- Epidemic propagation of state using "gossip"-style protocols (Demers et al., 1987).

Where data do not have to be communicated in real-time (i.e., noninteractive gathering or logging of data), they can, of course, be batched and exchanged according to some schedule or when the opportunity arises. It is often surprising how quickly persistent data or event debug logs and application output can grow to fill a (particularly embedded) device. For long-running systems, ensuring capacity by estimating growth based on the running system and considering housekeeping will help avoid unexpected problems when the device is full later on. Full disks lead to numerous problems with database integrity, virtual memory management, and consequent and typically unexpected system call failures.

2.3.3.7 Is It Working?

Debugging ubicomp systems is extremely challenging. Elements are often distributed and may not be available or remotely accessible for debugging. In many cases, embedded elements may not have much, if any, user interface. A common requirement is to monitor the system's output to check status messages or to be able to perform tests by injecting commands to emulate interactions and test components. Common strategies include

- Use of conventional mechanisms such as log files and network packet tracing to passively monitor running components

- Including status protocol messages that can be intercepted (often as periodic heartbeat messages)

- Adding status displays including use of hardware such as LED blink sequences, audible and visual feedback

- Including diagnostic interfaces such as embedded web servers that can be interrogated

- Enabling remote access to components such as remote shells, etc.

- Externalizing of state or communications by using a middleware such as a publish-subscribe event channel, Tuple Space, Message Oriented Middleware, etc.

For example, iROS (Ponnekanti et al., 2003) used the "EventHeap" a derivation of the Tuple Space to pass all communications between elements of the Stanford Interactive Workspaces project (Figure 2.3). All communication is observable, so liveness of a component is easy to establish. By injecting events manually, components could be tested. New applications and devices can be introduced that work with existing components by generating or using compatible events. Later, the behavior of the workspace could be changed at runtime by dynamically rewriting events to "replumb" the smart space. The design based around a central EventHeap contributed to the longevity and adaptability of the project, enabling a number of interesting extensions and projects to be built upon the system over time. Naturally, the need to communicate via the central entity meant that the performance of the system was bounded by the performance over

FIGURE 2.3 Stanford Interactive Workspaces Project (iWork).

the network to the EventHeap and load on this component, and dependent on its availability and robustness.

2.3.4 Summary

Designing good systems, by some metric such as elegance, robustness, extensibility, usability, or fitness for purpose, is extremely challenging and requires thoughtful design. This section stressed the importance of first setting and being cognizant of your objectives, but also early testing of your ideas. Our second, but not secondary, focus is on important boundaries and thresholds between the system, its environment, and its users; encouraging purposeful and intentional designs with respect to system knowledge of the world, accuracy and dependability of sensing, tolerance to ambiguity, and the role of the users and their interplay with the system. Finally, the section discusses important technical differences between ubicomp systems and many conventional system designs: volatility, transience of connectivity, handling of state and techniques for evolving, and debugging live ubicomp systems. The next section turns to the important business of implementation.

2.4 IMPLEMENTING UBICOMP SYSTEMS

2.4.1 Choosing "Off-the-Shelf" Components

As with any computer-based system, the design of your ubicomp system is just the first step in realizing it. The design is often refined as implementation choices are made and their limits tested. Given the richness and ambition of typical ubicomp systems and the typical development resources and timescales, pragmatic choices have to be made as to define what you will build and what you will appropriate to construct your system. It is natural to seek third party components from hardware and software vendors or, increasingly, from the public domain.

A key challenge is balancing this expedient use of off-the-shelf hardware and software against more bespoke solutions. Although the latter may offer a better fit to the problem domain or intended deployment environment than an off-the-shelf solution, it requires enormous effort to develop new technologies that meet the functional, aesthetic, reliability, or time constraints of the project. Again, this is a choice best made in the context of your objectives (which may aim to explore the creation of novel devices).

Building using proven components or implementations of standards may increase robustness or extend the range of functionality available to you more quickly, but there are also limitations to this approach that you should keep in mind when evaluating your choice:

1. You should not underestimate how much time can be spent in attempting to integrate disparate pieces of hardware and software.

2. Software perhaps successful or designed for one domain will not necessarily confer similar benefits to your domain.

3. The chosen software or hardware may place constraints on what you can build, or offer far more functionality than you require (with implications on software complexity and footprint).

4. Using proprietary hardware or software may imply the need to work around features and limitations that are outside of your control.

5. Versatile toolkits, for all their tempting power and flexibility, may introduce unneeded functionality and unwanted software bloat (particularly problematic when working with embedded and mobile devices).

Ask yourself critically whether the flexibility is really needed and compare to other strategies, such as just taking parts of the toolkit in question or simply coding the portion you need.

A hidden side effect of using third party libraries is that they may introduce dependencies that are not easy to understand or are too tightly integrated into the tool you have chosen to be removed or replaced. In longer-lived systems, the interdependencies between different versions of libraries and the level of skill and tacit knowledge required to update the system can become a particular burden. Do not forget that unknown systems may contain bugs or security vulnerabilities, or exhibit unwanted behavior. Because these are components that you do not necessarily fully understand, they may be difficult to detect and may take time to fix—a good justification for building with components you can get the source code for!

It is important not to "let the tail wag the dog," that is, to consider carefully whether the limitations or implications of accepting a constraint or technology are worth the compromise to your overall design. Recall the seamful design and role of the user design considerations discussed previously, and review carefully whether the perceived limitations can be embraced or taken advantage of in some way.

GAIA: Building on a Solid Foundation

In the GAIA, a meta-operating system for smart rooms (Figure 2.4) (Roman and Campbell, 2000), an industry quality CORBA middleware implementation was chosen as the core for the system. System components were implemented as distributed objects with CORBA IDL interfaces. This implementation choice enabled the project to build on a reliable core and focus on developing the higher-level GAIA OS services. As new services for CORBA matured (event channels, Lua scripting), these features could be exploited to enrich

FIGURE 2.4 GAIA meta-operating system integrated a wide range of situated and mobile devices to offer an interactive smart room.

the GAIA OS. This implementation choice enabled the project team to focus their development effort on higher-level services such as security and configuration management without worrying about object distribution and lifecycle management. An undesirable side effect of the choice was the proprietary dependency it introduced, potentially limiting uptake by other sites that might otherwise have wished to adopt GAIA but did not want to accept the licensing implications. There was also limited scope for optimizing the interconnection and performance of the many objects and communication channels underpinning each GAIA smart room application.

Cooltown: The Power of a Well-Chosen Paradigm

HP's Cooltown was a system to support nomadic computing by associating digital information and functionality with "people, places and things" (Kindberg et al., 2002). An extremely versatile system, allowing both access to information and access to services, Cooltown was based on straightforward and elegant technical choice: that people and artifacts could be tagged, and the tag resolved to a uniform resource identifier (URI) that linked to a web point of presence for the person or artifact in question. The flexible use of tags, decentralized resolvers, and the innate flexibility in the design of URIs made the entire system extremely lightweight and extensible. Infrared beacons (Figure 2.5) broadcast URIs to mobile devices to link physical artifacts to digital information.

In summary, paraphrasing Ockham's razor: the implementation choice that makes the fewest assumptions and introduces the fewest dependencies without making a difference to the observable behavior of the system is usually the best.

2.4.2 Deploying Ubicomp Systems

One of the most valuable lessons to take from looking at successful ubicomp systems is the need to mature the system through actual use. Colloquially, by "eating your own dog food," or rather, deploying and using the system initially yourself (but ideally also with other users, who are not necessarily the developers) can gain early feedback and highlight usability and interaction issues that may otherwise get missed until such decisions are too well entrenched to be easily reversed. This lends itself to an agile development process where simple prototypes are put out early and refined during the development cycle.

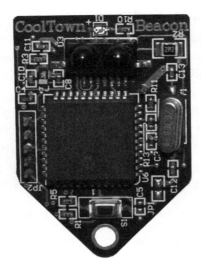

FIGURE 2.5 Cooltown beacon that enabled the physical environment to be augmented with digital information (top edge, just 3.3 centimeters wide).

With many systems, developers are also just another class of user; running training sessions with developers and having to explain the system and its application programming interfaces (APIs) to others can be a valuable source of insights. If your aim is to encourage adoption by others, and you plan to put the software in the public domain, then doing a "clean room install" helps "quantify the magic" and tacit knowledge that the systems' own developers are able to apply when using and installing the system. Documenting this type of information (e.g., as installation and maintenance guides) in a wiki associated with the software can help smooth adoption and also provide a resource for continuity if there are changes of personnel in the project team over the longer term.

Deploying systems for people to use is always a costly process. Designing a system that meets peoples' expectations, and indeed, helping set those expectations requires great care and expertise. The key is, of course, identifying the stakeholders and involving them in discussions from an early stage. How to design with users, known as participatory design, is a major topic for discussion in its own right; so we direct the interested reader to such texts as that published by Schuler and Namioka (1993). See also Chapter 6 for further discussion on participatory design.

Ubiquitously deploying technologies inevitably implies that, at some point, technologies must move out of the research laboratory and into the "real world." Experience has shown that with this comes a number of real-world constraints and practical concerns that may be unexpected and are certainly worth being highlighted (Fox et al., 2006; see Table 2.1).

There are many issues due to the real world and organizational settings that can catch the unwary developer by surprise (Hansen et al., 2006; Storz et al., 2006). For example,

1. The need to comply with health and safety or disabilities legislation, which can constrain the citing of equipment and place certain usability requirements for disabled users (for guidance on how to design inclusively for all users and design assistive technologies for those with disabilities in particular, the reader is referred to Clarkson et al., 2003).

2. To be sensitive to data protection legislation, which may impact what data you can store, whether users have the right to opt-in, opt-out, or declare (e.g., with notices) that the system is in operation. Public deployments are by their very nature public, so you should prepare to be accountable for your system and prepare yourself, your team, and your work for public scrutiny.

3. Environmental factors (including weather, pollution, etc.) can have a devastating effect on equipment that is not adequately protected. It is worth doing test installs before your main deployment to uncover unexpected issues due to environmental factors (particularly important for external and outdoor deployments).

4. Privacy and organizational sensitivity. The nature of putting technologies into real-world situations can potentially open vulnerabilities (perceived or actual) to expose private information or interfere with existing systems or processes. This is particularly true for organizations managing sensitive data or in high-pressure situations, such as healthcare and emergency services. It is always worth approaching such situations responsibly and involving and addressing the concerns of local experts.

With any system, there is an ongoing cost in supporting the system that is proportional to the length of the deployment. Robust engineering and clever design can help mitigate this cost, but it is a research challenge in itself to drive this to zero and make the system self-maintaining. To keep down the impact of remote maintenance and support, you should

TABLE 2.1 Self-Check Questions to Consider before Undertaking a Real-World Deployment

Category	Issues	Questions
Hardware	Cost, Security, Environment, Power, Network, Space, Safety issues	What will implementation cost? Will scaling up the system affect the price? Is special equipment needed? Is the equipment secure? Is there a risk of theft? Does the environment pose special requirements on the equipment? Is the system going to be used outdoors? Can it handle vandalism? Can it withstand being dropped or cleaned? Does the system require a power plug? How long can it run without being recharged? Do the batteries run flat if radio communication is used excessively? How do you recharge the system? How will the device communicate? Does it require an Ethernet connection? Is the wireless infrastructure in place? Do you need to transmit data in an external network? How much physical space does the system use? Is there space on the wall for large wall displays? Is there table space for another computer? Do the doctors have enough room in their pockets for another device? Is there space on the dashboard for another display? Will a system malfunction affect safety issues? What is the contingency plan in case of a full system crash? Will the system interfere with other systems? Can the system pose a threat to the user?
Software	Deployment and updates, Debugging, Security, Integration, Performance and scalability, Fault tolerance, Heterogeneity	How is the software transferred to the device? Does the deployment mechanism scale to a large number of devices? Can you update the system? How do you update the different devices? Are the devices accessible after deployment? If the system malfunctions, how do you find the error? Does the system store debugging information? How do you detect serious errors in the system? How is logging done? Does the system need to be secure? How does it keep information confidential and secure? Is there a concrete security risk? Is the deployed system stand-alone? Does it need to communicate with other deployed systems or integrate with third party systems? Is there a public API and converters for communicating between systems? How does the system perform? Is system performance acceptable in the real-world setting? How many devices are needed for deployment? Does the system scale? What happens when an error occurs? Can the system recover automatically? Can the daily system users bring the system back up to a running state? Is the developer team notified about errors? Is the system configured for remote support? Does the system run in a heterogeneous environment? Do heterogeneous elements need to communicate?

User setting	Usability, Learning, Politics, Privacy, Adaptation, Trust, Support	Will end users use the system? If so, how many? Can the average user use the system? Does the interface pose problems? Does the system's overall usability match the average user? How do the users learn to use the system? Is it individual instruction or group lessons? Does the system need superusers? Is a manual or help function needed? How does the user get support? Who controls the system? Does the system change the power balance in the user setting? Who benefits from the system? Is the person that benefits from the system the same as the person that provides data to the system? Does the system require extra work from users? Does the system reveal private information? What kind of personal information does the system distribute and to whom? Is the organization ready for the system? Is there organizational resistance? Will the system change formal or informal structures in the organization? Does the user trust the system? Is the information given to users reliable? Who sends the information? Will the developers support the system? Does the support organization have remote access to the system?

Source: Hansen et al. (2006), *IEEE Pervasive Computing*, 5(3): 24–31. With permission.

ensure that it is possible to remotely monitor it, ideally as the user perceives it (remote cameras and microphones can be extremely valuable, but are unpopular in many deployment settings). Remote access via the network is also important for resolving problems, especially if the system is inaccessible or far removed from the project team: it is easy to assume that the system will need less ongoing maintenance than perhaps it does, in fact, require, especially in the early phases of the project. If the system is physically inaccessible, then this is likely to cause problems going forward. Particularly in unsupervised deployments, there is always a chance of unexpected or accidental intervention. Equipment that is installed and left in working order can sometimes find itself unplugged unexpectedly (e.g., by cleaners looking for a power socket or due to a power outage). You should realize that you cannot mitigate against all eventualities, but if your system requires complex manual setup or cannot be diagnosed and maintained remotely, then you are asking for trouble!

Runtime Orchestration of the Ambient Wood

Ambient wood was an augmented "ubicomp woodland" (Figure 2.6) designed to promote learning about woodland environments (Rogers et al., 2004). Mobile sensor devices allowed children to

FIGURE 2.6 Children using a situated "periscope" to overlay augmented reality information onto the woodland.

collect geo-tagged light and moisture readings; installed information appliances allowed information concerning tagged objects to be explored. The system also used a mesh of wirelessly connected devices installed in the wood to generate ambient sounds based on sensed contextual triggers. Technically, all devices synchronized their data to a shared dataspace ("Equip" middleware), allowing interactions in the woodland to later be visualized during supervised teaching sessions. This system was not designed for unsupervised operation or for longitudinal deployment. To maintain the quality of the experience for end users, a degree of orchestration was required to address any problems that arose during each teaching session in the wood. To help make orchestration easier (e.g., to introduce a new object representing a sound or piece of information, inspect readings being sent from devices, etc.), the developers integrated a multiuser dungeon (MUD) into the system to provide another interface onto the Equip data. Each area of the augmented woodland was represented in the MUD as a "virtual room." The team could easily use this interface and MUD metaphors to interactively "walk around" the representation of the experience and to remotely control it by inspecting, picking up, and dropping virtual objects. Orchestration of the configuration of hardware and software beyond the game was still largely a laborious manual process.

2.4.2.1 Expect the Unexpected
In all deployments, the unexpected is the hardest thing to prepare for. Volatility is unfortunately endemic to the real world and hence to your ubicomp system (Coulouris et al., 2005). As a thought experiment and ideally during predeployment testing, consider how your system will react to

- Presence or use by unknown users
- Unrecognized devices (e.g., new phones and laptops)
- Changes to the wireless environment (new wireless networks)
- Devices being power cycled
- Batteries failing (particularly hard to tell on embedded devices with no moving parts or LEDs!)

- You not being there (it is easy to forget that the developer is present during development and testing, potentially impacting the sensing, wireless connectivity, etc.)
- Improper use (developers can quickly learn which interactions "break" the system and almost subconsciously adapt to avoid exercising these paths)

Anticipating these types of conditions, testing for them, and ideally having a strategy to deal with them will serve you well, particularly for longer-term deployments. Even something as simple as logging unexpected conditions to a persistent store can help with the posthoc diagnosis of "mysterious" system misbehavior.

2.4.3 Summary

Constructing ubicomp systems that not only meet the objectives and ambition of your project, but that are also sufficiently robust to be deployed for evaluation purposes, and in extremis, long-term use is very challenging. Off-the-shelf hardware and software can be an expedient means of building ambitious systems more quickly, but do not come without strings attached: careful, qualified, and dispassionate evaluation of the choices and implications of those choices is called for. The end game as we strive for ubicomp "for real," has to be moving toward daily and widespread use of ubicomp systems. Deployments, however, are not to be undertaken lightly; diligent preparation based on, for example, the above anticipated issues presented from past deployment-led projects will help you avoid many of the more common pitfalls. The next section concentrates on how to evaluate and learn from your built system.

2.5 EVALUATING AND DOCUMENTING UBICOMP SYSTEMS

In this section, we shall look more closely into how ubicomp systems can be evaluated and how the insight from your research can be documented and communicated.

2.5.1 Evaluating Ubicomp Systems

Evaluation of ubicomp systems and/or their smaller subcomponents needs to be carefully designed from the outset of a research project; different types of research contributions often need to be evaluated differently. For example, routing protocols and their applicability under given circumstances can often be evaluated using network simulation tools, whereas systems support for smart room technologies would often involve real-world testing with end users. It is particularly important to ensure a tight coupling between the *claims* you make about your system and the evaluation methods that you use to demonstrate that these claims hold. If, for example, you claim that your network protocol scales to many nodes, a simulation is a reasonable evaluation strategy; but if you claim that the protocol supports biologists to easily pair devices in a deployment situation, this claim needs to be evaluated with biologists using the nodes (and their protocol) in a deployment field study.

Now, the observations above may seem trivial and obvious. Unfortunately, however, our experience is that it is exactly the discrepancy between claims made by researchers and their evaluation strategy, approach, and methods that often leads to criticisms of the designed systems. Generally speaking, there are a number of approaches to evaluating ubicomp systems with varying degrees of ambition and required effort. A few important ones are introduced below.

2.5.1.1 Simulation
The design of a system can be modeled and subsequently simulated. For example, simulation is the research tool of choice for a majority of the MANET community. A survey of MANET research published in the premiere conference for the MANET community shows that 75% of all papers published used simulation to evaluate their research, and that the most widely used simulator is the Network Simulator (NS-2) (Kurkowski et al., 2005).

Once a system or a systems feature has been implemented, simulations can also be used to evaluate properties of the implementation. Simulations are typically used to evaluate nonfunctional systems qualities such as scalability, performance, and resource consumption. For example, the systems qualities of a ubicomp infrastructure for a smart room may be simulated by deploying it in a test setup where a number of test scripts are simulating the use of the infrastructure according to a set of evaluation scenarios. While running the test scripts, the technical behavior of the infrastructure is gauged with respect to responsiveness, load balancing, resource utility, and fault tolerance. Such a technical simulation of a complicated piece of system infrastructure is extremely valuable in systems research, and helps discover and analyze various technical issues to be mitigated in further research. It is, however, important to recognize that this type of technical simulation says absolutely nothing about the infrastructure's functional ability to, for example, support the creation of smart rooms, or about the usefulness and usability of the application built on top of it. To verify claims about the usefulness and usability of a system, you would need to make user-oriented evaluations.

2.5.1.2 Proof-of-Concept

Just as Marc Weiser coined the concept *ubiquitous computing*, he also described how these technologies were designed at Xerox Palo Alto Research Center (PARC) by building and experimenting with so-called *proof-of-concepts*. A proof-of-concept (PoC) was defined as

> The construction of working prototypes of the necessary infrastructure in sufficient quality to debug the viability of the system in daily use; ourselves and a few colleagues serving as guinea pigs.
>
> (WEISER, 1993)

A PoC is a rudimentary and/or incomplete realization of a certain technical concept or design to prove that it can actually be realized and built, while also to some degree demonstrating its feasibility in a real implementation. A PoC is not a theoretical (mathematical) proof of anything; it is merely a proof that the technical idea can actually be designed, implemented, and run. In analogy, even though Jules Verne introduced the concept of traveling to the moon in his famous 1865 novel, *From the Earth to the Moon*, the actual PoC was not designed, built, and run until a century later.

Creating PoCs is the most prevalent evaluation strategy in ubicomp systems research. The original work on the pad, tab, and wall sized ubicomp devices and their infrastructure at Xerox PARC is the classic example of this. But a wide range of other PoC examples exists, including ABC infrastructures for hospitals (Bardram and Christensen, 2007), different PoC for tour guiding systems such as GUIDE (Cheverst et al., 2000) and the San Francisco Museum Guide system (Fleck et al., 2002), home-based ubicomp systems such as EasyLiving (Brumitt et al., 2000), and ubicomp systems for smart rooms such as Gaia (Roman and Campbell, 2000), iRoom (Borchers et al., 2002), and iLand (Streitz et al., 1999).

However, looking at it from a scientific point of view, a PoC is a somewhat weak evaluation strategy. A PoC basically shows only that the technical concept or idea can be implemented and realized. Actually, however, a PoC tells us very little about how well this technical solution meets the overall goals and motivation of the research. For example, even if several PoCs of an ABC infrastructure have been implemented, this actually only tells us that it is possible to build and run a technical implementation of the underlying concepts and ideas. A PoC, however, does not tell us anything about whether it actually meets any of the functional and/or technical goals. For example, does the ABC framework support the highly mobile and collaborative work inside hospitals? Moreover, the PoC does not tell us anything about the nonfunctional aspects of the infrastructure: Does it scale to a whole hospital? Is the response time adequate for the life- and time-critical work in a hospital? Is it extensible in a manner that would allow clinical applications to be built and deployed on top of it? All of these questions can only be answered if the PoC is put under more rigorous evaluation.

2.5.1.3 Implementing and Evaluating Applications
A stronger evaluation approach is to build end user applications using ubicomp systems component and infrastructures, and then put these applications into subsequent evaluation. For example, the Context Toolkit was used to build several applications such as the In/Out Board and the DUMMBO Meeting Board. These applications can then be evaluated by end users in either a simulated environment or in a real-world deployment. For example, the ABC framework was used to implement a series of clinical applications, which was subsequently evaluated in a test setup where a hospital was simulated (Bardram and Christensen, 2007).

This evaluation approach is strong in several respects. First, using underlying systems technologies such as components, toolkits, or middleware infrastructures to build real applications, demonstrates that the systems components are indeed useful for building systems. Second, the act of building these applications helps the systems researcher to judge whether their building blocks actually help the application developer meet his or her application goals. For example, how easy is it to model, capture, and distribute context information using the Context Toolkit? Third, once the application is built and put into use, this provides a test bed for the underlying systems components and helps you answer more nonfunctional questions, such as: How well does the system scale, perform, and handle errors? For example, the implementation of clinical applications on top of the Java Context-Awareness Framework (JCAF)

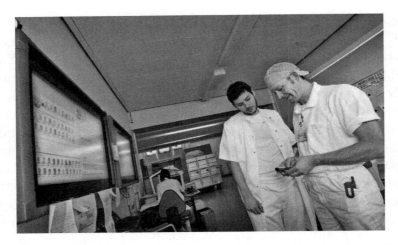

FIGURE 2.7 Context-aware technology deployed inside a hospital. (From Hansen et al., *IEEE Pervasive Computing*, 5(3), 24–31, 2006. With permission.)

framework and the subsequent evaluation sessions helped the creators inspect how well the context-aware technology scaled to multiple concurrent users, and what happened when clients lost network connectivity. The infrastructure and the application were put into pilot use in a hospital and evaluated over a 6-month period (Hansen et al., 2006). Figure 2.7 shows the use of context-aware public displays and smart phone in use.

However, as pointed out by Edwards et al. (2003), this evaluation strategy has its pitfalls and drawbacks. Essentially, if not carefully designed, the application and the subsequent evaluation may tell us little, if anything, about the systems aspect of the whole application. It is important to be absolutely clear about what your test application will tell you about your systems components, infrastructure, or toolkit. It is easy to get distracted by the demands of building useful and usable application in themselves and lose sight of the real purpose of the exercise, which may purely be to understand the pros and cons of the systems part. Moreover, whether or not the evaluation of an application turns out to be extremely successful may have very little to do with the systems properties of the application. For example, an application may fail simply because of poor usability, or because it is so novel that users have a hard time actually using it. This failure, on the other hand, may say very little about the usefulness of the underlying systems support.

2.5.1.4 Releasing and Maintaining Ubicomp Systems

The strongest evaluation of ubicomp systems components is to release them for third party use, for example, as open source. In this manner, the system research is used and evaluated by other than its original designers, and the degree to which the systems components helps the application programmers to achieve their goals directly reflects the qualities of the system components. One may even argue that there is a direct correlation between the number of application developers and researchers using the system in their work, and the value and merits of the work.

Releasing and maintaining systems software does, however, require a substantial and continuing effort. Releasing systems building blocks such as hardware platforms, operating systems, toolkits, infrastructures, middleware, and programming APIs entail a number of things such as a stable and well-tested code base, technical, and API documentation; tutorials helping programmer to get started; example code and applications; and setting up licensing policies. And once the system has been released for third party use, issues of bug reporting and fixing, support, general maintenance, and new system releases need to be considered.

This is a real dilemma in systems research and evaluation. On the one hand, the best way to evaluate your systems research is to implement the idea in sufficient quality for the rest of the world to use it, and then continuously document, support, maintain, and evolve the technology. The degree to which the world adopts your technology is a direct indicator of its usefulness. On the other hand, this limits the amount of systems research that you can do within your career to a few contributions, and there seems to be an internal and external pressure for continuously moving on to new systems research challenges. But—without a doubt—designing, implementing, documenting, releasing, and evolving systems contributions for third party use is the golden bar in systems research and is a goal pursued and reached by many researchers. Section 2.6 provides pointers on released ubicomp systems research, which can be used in further research.

2.5.2 Learning from What You Build

All ubicomp systems are complex and time consuming to design, implement, and deploy. It is easy to expend all efforts of the project on creating, deploying, and evaluating the system, while neglecting to dedicate sufficient resources to communicating your findings and experiences (both positive and negative) to others. As attributed to Plutarch between AD 46 and 120, "Research is the act of going up alleys to see if they are blind." If we do not communicate, then others will be doomed to repeat our mistakes and not learn from our innovations. It is important not to waste your efforts by sharing software, datasets, and knowledge for others to build on (or even contribute to).

2.5.2.1 Communicating Your Findings

There are many ways to communicate with the community at large, and comparatively recent innovations such as open-source software projects and contributory resources such as "wikis" make it easier to put work online and marshal interested parties around initiatives. Still, it takes work to engage with a community and provide the resources they will need to be able to work with and/or contribute to your system. For this purpose, there are several approaches to use:

1. Making your system available enables others to try, critique, compare, adopt, and potentially contribute to your project (e.g., the iRos interactive workspaces software (Borchers et al., 2006) and equip rapid prototyping toolkit (Greenhalgh and Egglestone, 2005) are both available in source form). If you do not just put materials online, but try to make the initiative open source, then it's important that you remain responsive and keep the information up-to-date, at least while the project is in active development.

2. Publishing datasets is another effective means of providing resources for the community to build on and also invites the scientific practice of experimental validation through repeatability and comparative analysis of approaches (the CRAWDAD wireless traces (Kotz and Henderson, 2009) and Massachusetts Institute of Technology (MIT) PlaceLab Datasets (Intille et al., 2006) are good exemplars of this approach).

3. Publishing (e.g., online) schematics, instructions, and documentation also provides critical insight into how to reconstruct experiments and follow on from your work [e.g., Multitouch table (Schmidt, 2009), Smart-Its (Smart-ITs, 2001)].

4. Traditional academic routes of dissemination (papers, magazine articles, demonstrations, workshop participation, etc.), which provide a means to obtain valuable peer feedback on your work.

As with any packaging of the work, be it open source or commercialization activities, any such effort should be undertaken advisedly. It takes effort to seed these initiatives, for example, creating documentation, putting up example code, instructions and tutorials, making public versions, choosing appropriate licenses, etc. However, one has to question whether it is valid to undertake the research without considering and budgeting for evaluating it and communicating your findings.

2.5.2.2 Rigor and Scientific Communication
Ubicomp systems are always difficult to describe due to their complexity and wide-ranging lessons that one accrues during a typical project. Not everything that becomes a time sink is worth communicating; conversely, it is easy to forget the many problems and compromises that have been overcome or bypassed that may hint at important research questions worth detailing. Keeping a laboratory notebook as you progress can be a valuable resource when writing papers and dissertations.

Academic forms of dissemination (e.g., papers) do not always value experience reports or negative results as much as they should. This, in turn, has a tendency to encourage some researchers to focus on positive contributions of their work and why it is new or different from existing approaches at the expense of objectivity; it is far more common in other disciplines to repeat experiments and validate the work of other scientists, rather than focus on novelty and differentiation. Your work should be grounded in the literature; it is definitely acceptable to learn and build on the work of others, and it is acceptable to stress the commonalities as well as the differences. If there is genuinely nothing new to learn from your proposed project, then you have to question whether your objectives are correct; an early search of the literature is particularly important for this reason. For a thought-provoking discussion of experimental methods in Computer Science, the interested reader is directed to Feitelson (2005).

2.5.3 Documenting Ubicomp Systems

This last section shall present what, in our experience, is the best way to document ubicomp systems research and what needs to be addressed. This may work as an outline of your technical documentation as well as some basic directions for writing good ubicomp systems papers for the research community.

- Explain the specific (systems) *question and challenge* that you are addressing.

- Enumerate and explain the *assumptions* you make—both technical as well as any assumptions on the developers' and users' behalf.

- Carefully *relate your work* to others, paying special attention to where your work extends the work of others, and where it differs. Your work may differ in several areas, but it is important to highlight a few significant differences that constitute your main contributions.

- Divide documentation into

  - Technical documentation—contains all the technical details on the system, its implementation, and evaluation.

  - Research paper describing the overall research approach, questions, contribution, concepts, and technical innovation—always be careful not to include trivia or irrelevant implementation details; refer to the technical document if necessary.

- Describe your evaluation—especially why the system was evaluated in the manner with which it was conducted—addressing the following issues:

  - Evaluation strategy and overall approach.

  - The aim of the evaluation, including a description of how to measure it. Outline evaluation criteria and how to measure success.

  - Evaluation setup, including technical setup, configuration, runtime environment, simulation parameters, users, their background, the physical setup, etc.

  - Results of the evaluation, including measurable results such as time measurements of performance, throughput, and resource contribution, as well as qualitative results such as user feedback based on interviews, observations, and questionnaires.

- Discuss the contribution of the system as related both to the results of the evaluation as well as to the results from others.

In technical and scientific documentation, it is important to maintain objectivity and honesty when reporting results and findings. Try to avoid unnecessary adjectives and provide a prosaic description. Carefully present and discuss what can be learned from your research and the results you have obtained. Documenting and reporting on apparently negative results may entail a contribution in itself; it may be an inspiration for others to try to address this particular challenge or it may be associated with a flaw in the evaluation setup, which can be fixed once discovered.

2.5.4 Corollary

On a final note, it is worth keeping the scientific mindset to the fore in order to cultivate a scientific and balanced approach in describing your work. Scientists should be uncertain, open-minded, skeptical, cautious, and ethical—readers will question your work and will be cautious to accept your claims without appropriate evidence and grounding with respect to other approaches. Balanced and objective self-reflective analysis and evaluation of your work is crucial to its acceptability by others and particularly by the best quality conferences and journals. Evaluations must be methodologically sound and include adequate explanation of how they were conducted, because this is important for confidence in the quality of the results and trustworthiness of the inferences drawn from them. Results and lessons should also be clearly and concisely presented. Try, if possible, to "quantify the magic" (Barton and Pierce, 2006) that made your system work for the setting and users you chose; understanding the scope and limitations of your system and how seriously these might impact the generalization of your work is important for setting the boundaries and research questions for further work in that area.

A simple guiding principle is: "What does the reader learn from reading my paper?" If a paper lacks useful insight, lessons, or results, then it is highly likely to be rejected.

2.6 GETTING STARTED

> If I have seen further than others, it is by standing upon the shoulders of giants.
>
> ISAAC NEWTON, LETTER TO ROBERT HOOKE
>
> FEBRUARY 5, 1675

Many ubicomp systems research projects have put tools, toolkits, and datasets into the public domain. Here are a few examples of tools that we, as experimental scientists and designers of next-generation ubicomp systems, can download and evaluate. These can provide a quick route to getting your ideas up and running and allowing low-cost experimentation with ubicomp systems. You should feel positively encouraged to offer feedback to the creators, contribute to projects and dataset archives, and objectively compare your work with others in the domain; as a matter of principle, we can only benefit as a community from trying out each other's systems and paradigms, and working together to address the many challenges ubicomp poses. In general, there are different types of technology that can help you realize your system and prototype your ideas:

1. Rapid prototyping tools for creating situated or mobile ubicomp systems

2. Libraries that can form components of your system, for example, handling computer vision, gesture recognition, processing sensor data, handling context

3. Hardware components including wireless sensors for augmenting artifacts or forming sensor networks

Given the typical lifetime of the average research project or these types of technology, this section merely aims to serve as an indicator of the types of systems available to you. More up-to-date resources should be kept in the public domain where they can be added to by active researchers, such as yourself (e.g., see http://ubisys.org).

2.6.1 Prototyping Your Ideas

There are many hardware and software platforms available to assist with deploying test ubicomp infrastructures to test your ideas and novel forms of interaction. Tools such as ActivityStudio (Li and Landay, 2008), exemplar (Hartmann et al., 2007), and iStuff Mobile (Ballagas et al., 2007) support the creation of low-fidelity functional prototypes that can be used to experiment with different ubicomp application designs. Each has a different focus: ActivityStudio provides an environment for moving from field notes, through a storyboarding and visual programming step through to simulation and in situ deployment of a Web-based prototype; exemplar encourages demonstration of sensor-based interactions (e.g., gestures) that are then filtered and transformed using a visual development environment to trigger other applications; and iStuff Mobile provides a visual programming interface for novel mobile phone–based interfaces (an otherwise notoriously difficult platform to develop for).

Systems such as the EQUATOR Component Toolkit (ECT) (Greenhalgh and Egglestone, 2005) and Wiring (Barragán, 2006) provide programmatic glue for constructing ubicomp systems that integrate sensing (input from sensors such as phidgets, motes, and d.tools boards), actuation (of physical actuators including X10 modules, output to Internet applications, etc.). ECT uses a visual graph-based editor to allow runtime interconnection of modules. The underlying EQUIP instances can support multimachine and distributed configurations. Wiring offers a high-level language based on the popular open-source visualization language Processing (Fry and Reas, 2001). Similar dataflow-like graphing metaphors are also exploited in Max/MSP (Zicarelli, 1997), a commercial system used by artists and designers to create interactive installations. Max uses a powerful graphical wiring metaphor (an interesting and flexible design in its own right) for connecting input and output components with channels that communicate messages. PureData (Puckette, 1996) and jMax (Cecco et al., 2008) are open-source derivatives of Max. Extensions to these (e.g., Digital Image Processing with Sound for jMax) allow real-time processing and transformation of video suitable for use in video-based installations and art pieces.

The Context Toolkit (Salber et al., 1999), Java Context-Awareness Framework (Bardram, 2005), and Context Aware Toolkit (CAT) (Prideau, 2002) allow sensors of context to be decoupled from higher level context reasoning in applications. Whereas the Context Toolkit and JCAF support ubicomp applications based on the integration of distributed context sensors, CAT does a similar job for embedded wearable devices.

Topiary (Li et al., 2007) and MyExperience (Froehlich, 2009) use context in a mobile environment to trigger interactions with the user. However, the aim of the two systems is quite different: Topiary's focus is low-fidelity contextual presentation of interactive design sketches to the user, whereas MyExperience is designed to ask the user a contextually relevant set of questions to survey them in situ (a methodology known as the experience sampling method).

2.6.2 Smart Room in a Box

If your aim is to create a smart environment populated with multiple displays and interaction devices, then Stanford's interactive workspaces spin off iROS (Borchers et al., 2006) is available as open source.

A meta-operating system for creating interactive rooms, iROS includes a set of core middleware (Event Heap, DataHeap, iCrafter) for unifying machines and displays together to form larger interactive surfaces. The MeetingMachine (Barton et al., 2003; Barton, 2003) repackages iROS to create a shared networked appliance supporting the exchange, discussion, and collation of electronic documents in a meeting setting. Radically different approaches are taken by Plan B (Ballesteros et al., 2008), where smart environments are built on the Bell Labs Plan 9 operating system (Pike et al., 2003) and PCOM (Rothermel et al., 2006), a peer-to-peer component middleware for constructing pervasive applications.

2.6.3 Public Domain Toolkits

There are many useful libraries that can provide solutions to well-known algorithmic or integration problems, for example, integrating computer vision or detecting human activities.

2.6.3.1 Vision and Augmented Reality

A very common requirement is to integrate computer vision systems. OpenCV (OpenCV, 2009) provides over 500 algorithms for real-time vision processing. CANTag (DTG Research Group, 2005) supports the tracking of fiducial tags including their orientation and rotation, enabling a range of possible interaction gestures. Using a similar technique, the popular augmented and mixed-reality ARToolkit (Lamb et al., 2007) has been used to great effect for overlaying 3-D graphics onto similar fiducials. Building on this, the Designers Augmented Reality Toolkit (Macintyre et al., 2005) integrates this into an experience design environment based on Macromedia Director, simplifying augmented reality experience design.

2.6.3.2 Sensing

Ubicomp systems are often required to interpret sensor data to identify user interactions or human activities. Weka (Frank et al., 2008) is a collection of machine learning algorithms for preprocessing, classification, processing, and visualizing of data. Sensor networks increasingly underlie many ubicomp installations, particularly in the healthcare and emergency services domains. For example, DexterNet (Kuryloski et al., 2008) is an open framework for integrating wearable sensors for medical applications. It provides support for communicating with medical sensors (e.g., ECG), network support for communicating readings from the device, and a higher layer toolkit called SPINE (Giannantonio et al., 2008; Gravina et al., 2008), which helps simplify code development for embedded wearable sensors and deployment on arrays of sensor nodes.

2.6.3.3 Hardware
Many of the toolkits described above have been adapted to allow the integration of commodity hardware and tangible prototyping tools. Common ones include

- Wireless sensor nodes such as Motes (Crossbow, 2008), SunSpots (Sun Microsystems Laboratories, 2004), jStamps (Systronix, 2009), and µParts (Beigl et al., 2005)

- Interface prototyping boards such as the popular Phidgets (Phidgets, 2009), Arduino (Arduino, 2009), and d.tools (Hartmann et al., 2006) kits

- Wearable sensor boards for medical applications, for example, Harvard's CodeBlue (Welsh et al., 2008) and University of Alabama in Huntsville's Wearable Health Monitoring Systems (WHMS) (Otto et al., 2008)

Most of these have active communities developing tutorial materials and examples.

2.6.4 Datasets

A positive side effect of the standardization of some of these components (particularly sensor platforms) is that it becomes possible to repeat experiments and validate other people's findings. Datasets from such platforms are being increasingly collected online and are often open to contributions from other researchers. Examples of useful datasets already in the public domain include:

1. MIT's House_n PlaceLab (Intille et al., 2006) includes traces of human activity that have been used by the community to develop activity detection algorithms.

2. Intel's Place Lab (Hightower et al., 2006; Lamarca et al., 2005) provides a freely available system for mobile localization, together with contributed location traces, which have been used to look at destination prediction, context-aware assistive technologies, and privacy.

3. Dartmouth's CRAWDAD (Kotz and Henderson, 2009) is a community archive of wireless network traces that have been used for a wide range of uses including developing improved MAC layer protocols and location prediction.
4. Berkeley's Wearable Action Recognition Database (Yang et al., 2008) and WHMS (Otto et al., 2008) activity traces for developing human action recognition systems based on wearable motion sensors.

The appearance and growth of initiatives such as these can help us collectively identify and solve common systems problems in ubicomp, enabling the field to move forward more rapidly.

2.6.5 Summary

Prototyping your ideas using available prototyping and smart room tools is a laudable approach for exploring the ubicomp design space and soliciting feedback. These types of tools help us commodify ubicomp systems, simplifying rapid creation of prototypes and broadening ubicomp experience. Ubicomp systems researchers are to be encouraged to use, refine, contribute to these initiatives, and start new ones as needed, to continue the technological dialogue that helps support our community. Exploiting and contributing new tools and datasets in the public domain can only serve to stimulate further research activity and promote increased adoption of scientific practices such as repeatability and comparison.

2.7 CONCLUSION

Systems research is central to ubicomp research and provides the fundamental building blocks for moving the field forward in terms of new applications and user experiences. As a research field, ubicomp must continue to build and evaluate systems components that ease the design, implementation, deployment, and maintenance of real-world ubicomp applications.

This chapter has outlined the special challenges pertaining to ubicomp systems and applications, including issues of designing systems that have to run in resource-constrained, volatile, and heterogeneous execution environments. But, in addition to these technical challenges to ubicomp systems design, the chapter has also tried to highlight that the special characteristics of ubicomp applications force systems researchers to address a whole new set of systems challenges, including the need to design for fluctuating environments and circumstances, and invisible computing. To a large degree, the assumptions that contemporary personal and client-server computing relies on, breaks down in a ubicomp environment.

The chapter then moved on to discuss how to create ubicomp systems, putting emphasis on the experimental nature of systems design and implementation. Advice on how to implement and deploy ubicomp systems was given with reference to concrete ubicomp technologies and projects. Special emphasis was placed on evaluating and documenting ubicomp systems research; it is essential for the research community that ubicomp systems research is properly evaluated and documented in order to move the field forward. Evaluation of ubicomp systems is far from easy, and the chapter offers advice on how to conduct evaluation under specific conditions, including the use of simulation, proof-of-concepts, end user application building and evaluation, and technology releases to the research community. Similar advice is offered on how to document ubicomp systems research, both technically and scientifically.

With this chapter, we hope that researchers are motivated to engage in creating systems support for the ubicomp application area and, with the chapter in hand, have some specific pointers and tools for engaging in this research. After all, the ubicomp systems research field is still in its infancy, and there is ample space for new exciting systems innovations.

REFERENCES

Balan, R., Flinn, J., Satyanarayanan, M., Sinnamohideen, S., and Yang, H., 2002, The case for cyber foraging, in *Proceedings of the 10th Workshop on ACM SIGOPS European Workshop*, ACM, New York, NY, pp. 87–92.

Bardram, J. E., and Christensen, H. B., 2007, Pervasive computing support for hospitals: An overview of the activity-based computing project. *IEEE Pervasive Computing* 6(1): 44–51.

Bardram, J. E., Baldus, H., and Favela, J., 2006, Pervasive computing in hospitals, in *Pervasive Healthcare: Research and Applications of Pervasive Computing in Healthcare*, CRC Press, Boca Raton, FL, pp. 49–78.

Bardram, J. E., and Schultz, U. P., 2004, Contingency management, Palcom Working Note #30, Technical report, Palcom Project IST-002057.

Bardram, J. E., 2004, Applications of context-aware computing in hospital work—examples and design principles, in *Proceedings of the 2004 ACM Symposium on Applied Computing*, pp. 1574–1579.

Barton, J., and Pierce, J., 2006, Quantifying magic in ubicomp system scenarios, in *Ubisys 2006: System Support for Ubiquitous Computing Workshop*, 8th Annual Conference on Ubiquitous Computing (Ubicomp 2006), Orange County, CA, USA, September 17–21.

Benford, S., Seagar, W., Flintham, M., Anastasi, R., Rowland, D., Humble, J., Stanton, D., Bowers, J., Tandavanitj, N., Adams, M., Farr, R. J., Oldroyd, A., and Sutton, J., 2004, The error of our ways: The experience of self-reported position in a location-based game, in *Proceedings of the 6th International Conference on Ubiquitous Computing (UbiComp 2004)*, Nottingham, September, pp. 70–87.

Borchers, J., Ringel, M., Tyler, J., and Fox, A., 2002, Stanford interactive workspaces: A framework for physical and graphical user interface prototyping, *Wireless Communications, IEEE* (see also *IEEE Personal Communications*) 9(6): 64–69.

Boucher, A., Steed, A., Anastasi, R., Greenhalgh, C., Rodden, T., and Gellersen, H., 2003, Sensible, sensable and desirable: A framework for designing physical interfaces, Technical report, Technical Report Equator-03-003.

Brumitt, B., Meyers, B., Krumm, J., Kern, A., and Shafer, S., 2000, EasyLiving: Technologies for intelligent environments, in *Proceedings of the Second International Symposium on Handheld and Ubiquitous Computing*, Bristol, UK, 25–27 September, pp. 12–29.

Chalmers, M., MacColl, I., and Bell, M., 2003, Seamful design: Showing the seams in wearable computing, *IEEE Seminar Digests* (10350), 11–16.

Cheverst, K., Davies, N., Mitchell, K., Friday, A., and Efstratiou, C., 2000, Experiences of developing and deploying a context-aware tourist guide: The GUIDE project, 6th Annual International Conference on Mobile Computing and Networking (MobiCom 2000), Boston, MA, August, ACM Press, New York, NY, pp. 20–31.

Clarkson, P. J., Coleman, R., Keates, S., and Lebbon, C., 2003, *Inclusive Design: Design for the Whole Population*, Springer, London.

Coulouris, G., Dollimore, J., and Kindberg, T., 2005, *Distributed Systems: Concepts and Design*, 4th ed., Addison-Wesley, Reading, MA.

Demers, A., Greene, D., Hauser, C., Irish, W., Larson, J., Shenker, S., Sturgis, H., Swinehart, D., and Terry, D., 1987, Epidemic algorithms for replicated database maintenance, in *PODC '87: Proceedings of the 6th annual ACM Symposium on Principles of Distributed Computing*, ACM Press, New York, NY, pp. 1–12.

Edwards, W. K., Bellotti, V., Dey, A. K., and Newman, M. W., 2003, Stuck in the middle—The challenges of user-centered design and evaluation for infrastructure, in *CHI '03: Proceedings of the SIGCHI Conference on Human Factors in Computing Systems*, ACM Press, New York, NY, pp. 297–304.

Feitelson, D. G., 2005, Experimental computer science: The need for a cultural change, Technical report, School of Computer Science and Engineering, Hebrew University, Jerusalem.

Fleck, M., Frid, M., Kindberg, T., O'Brien-Strain, E., Rajani, R., and Spasojevic, M., 2002, Rememberer: A tool for capturing museum visits, in *Proceedings of UbiComp 2002: Ubiquitous Computing*, pp. 379–385.

Flintham, M., Anastasi, R., Hemmings, T., Crabtree, A., Greenhalgh, C., and Rodden, T., 2003, *Where On-Line Meets on-the-Streets: Experiences with Mobile Mixed Reality Games*, ACM Press, New York, NY, pp. 569–576.

Floyd, S., Jacobson, V., Liu, C.-G., McCanne, S., and Zhang, L., 1997, A reliable multicast framework for light-weight sessions and application level framing, *IEEE/ACM Transactions on Networking* 5(6): 784–803.

Fox, A., Davies, N., de Lara, E., Spasojevic, M., and Griswold, W., 2006, Real-world ubicomp deployments: Lessons learned, *IEEE Pervasive Computing* 5(3): 21–23.

Friday, A., Roman, M., Becker, C., and Al-Muhtadi, J., 2005, Guidelines and open issues in systems support for ubicomp: Reflections on ubisys 2003 and 2004, *Personal and Ubiquitous Computing* 10: 1–3.

Friday, A., Davies, N., Seitz, J., and Wade, S., 1999, Experiences of using generative communications to support adaptive mobile applications, *Kluwer Distributed and Parallel Databases Special Issue on Mobile Data Management and Applications* 7(3): 319–342.

Garlan, D., Siewiorek, D. P., Smailagic, A., and Steenkiste, P., 2002, Project Aura: Toward distraction-free pervasive computing, *IEEE Pervasive Computing* 2(1): 22–31.

Gaver, W. W., Beaver, J., and Benford, S., 2003, Ambiguity as a resource for design, in *CHI '03: Proceedings of the SIGCHI Conference on Human Factors in Computing Systems*, ACM Press, New York, NY, pp. 233–240.

Hansen, T. R., Bardram, J. E., and Soegaard, M., 2006, Moving out of the lab: Deploying pervasive technologies in a hospital, *IEEE Pervasive Computing* 5(3): 24–31.

Hightower, J., and Borriello, G., 2001, Location systems for ubiquitous computing, *Computer* 34(8): 51–66.

Kephart, J. O., and Chess, D. M., 2003, The vision of autonomic computing, *IEEE Computer* 36(1): 41–50.

Kindberg, T., and Fox, A., 2002, System software for ubiquitous computing, *Pervasive Computing IEEE* 1(1): 70–81.

Kindberg, T., Barton, J., Morgan, J., Becker, G., Caswell, D., Debaty, P., Gopal, G., Frid, M., Krishnan, V., Morris, H., Schettino, J., Serra, B., and Spasojevic, M., 2002, People, places, things: Web presence for the real world, *Mobile Networks and Applications* 7: 365–376.

Kistler, J. J., and Satyanarayanan, M., 1992, Disconnected operation in the coda file system, *ACM Transactions on Computer Systems* 10(1): 3–25.

Kurkowski, S., Camp, T., and Colagrosso, M., 2005, MANET simulation scenarios: The incredibles, *ACM Mobile Computing and Communications Review (MC2R)* 9(4): 50–61.

Lamarca, A., Chawathe, Y., Consolvo, S., Hightower, J., Smith, I., Scott, J., Sohn, T., Howard, J., Hughes, J., Potter, F., Tabert, J., Powledge, P., Borriello, G., and Schilit, B., 2005, Place Lab: Device positioning using radio beacons in the wild, in *Proceedings of the 3rd International Conference on Pervasive Computing*, May.

Langheinrich, M., 2001, Privacy by design—principles of privacy-aware ubiquitous systems, in *Ubicomp 2001: Ubiquitous Computing*, vol. 2201, *Lecture Notes in Computer Science*, pp. 273–291, Springer Verlag, Berlin.

Ponnekanti, S., Johanson, B., Kiciman, E., and Fox, A., 2003, Portability, extensibility and robustness in iROS, in *Proceedings of the First IEEE International Conference on Pervasive Computing and Communications (PerCom 2003)*, March, pp. 11–19.

Rogers, Y., 2006, Moving on from Weiser's vision of calm computing: Engaging ubicomp experiences, *UbiComp 2006: Ubiquitous Computing* 4206: 404–421.

Rogers, Y., Price, S., Fitzpatrick, G., Fleck, R., Harris, E., Smith, H., Randell, C., Muller, H., O'Malley, C., Stanton, D., Thompson, M., and Weal, M. J., 2004, Ambient wood: Designing new forms of digital augmentation for learning outdoors, Third International Conference for Interaction Design and Children (IDC 2004), ACM Press, New York, NY, pp. 1–9.

Roman, M., and Campbell, R. H., 2000, Gaia: Enabling active spaces, in *EW 9: Proceedings of the 9th Workshop on ACM SIGOPS European Workshop*, ACM Press, New York, NY, pp. 229–234.

Roman, R., Hess, C., Cerqueira, R., Ranganathan, A., Campbell, R. H., and Nahrstedt, K., 2002, A middleware infrastructure for active spaces, *IEEE Pervasive Computing* 1(4): 74–83.

Satyanarayanan, M., 2001, Pervasive computing: Vision and challenges, *Personal Communications IEEE* 8(4): 10–17.

Salber, D., Dey. A. K., and Abowd, G. D., 1999, The context toolkit: Aiding the development of context-enabled applications, *CHI'99: Proceedings of the SIGCHI Conference on Human Factors in Computing Systems*, ACM Press, New York, NY, pp. 434–441.

Schuler, D., and Namioka, A. (Eds.), 1993, *Participatory Design: Principles and Practices*, Erlbaum Associates, Hillsdale, NJ.

Storz, O., Friday, A., Davies, N., Finney, J., Sas, C., and Sheridan, J., 2006, Public ubiquitous computing systems: Lessons from the e-campus display deployments, *IEEE Pervasive Computing* 5: 40–47.

Streitz, N. A., Geissler, J., Holmer, T., and Konomi, S., 1999, ILand: An interactive landscape for creativity and innovation, in *Proceedings of the ACM Conference on Human Factors in Computing Systems: CHI 1999*, ACM Press, New York, NY, pp. 120–127.

Terry, D. B., Theimer, M. M., Petersen, K., Demers, A. J., Spreitzer, M. J., and Hauser, C. H., 1995, Managing update conflicts in bayou, a weakly connected replicated storage system, in *SOSP '95: Proceedings of the 15th ACM Symposium on Operating Systems Principles*, ACM Press, New York, NY, pp. 172–182.

Weiser, M., 1991, The computer for the 21st century, *Scientific American* 265(3): 66–75.

Weiser, M., 1993, Some computer science issues in ubiquitous computing, *Communications of the ACM* 36(7): 74–84.

Zambonelli, F., Jennings, N. R., and Wooldridge, M., 2003, Developing multiagent systems: The Gaia methodology, *ACM Transactions on Software Engineering and Methodology* 12(3): 317–370.

URLs

Arduino Project, 2009, Arduino: An Open-Source Electronics Prototyping Platform, http://arduino.cc/.

Ballagas, R., Memon, F., Reiners, R., and Borchers, J., 2007, iStuff mobile: Rapidly Prototyping New Mobile Phone Interfaces for Ubiquitous Computing, http://research.nokia.com/people/tico_ballagas/istuff_mobile.html.

Ballesteros, F. J., Soriano, E., Guardiola, G., de las Heras, P., de Mingo Gil, S., Higuera, O., Lalis, S., and Garcia, R., 2008, Plan B: An Operating System for Distributed Environments, http://lsub.org/ls/planb.html.

Bardram, J. E., 2005, JCAF: The java context-awareness framework, http://www.daimi.au.dk/~bardram/jcaf/.
Barragán, H., 2006, Wiring, http://wiring.org.co/.
Barton, J. J., 2003, The MeetingMachine: Interactive Workspace Support for Nomadic Users, http://home.comcast.net/~johnjbarton/ubicomp/mm/.
Barton, J. J., Hsieh, T., Johanson, B., Vijayaraghavan, V., Fox, A., and Shimizu, T., 2003, The MeetingMachine: Interactive workspace support for nomadic users, in 5th IEEE Workshop on Mobile Computing Systems and Applications (WMCSA 2003), Monterey, CA, USA, pp. 2–12, IEEE Computer Society, October.
Beigl, M., Decker, C., Krohn, A., Riedel, T., and Zimmer, T., 2005, µParts: Low Cost Sensor Networks at Scale, http://particle.teco.edu/upart/.
Borchers, J., Ringel, M., Tyler, J., and Fox, A., 2006, iROS Meta-Operating system for Interactive Rooms, http://sourceforge.net/projects/iros/.
Cecco, M. D., Dechelle, F., and Maggi, E., 2008, jmax phoenix, http://sourceforge.net/projects/jmax-phoenix/.
Crossbow, 2008, Motes Wireless Modules, http://www.xbow.com.
DTG Research Group, 2005, Cantag Machine Vision Framework, http://www.cl.cam.ac.uk/research/dtg/research/wiki/Cantag.
Frank, E., Hall, M., and Trigg, L., 2008, Weka 3: Data Mining Software in Java, http://www.cs.waikato.ac.nz/ml/weka/.
Froehlich, J., 2009, MyExperience: A Context-Aware Data Collection Platform, http://myexperience.sourceforge.net/.
Fry, B., and Reas, C., 2001, Processing, http://processing.org/.
Giannantonio, R., Bellifemine, F., and Sgroi, M., 2008, "SPINE (Signal Processing in Node Environment), http://spine.tilab.com/.
Gravina, R., Guerrieri, A., Iyengar, S., Bonda, F. T., Giannantonio, R., Bellifemine, F., Pering, T., Sgroi, M., Fortino, G., and Sangiovanni-Vincentelli, A., 2008, Demo: Spine (Signal Processing in Node Environment) Framework for Healthcare Monitoring Applications in Body Sensor Networks, in 5th European Conference on Wireless Sensor Networks 2008 (EWSN '08), Bologna, Italy, Jan 30–Feb 1.
Greenhalgh, C., and Egglestone, S. R., 2005, Equip and Equator Component Toolkit (ECT), http://equip.sourceforge.net.
Hartman, B., Abdulla, L., Klemmer, S., and Mittal, M., 2007 Exemplar: Authoring Sensor Based Interactions, http://hci.stanford.edu/exemplar/.
Hartmann, B., Klemmer, S., Abdulla, L., Mehta, N., Bernstein, M., Burr, B., Robinson-Mosher, A. L., and Gee, J., 2006, d.tools: Enabling Rapid Prototyping for Physical Interaction Design, http://hci.stanford.edu/dtools/.
Hightower, J., LaMarca, A., and Smith, I., PlaceLab: A Privacy-Observant Location System, http://www.placelab.org/.
Intille, S. S., Larson, K., Munguia Tapia, E., Beaudin, J., Kaushik, P., Nawyn, J., and Rockinson, R., 2006, Using a Live-in Laboratory for Ubiquitous Computing Research, in *Proceedings of PERVASIVE 2006*, vol. LNCS 3968, Fishkin, K. P., Schiele, B., Nixon, P., and Quigley, A. (Eds.), Springer-Verlag, Berlin, pp. 349–365, http://architecture.mit.edu/house_n/data/PlaceLab/PlaceLab.htm.

Kotz, D., and Henderson, T., 2009, CRAWDAD: Community Resource for Archiving Wireless Data at Dartmouth, http://crawdad.cs.dartmouth.edu/.

Kuryloski, P., Giani, A., Giannantonio, R., Gilani, K., Gravina, R., Seppa, V.-P., Seto, E., Shia, V., Wang, C., Yan, P., Yang, A., Hyttinen, J., Sastry, S. S., Wicker, S., and Bajcsy, R., 2008, Dexternet: An Open Platform for Heterogeneous Body Sensor Networks and Its Applications, Tech. Rep. UCB/EECS-2008-174, EECS Department, University of California, Berkeley, December, http://www.eecs.berkeley.edu/Pubs/TechRpts/2008/EECS-2008-174.html.

Lamb, P., Looser, J., Grasset, R., Pintaric, T., Woessner, U., Piekarski, W., and Seichter, H., ARToolKit: A Software Library for Building Augmented Reality (AR) Applications, http://www.hitl.washington.edu/artoolkit/.

Li, Y., and Landay, J., 2008, ActivityStudio: Design and Testing Tools for Ubicomp Applications, http://activitystudio.sourceforge.net/.

Li, Y., Hong, J., and Landay, J., 2007, Topiary: A Tool for Prototyping Location-Enhanced Applications, http://dub.washington.edu:2007/topiary/.

Macintyre, B., Bolter, J. D., Gandy, M., and Dow, S., 2005, DART: The Designers Augmented Reality Toolkit, http://www.cc.gatech.edu/dart/.

OpenCV Project, 2009, OpenCV (Open Source Computer Vision), http://opencv.willowgarage.com/wiki/.

Otto, C., Milenkovic, A., Sanders, C., and Jovanov, E., 2008, WHMS: Wearable Health Monitoring Systems, http://www.ece.uah.edu/~jovanov/whrms/.

Phidgets Inc., 2009, Phidgets: Products for USB Sensing and Control, http://www.phidgets.com/.

Pike, R., Presotto, D., Dorward, S., Flandrena, B., Thompson, K., Trickey, H., and Winterbottom, P., 2003, Plan 9 from Bell Labs., http://plan9.bell-labs.com/plan9/.

Prideau, J., 2002, CAT: Context Aware Toolkit, http://www.cs.uoregon.edu/research/wearables/CAT/.

Puckette, M. S., 1996, Puredata (PD), http://puredata.info/.

Rothermel, K., Becker, C., Schiele, G., Handte, M., Wacker, A., and Urbanski, S., 2006, 3PC: Peer 2 Peer Pervasive Computing, http://3pc.info.

Schmidt, D., 2009, Multi-Touch Table, http://eis.comp.lancs.ac.uk/~dominik/cms/, accessed Feb. 27, 2009.

Smart-ITs, 2001, http://www.smart-its.org/, accessed Feb. 27, 2009.

Sun Microsystems Laboratories, 2004, Sun SPOT: Sun Small Programmable Object Technology, http://www.sunspotworld.com/.

Systronix Inc., 2009, JStamp: Java Embedded Processor, http://jstamp.systronix.com/.

Welsh, M., Wei, P. G.-Y., Moulton, S., Bonato, P., and Anderson, P., 2008, CodeBlue: Wireless Sensors for Medical Care, http://fiji.eecs.harvard.edu/CodeBlue.

Yang, A., Kuryloski, P., and Bajcsy, R., 2008, WARD: A Wearable Action Recognition Database, http://www.eecs.berkeley.edu/~yang/software/WAR/.

Zicarelli, D., 1997, Max/msp/jitter, http://www.cycling74.com/.

CHAPTER 3

Privacy in Ubiquitous Computing

Marc Langheinrich

CONTENTS

3.1	Introduction	96
	3.1.1 Why a Privacy Chapter in a Ubicomp Book?	97
	3.1.2 Isn't Privacy the Same as Security?	97
	3.1.3 What Is in This Chapter?	98
	3.1.3.1 Conclusions	98
3.2	Understanding Privacy	98
	3.2.1 Defining Privacy	101
	3.2.1.1 Conclusions	107
	3.2.1.2 Further Reading	108
	3.2.2 Motivating Privacy: Do People Care about Privacy?	108
	3.2.2.1 Conclusions	113
	3.2.2.2 Further Reading	114
	3.2.3 Legal Background	114
	3.2.3.1 Conclusions	118
	3.2.3.2 Further Reading	118
	3.2.4 Interpersonal Privacy	119
	3.2.4.1 Conclusions	121
	3.2.4.2 Further Reading	121
3.3	Technical Solutions for Ubicomp Privacy	121
	3.3.1 Novel Ubicomp Challenges to Privacy	122
	3.3.1.1 Collection Scale	123
	3.3.1.2 Collection Manner	124

		3.3.1.3	Data Types	125
		3.3.1.4	Collection Motivations	126
		3.3.1.5	Data Accessibility	127
	3.3.2	The Basics: Privacy Enhancing Technologies		128
	3.3.3	Example: Protecting Smart Spaces		131
	3.3.4	Example: Protecting RFID Tags		134
		3.3.4.1	Communication Confidentiality and Anticollision Protocols	138
		3.3.4.2	Access Control/Tag Deactivation	140
		3.3.4.3	Proxies	142
	3.3.5	Example: Protecting Location Information		145
		3.3.5.1	Conclusions	149
		3.3.5.2	Further Reading	150
3.4	How to Address Privacy in Your Ubicomp Work			151
	3.4.1	Understand Your Application (Consider Users and Use)		151
	3.4.2	Define the Problem (Think Attacker Model in Security)		152
	3.4.3	Know Your Tools (Get the Technical Details Right)		154
		3.4.3.1	Conclusions	155
		3.4.3.2	Further Reading	155
Bibliography				156

3.1 INTRODUCTION

Privacy is by no means a recent addition to the ubiquitous computing (ubicomp) research curriculum. In his 1991 *Scientific American* article, Mark Weiser already identified it as one of its biggest challenges: "Perhaps key among [the social issues that embodied virtuality will engender] is privacy: hundreds of computers in every room, all capable of sensing people near them and linked by high-speed networks, have the potential to make totalitarianism up to now seem like sheerest anarchy." (Weiser, 1991). It would be nice if by now, almost two decades later, we would have a standard set of solutions that we could easily prescribe for any ubicomp system (or any computer system in general): "in order to protect privacy, implement subroutines A, B, and C."

Unfortunately, privacy is such a complex issue that there is no single solution, no recipe for success, no silver bullet (or set of silver bullets) that will fix a system for us so that it is "privacy-safe." What exactly does it mean anyway, for a system to be privacy-safe? Whose privacy does it protect, when, and to what extent? None of these questions can be answered in general. Instead of a simple one-size-fits-all recipe, one needs to look into each single system and application in great detail, first understanding what the system does and what the implications of this are, and then working out how (and why) this needs to be changed in order to reach the right behavior.

This chapter attempts to provide some guidance for this process: First, by explaining the concept of privacy in more detail, so one understands what it is that should be protected; second, by giving some examples on how technology can safeguard personal information in ubicomp systems. This chapter is not meant as a cookbook that allows one to quickly find a solution for a particular problem, but rather as a starting point for recognizing and approaching privacy issues in one's own design, development, or use of ubicomp systems and applications.

3.1.1 Why a Privacy Chapter in a Ubicomp Book?

Privacy might indeed be an important topic, but one could argue that it would be much better suited for a text on legal or social issues. Why include it in book about ubicomp technology? Privacy and technology are closely intertwined. Shifts in technology require us to rethink our attitude toward privacy, as suddenly our abilities to see, hear, detect, record, find, and manipulate others and their lives is greatly enhanced. Ubicomp represents such a technology shift, and its widespread adoption will significantly influence the way we handle personal information.

One could still suggest putting this topic in front of people interested in databases or information retrieval in particular, rather than systems or usability evaluation in general. Is privacy not about the storage and processing of data, which in turn is really the job of database and information retrieval specialists? The answer is yes and no. Clearly, the ability to store, process, and analyze information is at the heart of the privacy debate. However, in order to make meaningful choices within any system parameters, one needs to understand the entirety of the system and its applications: What type of information is collected and in what manner? Who needs to have access to such information and for what purpose? How long should this information be stored and in what format, with what levels of accuracy and precision? Effective privacy protection can only work if it addresses the entirety of the information life cycle in each individual ubicomp application.

3.1.2 Isn't Privacy the Same as Security?

Security—that is, the confidentiality, integrity, and authenticity of information—is often a necessary ingredient to privacy, as it facilitates the control of information flows (i.e., who gets to know what when?) and helps to ensure the correctness of data. However, it is possible to have high levels of security but no privacy (think surveillance state), or even some sort of privacy without security (e.g., a private table conversation in a busy restaurant). The important insight is that simply implementing some form of security is not enough to ensure privacy. Ensuring the confidentiality and authenticity of a particular information does not say anything about how and when this particular piece of information will be *used* by its designated recipient.

3.1.3 What Is in This Chapter?

The main focus of this chapter is how ubicomp affects privacy and what technical methods can be used to counter or mitigate this influence. However, this requires a clear understanding of what exactly should be protected. A large part of this chapter is therefore dedicated to understanding the concept of privacy first. Only then can one discuss the particular effects of ubicomp on privacy and the required technological countermeasures. Of course, privacy issues of "regular" computer systems such as databases often apply equally to ubicomp system, simply because most ubicomp applications are built with such standard components. Including technical solutions for such computing systems in general would be beyond the scope of this book, so the discussion in this chapter is strictly limited to examples on how particular threats induced by ubicomp technology can be addressed.

3.1.3.1 Conclusions
1. Data collection and processing are core components of ubicomp technology. Privacy issues are thus of utmost importance to ubicomp researchers, designers, service providers, and users.

2. Simply offering strong security does not solve privacy issues. Although security is an integral part of any privacy solution, it fails to address questions such as scope, purpose and use, adequacy, lifetime, or access.

3. To build privacy-aware or privacy-compliant ubicomp systems, we need to understand the nature of privacy, its social and legal realities, and the technical tools at our disposal. This chapter attempts to introduce those three aspects in detail.

3.2 UNDERSTANDING PRIVACY

Imagine your kitchen of the future (it might look like the one shown in Figure 3.1). All appliances—your refrigerator, freezer, stove, microwave, but also cabinets, utensils, faucets, and lamps—are now "smart," that is,

FIGURE 3.1 The smart kitchen of the future: spy in your home or ultra convenience? (Copyright Anton Volgger. Design by Meier + Steinauer Partner AG, Zurich, Switzerland. Image reprinted with permission.)

they are able to sense their environment and communicate: among themselves, with you, and with other things and people via the Internet. The famous *Internet-fridge* monitors its contents and orders milk and other ingredients before they run out. Your microwave-grill combo interrogates the pizza package to make sure it properly heats and bakes your TV food to perfection. And your faucets, freezer, and stove coordinate the use of warm water and electricity with the local power plant to minimize your energy costs. Ubicomp heaven!

Now imagine a few extras: your fridge not only orders milk and other staples if you run low, but also scouts for offers and coupons from the supermarket. Interestingly enough, the offers you receive are often very different from what your friend's fridge receives, who regularly gets discounts for expensive organic products (you don't). For your convenience, you have allowed your home insurer to periodically query your belongings in order to verify that you are sufficiently covered. However, after having had a number of visitors during the past few months, your insurer yesterday suddenly doubled your premium, claiming that the contents and activities in your kitchen indicated that you no longer have a single-person household. And just now, a police officer stopped by, asking you to explain the large quantities of hydrogen peroxide stored in your shed (sensed by the chemical sensors that monitor its contents for fire safety). Ever since the government passed the Preventing Irregularities through Smart Appliances Act, all household appliances are required by law to report suspicious items and activities directly to law enforcement agencies. Even though you showed the officer the antique wooden sailboat in your backyard that you are planning to bleach with the hydrogen peroxide, he informed you that your name will be kept on a list used by pharmacies around the country, alerting the authorities of any additional products you buy that could be used for bomb making. Well, at least your appliances have not been broken into yet: Your chaotic neighbor forgot to renew the firewall update for his fridge and promptly had a hacker monitor its use in order to detect longer absences. As soon as your neighbor had left for a business trip, his apartment address was traded on a local underground bulletin board that, for a small fee, would show "inactive" households on a Google Maps mash-up. Ubicomp heaven?

Whether any of the above examples make you question this brave new world of ubiquitous computing depends largely on your personal conception of privacy. Is the automated creation of profiles for the purpose of offering you discounts a good thing, even if it might offer you worse deals than others? Is the detection of inaccurate statements on insurance applications (and thus savings of several millions of dollars) a good thing, even if the system *occasionally* gets it wrong? And what is so bad about smart appliances that report unlawful behavior to the police, if they stay silent otherwise? If you don't do anything wrong, what have you got to hide? And if your data got stolen, maybe you did not protect it well enough?

To build ubicomp systems that are privacy-aware or privacy-respecting, one obviously has to first define what exactly is meant by *privacy*. There certainly is no shortage of definitions for privacy, yet no single one seems to work for all cases and disciplines. Section 3.2.1 briefly summarizes some of them and illustrate the changes the concept of privacy has undergone over time. Motivating privacy is an important part of such a definition, and Section 3.2.2 looks into current trends and surveys in order to understand what consumers and citizens expect in terms of privacy protection. Section 3.2.3 then gives a very short summary of the legal status of privacy protection, as any technical solution will most likely need to conform to local laws and regulations. However, not all interactions are governed by laws, in particular, if they do not involve companies or government agencies, but friends and families. Section 3.2.4 introduces some concepts from social sciences and psychology that can help understand how people "regulate" access to themselves.

3.2.1 Defining Privacy

Privacy has been on people's minds even before credit cards or the Internet came along. Everybody seems to have some sort of intuitive understanding of what privacy means on a case-by-case basis, yet it is difficult to *objectively* define what exactly it is. Legal scholars have long since grappled with such a definition, in order to write laws and regulations that describe what type of privacy protections should be granted, and they provide a good start for analyzing this complex topic.

One of the earliest legal references to privacy is found in the 1361 *Justices of the Peace Act* in England, which laid down sentences for peeping Toms and eavesdroppers. The famous saying "My home is my castle" dates back to the eighteenth century, where the British parliamentarian William Pitt said in a 1763 speech in Parliament: "The poorest man may in his cottage bid defiance to all the forces of the Crown. It may be frail—its roof may shake—the wind may blow through it—the storm may enter—the rain may enter—but the King of England cannot enter—all his force dares not cross the threshold of the ruined tenement!" However, as old as the concept of privacy is, it stills seems unclear exactly what it means today.

One of the most popular definitions of privacy comes from two U.S. lawyers, Samuel Warren and Louis Brandeis (see Figure 3.2), who wrote

FIGURE 3.2 Louis Brandeis, coauthor of *The Right to Privacy*. (Harris & Ewing, Collection of the Supreme Court of the United States. Image reprinted with permission.)

the first legal article that framed privacy as a tort action, that is, a civil wrong that one could sue for compensation of injuries. In their 1890 *Harvard Law Review* paper, Warren and Brandeis (1890) described privacy as "the right to be let alone," a state of solitude and seclusion that would ensure a "general right to the immunity of the person, the right to one's personality."

It is interesting to look into the particular circumstances that prompted Warren and Brandeis to write that article. In their introduction, they write "numerous mechanical devices threaten to make good the prediction that 'what is whispered in the closet shall be proclaimed from the housetops'." What may sound like an accurate description of the new possibilities of ubiquitous computing systems is actually a reference to the technical progress in the field of photography at that time. Before 1890, getting one's picture taken usually required visiting a photographer in his studio and sitting still for a considerable amount of time, otherwise the picture would be blurred. But on October 18, 1884, George Eastman, the founder of the Eastman Kodak Company, received U.S. Patent #306 594 for his invention of the modern photographic film (Figure 3.3). Instead of having to use the heavy glass plates in the studio, everybody could now take Kodak's Camera out on the streets and take a snapshot of just about anybody without their consent. It was this rise of unsolicited pictures—pictures that started to appear more and more often in the ever-expanding tabloid newspapers—that prompted Warren and Brandeis to paint this dark picture of a world without privacy.

Today's developments of "Smart Labels," "Memory Amplifiers," and "Smart Dust" seem to mirror the sudden technology shifts experienced by Warren and Brandeis, opening up new forms of social interactions that change one's expectation of privacy. However, even the strong resemblance of technological progress cannot ignore the fact that their "right to be let alone" looks hardly practicable today: With the multitude of interactions in today's world, consumers find themselves constantly in need of dealing with people that do not know them in person, hence require some form of information from them in order to judge whether such an interaction would be beneficial. From opening bank accounts, applying for credit, obtaining a personal yearly train pass, or buying books online—one constantly has to disclose part of one's personal information in order to participate in today's modern society. Preserving privacy through isolation is just not as much an option anymore as it was 100 years ago.

Privacy in Ubiquitous Computing ■ 103

FIGURE 3.3   The Kodak camera suddenly allowed anybody to take and instant photograph. (Eastman Kodak Company. Images printed with permission.)

A more up-to-date definition thus comes from the 1960s, when automated data processing first took place on a national scale. Alan Westin, professor emeritus of public law and government at Columbia University, defined privacy in his groundbreaking book *Privacy and Freedom* (Westin, 1967) as follows:

> Privacy is the claim of individuals, groups, or institutions to determine for themselves when, how, and to what extent information about them is communicated to others.

This definition is often described as *information privacy*, contrasting it to Warren and Brandeis' definition of privacy as solitude, of being "let alone." Whereas solitude might be an effect of information privacy, Westin stressed the fact that "the individual's desire for privacy is never absolute, since participation in society is an equally powerful desire." However, as

Warren and Brandeis' definition suggests, being in control of one's personal data is only one facet of privacy. Back in the nineteenth century, the protection of the home—or *territorial privacy*—was the most prevalent aspect of privacy protection. Equally important was the idea of *bodily privacy*, the protection from unjustified strip searches or medical tests or experiments (e.g., drug testing). These two facets are also often called local privacy or physical privacy. And with the invention of the telegraph and telephone in the late nineteenth century, the rise of modern telecommunication required reevaluation of the well-known concept of *communication privacy*, previously manifested in the secrecy of sealed letters.

Over the past 200 years, the focus of privacy has thus shifted from things that one could perceive directly with one's own eyes and ears (bodily privacy and territorial privacy) to more "remote" forms of privacy such as communication privacy and information privacy, where the privacy violations are undertaken at a distance. It is interesting to note, however, that ubicomp has made those seemingly long-solved issues of bodily and territorial privacy become highly relevant again: Smart appliances, wearable computers, and activity recognition algorithms allow one to invade the bodily and territorial privacy of another person—not with one's own eyes and ears, but from a distance that had previously constituted the realm of communication and information privacy. A smart fridge might disclose the activities in a home to a grocery distributor, whereas a smart shirt would send a stream of vital signs to a remote health center or insurance provider.

The limitation of both Warren and Brandeis' and Westin's definition of privacy is that they do not specify exactly how one's privacy should be protected. In order to better understand how to build technology that safeguards privacy, one needs to look at how one's privacy can be violated. This is important because privacy, just as security, is often not a goal in itself, not a service that people want to subscribe to, but rather an expectation of being in a state of protection without having to actively pursue it. For example, most users would prefer systems without passwords or similar access control mechanisms, as long as they would not suffer any disadvantages from this. Only if any of their data are maliciously deleted or illegally copied, users will regret not having any security precautions in place. So what would be the analogy to a break-in from a privacy point of view?

Gary T. Marx, professor emeritus for sociology at the Massachusetts Institute of Technology, has done extensive research in the areas of privacy and surveillance, identifying *personal border crossings* as a core concept: "Central to our acceptance or sense of outrage with respect to surveillance… are the implications for crossing personal borders" (Marx, 2001). Marx differentiates between four such border crossings that are perceived as privacy violations:

- Natural borders—Physical limitations of observations, such as walls and doors, clothing, darkness, but also sealed letters and telephone calls. Even facial expressions can form a natural border against the true feelings of a person.

- Social borders—Expectations about confidentiality for members of certain social roles, such as family members, doctors, or lawyers. This also includes expectations that your colleagues will not read personal fax messages addressed to you, or material that you left lying around the photocopy machine.

- Spatial or temporal borders—The usual expectations of people that parts of their life, both in time and social space, can remain separated from each other. This would include a wild adolescent time that should not interfere with today's life as a father of four, or different social groups, such as your work colleagues and friends in your favorite bar.

- Borders due to ephemeral or transitory effects—This describes what is best known as a fleeting moment, an unreflected utterance or action that one hopes gets forgotten soon, or old pictures and letters that one puts out in the trash. Seeing audio or video recordings of such events later, or observing someone sifting through our trash, will violate one's expectations of being able to have information simply pass away unnoticed or hopefully forgotten.

Whenever your personal information crosses any of these borders without your knowledge, your potential for possible actions—your decisional privacy—gets affected. When someone at the office suddenly mentions family problems that you have at home, or if circumstances of your youth suddenly are being brought up again even though you assumed that they were long forgotten, you perceive a violation of your zonal, informational, or communication privacy. This violation is by no means an absolute measure, but instead depends greatly on the individual circumstances, such as the type of information transgressed, or the specific situation under which

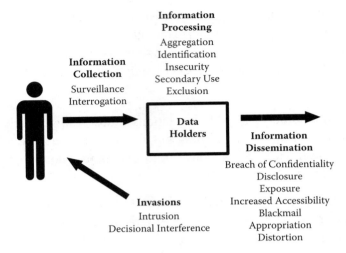

FIGURE 3.4 Solove's privacy taxonomy. (From D.J. Solove. *University of Pennsylvania Law Review 154*(3), 477–560. With permission of the author.)

the information is disclosed. The effects such border crossings have on people's lives, as well as the chances that they actually happen, are therefore a highly individual assertion.

In a similar fashion, Solove (2006) has attempted to create a *privacy taxonomy*, that is, an overview of the activities that might lead to privacy problems. He groups such activities into four sets (see Figure 3.4): information collection, information processing, information dissemination, and invasion. Starting from the affected individual, the data subject, various entities collect information. Most of the time this will be voluntary, but hidden or forced collections lead to *surveillance* or *interrogation* activities that violate the data subject's privacy. Once data holders process the data, that is, store, combine, search, or otherwise use it, they might engage in a number of activities that directly threaten the subject's privacy: through *aggregation* (multiple information sources might be linked that the subject might prefer to be separated); *identification* (which ties a particular information or activity to a person); acts of *insecurity* (a failure to properly protect the stored information that leads to improper access); *secondary use* activities (using the collected data for a purpose other than what was agreed with the data subject); and *exclusion* (the lack of letting the data subject know what data the holder has on file about her and how it is used). Information dissemination activities then propagate the information outward from the data holder: a *breach of confidentiality* is breaking a promise of keeping information confidential; *disclosure* means the publication of truthful facts about a person that might affect the person's reputation; *exposure* involves revealing private details, for example, nude pictures or bodily disabilities; *increased accessibility* concerns the publication of, for example, telephone numbers or email addresses; *blackmail* is the threat of disclosing information; *appropriation* is the use of the data subject's identity to serve someone else's interest; and *distortion* activities concern the dissemination of false or misleading information about the data subject. Solove also lists *intrusion* into one's life and *decisional interference* as privacy problems, even though they do not necessarily involve the use of personal information (but they often do).

Although Solove has drawn up the framework primarily to aid in discussing legal protections in the domain of privacy and tort law, his taxonomy is nevertheless also useful for technologists. Given his comprehensive list of privacy-related wrongdoings, one can systematically analyze whether a particular piece of software and/or technology might increase the chances of such problem occurring, and how to mitigate it.

3.2.1.1 Conclusions

1. Privacy has a long history and is by no means a fad of modern times.

2. Technological shifts often open up new ways of how privacy can be affected, thus prompting the need to reassess one's understanding of what privacy is and how it should be protected. With the advent of modern telecommunication and computers, society's focus changed from bodily and territorial privacy to communication and information privacy. Ubicomp's embedded and ubiquitous sensors now reassert the importance of bodily and territorial privacy.

3. A final definition of privacy is difficult. Privacy is related to, but not identical with, secrecy, solitude, liberty, autonomy, freedom, intimacy, and personhood.

4. Privacy violations can be seen as involuntary border crossings, that is, whenever information permeates barriers without our help (or contrary to our efforts)—barriers such as sealed letters, closed doors, the trust of confidentiality with a close friend, or the passage of time.

3.2.1.2 Further Reading
Ken Gormley: One hundred years of privacy. *Wisconsin Law Review*, 1335, 1992.
> Examines the evolution of privacy law in the United States, with a particular focus on the varying definitions of privacy over time. Available from: http://cyber.law.harvard.edu/privacy/Gormley--100%20Years%20of%20Privacy.htm

Paul Sieghart: *Privacy and Computers*. Latimer, London, UK, 1976.
> One of the earliest books on modern information privacy and still an insightful read about the consequences of computerized data processing.

Daniel Solove: *Understanding Privacy*. Harvard University Press, Cambridge, MA, 2008.
> Solove offers detailed legal analyses of various privacy problems, based on his taxonomy (cf. Figure 3.4). The book is based on an earlier journal article (Solove, 2006), which is freely available for download: http://papers.ssrn.com/sol3/papers.cfm?abstract_id=667622

3.2.2 Motivating Privacy: Do People Care about Privacy?

Defining privacy might be a fruitless endeavor if its goals remain unclear. Why is it important to protect one's privacy? What can be gained by providing privacy, what would be at stake if it was lost?

Another way of differentiating the various conceptions of privacy can be found by distinguishing the various effects privacy has on people's lives, grouping them around the three functional concepts of zonal, relational, and decisional privacy. Zonal privacy protects certain spaces, such as one's home, workplace, or car. Relational privacy protects the relationships in an individual's life, such as intimate family relations between husband and wife, or between mother and child. Decisional privacy is what Beate Rössler, professor for philosophy at the University of Amsterdam, calls "securing the interpretational powers over one's life," the freedom to decide for oneself "who do I want to live with; which job to take; but also: what clothes do I want to wear" (Rössler, 2001).[*]

Privacy is thus needed for the autonomy of the individual, to protect one's independence in making choices central to personhood. Without privacy, such personal choices would inevitably have to be done in the open, with society at large (e.g., friends, family, neighbors, colleagues, superiors, and subordinates) scrutinizing and judging them, thus leaving no margin for error. Privacy provides us with room to experiment, a space to explore choices and values, in order to find the right balance between

---

[*] Quote translated from the original German text by the author.

our own goals and society's expectations. Westin (1967) describes this as follows:

> Each person is aware of the gap between what he wants to be and what he actually is, between what the world sees of him and what he knows to be his much more complex reality. In addition, there are aspects of himself that the individual does not fully understand but is slowly exploring and shaping as he develops.

Westin's allegory of playing roles in everyday life has its roots in the interaction theory of sociologist Erving Goffman, who described the fact that people routinely disclose and withhold information about themselves in a very selective fashion in order to maintain different fronts, or *faces*, for different audiences. This also connects with what Westin calls the emotional release functionality of privacy, moments off stage, where an individual can be himself, finding relief from the various roles he plays on any given day (Westin, 1967): "stern father, loving husband, carpool comedian, skilled lathe operator, union steward, water-cooler flirt, and American Legion committee chairman." Equally important in this respect is the "safety-value" function of privacy, for example, the "minor non-compliance with social norms," and to "give vent to their anger at 'the system,' 'city hall,' the boss":

> The firm expectation of having privacy for permissible deviations is a distinguishing characteristic of life in a free society.
>
> (WESTIN, 1967)

One important aspect of motivating privacy is, of course, to ask the data subjects directly. What type of information about yourself do you consider private? What type of actions would you consider to be privacy invasive? There are and have been many such surveys, the most prominent ones perhaps by Alan Westin, who has conducted more than 30 such surveys in the United States since 1978. In most of these surveys, Westin provided a summary that classified the respondents into three categories—privacy fundamentalists, privacy pragmatists, and privacy unconcerned—based on their replies. Westin described privacy fundamentalists as "generally distrustful of organizations that ask for their personal information" and "worried about the accuracy of computerized information and additional uses made of it."

Pragmatists instead would "weigh the benefits to them of various consumer opportunities and services, protections of public safety or enforcement of personal morality against the degree of intrusiveness of personal information sought and the increase in government power involved," whereas unconcerned would be "generally trustful of organizations collecting their personal information, comfortable with existing organizational procedures and uses, and ready to forego privacy claims to secure consumer service benefits or public order values." Westin measured the distribution of these three types over the years, typically finding about 55–60% pragmatists, 25–30% fundamentalists, and 10–20% unconcerned (see Figure 3.5).

As to the actual data that are considered private, answers similarly differ. In a November 1998 survey (see Figure 3.6), administered among 381 Internet users in the United States on comfort levels for disclosing personal information online, Cranor et al. (1999) found that large numbers

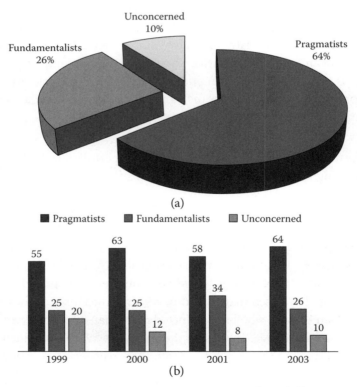

FIGURE 3.5 Westin Core Privacy Orientation Index, differentiating between privacy pragmatists, fundamentalists, and unconcerned. (Data from Westin, as cited by Kumaraguru and Cranor, 2005.)

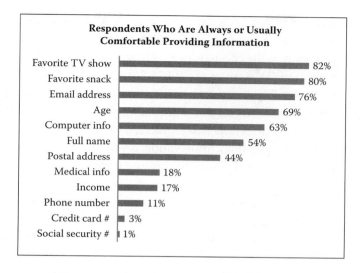

FIGURE 3.6 Results of a 1998 survey among 381 U.S. Internet users regarding the disclosure of personal information online. (Data from Cranor, Reagle, and Ackerman, 1999.)

were always or usually comfortable with disclosing their email address (76%), but only half would be comfortable giving out their full name (54%) or postal address (44%). Few said they would be comfortable giving out their telephone number (11%). A 2007 poll[*] among 1200 U.S. adults also found strong differences in privacy perception between different age groups. For example, only 35.6% of 18- to 24-year-olds consider someone posting a picture of them in a swimsuit to be an invasion of their privacy, compared to 65.5% of other respondents, and only 19.6% of 18- to 24-year-olds consider the publication of their dating profiles to be an invasion of their privacy, compared to 54.6% of other respondents.

At the same time, however, there is ample evidence that people value their privacy far less than they indicate in their survey replies. In Germany, recent polls show that some 64% of all citizens carry at least one consumer loyalty card with them at all times, with 34% of respondents using it for almost every purchase (TNS Emnid, 2006). Obviously, many of today's consumers are willing to disclose detailed shopping records in return for

---

[*] What is Privacy? Poll Exposes Generational Divide on Expectations of Privacy, Zogby/Congressional Internet Caucus Advisory Committee Survey. January 31, 2007 (http://www.zogby.com/search/ReadNews.dbm?ID=1244).

savings of often less than 0.5%. Several surveys found that users of (imaginary or prototypical) location-based services would not mind sharing this information widely with companies or other interested parties, given a small remuneration (Danezis et al., 2005). Some scholars have called for a "privacy as property" approach, where personal data becomes a commodity that people can sell, arguing that market forces might do a better job in protecting privacy than laws and regulations alone.* And the recent rise of publicly available personal information on blogs and social networking sites continues to astonish sociologists and legal scholars alike (Solove, 2007).†

Should people be allowed to sell large parts of their personal data, maybe even on a lifetime contract? Is there a benefit to having privacy, even if the data subject does not want it? One important alternative motivation for privacy goes beyond the value for the individual, and sees privacy as a social good necessary for the functioning of a democratic society. Representative of this paradigm shift was the so-called "census-verdict" of the German federal constitutional court (Bundesverfassungsgericht) in 1983, which extended the existing right to privacy of the individual (Persönlichkeitsrecht) with the right of *self-determination over personal data* (informationelle Selbstbestimmung). The judgment reads as follows:

> Those who cannot with sufficient surety be aware of the personal information about them that is known in certain parts of their social environment, and who cannot judge the amount of information that potential communication partners may know about them, can be seriously inhibited in their freedom to plan and decide in a self-determined manner. A society, in which the individual citizen would not be able to find out who knows what when about them, would not be reconcilable with the right of self-determination over personal data. Those who are unsure if deviant attitudes and actions are ubiquitously noted and permanently stored, processed, or distributed, will try not to stand out with their behavior. [...] This would not only limit the chances for individual development, but also affect public welfare, since self-determination is an

---

* See Litman (2000) for an overview and a critique.
† See, for example, http://www.davidhenderson.com/2009/01/21/key-online-influencer/ or http://www.seo-pr-tips.com/2009/01/26/facebook-new-big-brother/ for recent examples of online self-disclosure of personal information gone wrong.

essential requirement for a democratic society that is built on the participatory powers of its citizens.*

The concept of self-determination over personal data constitutes an important part of modern privacy legislation with respect to ensuring the autonomy of the individual. First, it extended classical data protection principles with a participatory approach, which would allow the individual to decide beyond a "take it or leave it" choice over the collection and use of his or her personal information. Second, it frames privacy protection no longer as only an individual right, but emphasizes its positive societal role. Privacy is thus seen not as an individual fancy, but as an obligation of a democratic society, as Julie Cohen (2000) notes†:

> Prevailing market-based approaches to data privacy policy...treat preferences for informational privacy as a matter of individual taste, entitled to no more (and often much less) weight than preferences for black shoes over brown, or red wine over white. But the values of informational privacy are far more fundamental. A degree of freedom from scrutiny and categorization by others promotes important noninstrumental values, and serves vital individual and collective ends.

3.2.2.1 Conclusions

1. Privacy plays an important role in human relationships, enabling us to create intimacy, as well as in personal development, supporting decisional autonomy.

2. Many people wish to control the flow of information about themselves, but they often differ widely about what types of information they want to control.

3. Although public interest often conflicts with the privacy of the individual, there is a strong societal benefit in offering its members at least some degree of freedom from scrutiny and categorization.

---

* BVerfGE 65, 154. Translation by the author. See http://www.servat.unibe.ch/law/dfr/bv065001.html for the full text in German. The quote comes from paragraph 154 (as indicated on the right-hand side of the page).
† As reprinted in the work of Solove, Rotenberg, and Schwartz (2006).

3.2.2.2 Further Reading
Colin J. Bennett and Charles D. Raab: *The Governance of Privacy*. MIT Press, Cambridge, MA, 2006.

> Gives an excellent overview on international privacy protection policy and examines regulatory instruments, particularly in light of technological change and globalization. Argues that privacy related problems are as much public policy issues as they are legal and technological ones.

Simson Garfinkel: *Database Nation*. O'Reilly, Sebastopol, CA, 2000.

> A vivid illustration of the myriad data collections taking place today, and the danger these might pose to the individual.

Daniel Solove: *The Future or Reputation*. Yale University Press, New Haven, CT. 2008.

> Discusses how recent trends in publishing personal information online—both by ourselves and by our friends or enemies—can seriously affect personal freedom and self-development. The full text is licensed under a creative commons and is freely available at http://docs.law.gwu.edu/facweb/dsolove/Future-of-Reputation/text.htm

3.2.3 Legal Background

Data protection and privacy laws provide an important aspect for understanding both the "Why?" as well as the "How?" of privacy protection. Laws and regulation mirror a social process that defines, for a particular culture, a set of practices that are acceptable and unacceptable. Looking at privacy laws thus helps us understand what society envisions privacy to be. At the same time, laws are also tools that may complement, support, or in turn rely on technical privacy tools.

The work of Warren and Brandeis in 1890 argued for a "right to privacy" in the realm of *tort law*, a part of the law that provides remedies for civil wrongs. In the 1960s, William L. Prosser described a set of four *privacy torts* that have since become established legal concepts (cf. with Solove's extended taxonomy in Figure 3.3):

1. Intrusion upon seclusion or solitude, or into private affairs

2. Public disclosure of embarrassing private facts

3. Publicity that places a person in a false light in the public eye

4. Appropriation of name or likeness

Whereas tort law addresses conflicts between two private parties, *privacy law* or *data protection law* regulates how the government and companies can collect and process personal information. The basis for all modern privacy laws are the Fair Information Principles, which were drawn up in the early 1980s by the Organization for Economic Cooperation and Development (OECD) and which describe eight practical measures aimed at harmonizing the processing of personal data in its member countries. By setting out core principles, the organization hoped to "obviate unnecessary restrictions to trans-border data flows, both on- and off-line." The eight principles are as follows:[*]

1. *Collection Limitation Principle.* There should be limits to the collection of personal data and any such data should be obtained by lawful and fair means and, where appropriate, with the knowledge or consent of the data subject.

2. *Data Quality Principle.* Personal data should be relevant to the purposes for which they are to be used, and, to the extent necessary for those purposes, should be accurate, complete, and kept up-to-date.

3. *Purpose Specification Principle.* The purposes for which personal data are collected should be specified no later than at the time of data collection and the subsequent use limited to the fulfillment of those purposes or such others as are not incompatible with those purposes and as are specified on each occasion of change of purpose.

4. *Use Limitation Principle.* Personal data should not be disclosed, made available, or otherwise used for purposes other than those specified in accordance with the *Purpose Specification* principle except

    (a) With the consent of the data subject or

    (b) By the authority of law

5. *Security Safeguards Principle.* Personal data should be protected by reasonable security safeguards against such risks as loss or unauthorized access, destruction, use, modification, or disclosure of data.

6. *Openness Principle.* There should be a general policy of openness about developments, practices, and policies with respect to personal data. Means should be readily available of establishing the existence and nature of personal data, and the main purposes of their use, as well as the identity about usual residence of the data controller.

---

[*] See http://www.oecd.org/document/18/0,2340,en_2649_34255_1815186_1_1_1_1,00.html.

7. *Individual Participation Principle.* An individual should have the right to

    (a) Obtain from a data controller, or otherwise, confirmation of whether the data controller has data relating to him

    (b) Have communicated to him, data relating to him

        i. Within a reasonable time

        ii. At a charge, if any, that is not excessive

        iii. In a reasonable manner

        iv. In a form that is readily intelligible to him

    (c) Be given reasons if a request made under subparagraphs (a) and (b) is denied, and to be able to challenge such denial

    (d) Challenge data relating to him and, if the challenge is successful, to have the data erased, rectified, completed, or amended

8. *Accountability Principle.* A data controller should be accountable for complying with measures that give effect to the principles stated above.

Even though the OECD principles carry no legal obligation, they nevertheless constitute an important international consensus that has substantially influenced national privacy legislation in the years since. Consequently, they provide a good starting point to measure any ubicomp application against, in order to assess its privacy compliance: Is the collection limitation principle adequately addressed? Can data quality be ensured? Are reasonable security safeguards in place? And how are data subject participation and data controller accountability supported in the system?

In practice, it depends on national legislation how these principles are incorporated into actual regulations. In today's legal landscape, two main approaches to privacy legislation exist: the sectorial approach, favored in the United States, and the omnibus approach in Europe. U.S. legislation features strong, overarching privacy laws only for the federal government, whereas state governments and private organizations are regulated on an "as-needed" basis with a variety of highly focused laws, such as the Driver's Privacy Protection Act of 1994, which safeguards an individual's motor vehicle record, or the Video Privacy Protection Act of 1988, which limits access to customers' video rental records. Laws protecting financial (Financial Modernization Act, also known as Gramm–Leach–Bliley Act) and health records (Health Insurance Portability and Accountability Act) are limited and only went into effect in 2001 and 2003, respectively. Probably most relevant to ubicomp applications are recent efforts in several U.S. states to address radio frequency identification (RFID) technology* and several rulings involving location privacy,† although traditional privacy law that addresses privacy at home, at school, or at work, is of course equally relevant (see, e.g., Solove and Schwartz, 2009, for an overview).

On the other side of the Atlantic, Europe has long since favored overarching frameworks that apply to both governments and commercial entities. Probably the most influential pieces of privacy legislation in recent years is the 1995 "Directive 95/46/EC of the European Parliament and of the Council of 24 October 1995 on the protection of individuals with regard to the processing of personal data and on the free movement of such data"‡ (often called the Data Protection Directive or "the Directive" for short). It requires all member states of the European Union (EU)§ to enact national law that offers the same level of privacy protections as those set forth in the Directive, in effect harmonizing privacy law across all EU member states and thus ensuring the free flow of information that the OECD had in mind with their fair information principles. The Directive was complemented by Directive 1997/66/EC¶ and later Directive 2002/58/EC** (which replaced 1997/66/EC) to offer specific guidelines on "the processing of personal data and the protection of privacy in the electronic communications sector" (called the "e-Privacy Directive" for short). Although directive 1995/46 sets

---

* See, for example, the Identity Information Protection Act of 2007 (SB30) in California (www.eff.org/issues/rfid/sb30facts), the New Hampshire House Bill 478 (rfidlawblog.mckennalong.com/2009%20NH%20H%20478.pdf) and the recently proposed New York Assembly Bill A275 (rfidlawblog.mckennalong.com/archives/privacy-new-york-assembly-bill-a275.html). At least 12 U.S. states have already introduced RFID legislation, see www.ncsl.org/programs/lis/privacy/rfid05.htm for an overview.
† See, for example, the recent ruling in Pennsylvania on mobile phone tracking (www.eff.org/press/archives/2008/09/11) or GPS tracking in *US vs. Jones* (www.eff.org/cases/us-v-jones).
‡ See http://eur-lex.europa.eu/LexUriServ/LexUriServ.do?uri=CELEX:31995L0046:EN:HTML
§ As of 2009, the EU has 27 member states: Austria, Belgium, Bulgaria, Cyprus, the Czech Republic, Denmark, Estonia, Finland, France, Germany, Greece, Hungary, Ireland, Italy, Latvia, Lithuania, Luxembourg, Malta, the Netherlands, Poland, Portugal, Romania, Slovakia, Slovenia, Spain, Sweden, and the United Kingdom.
¶ See http://eur-lex.europa.eu/LexUriServ/LexUriServ.do?uri=CELEX:31997L0066:EN:HTML.
** See http://eur-lex.europa.eu/LexUriServ/LexUriServ.do?uri=CELEX:32002L0058:EN:HTML.

out general guidelines, it is Directive 2002/58 that details their implementation in the telecommunications and networking sector, containing, for example, provisions for calls, communications, traffic data, and location data. By 2009, the 2002 e-Privacy Directive has again become dated and a review of the e-Privacy Directive is currently underway that attempts to take recent technological developments into account, for example, the rise of semipublic and private networks such as WiFi hotspots, which are not covered by the 2002 directive. In addition, the European Commission is also evaluating whether specific regulations for the "Internet of Things," in particular the rise of RFID technology, will be required.*

3.2.3.1 Conclusions

1. Data collection and processing does not happen in a legal vacuum. Most countries have laws in place that specifically address privacy issues, both for institutions (i.e., government or industry) and private persons. Ubicomp systems must at least be compliant with these national laws.

2. Laws also help to further understand privacy issues as they provide an overview of negative consequences of data collection practices (see, e.g., privacy torts).

3. The fair information principles are at the core of most modern privacy laws. They require that data subjects be informed of a data collection taking place, give their consent, and have access to the collected data. Security safeguards must be in place and data must only be used for the purpose it was collected under.

3.2.3.2 Further Reading

Electronic Privacy Information Center & Privacy International (Eds.): *Privacy and Human Rights Report 2006: An International Survey of Privacy Laws and Developments*. Electronic Privacy Information Center, Washington, D.C., 2006.

    An annual report that provides an overview of key privacy topics and reviews the corresponding legislation and state of privacy around the world. Full text of latest available edition should be online at the Privacy International Web site (see "Key PI Resources"): http://www.privacyinternational.org/.

Christopher Kuner: *European Data Protection Law: Corporate Regulation and Compliance*, 2nd Edition. Oxford University Press, Oxford, UK, 2007.

---

* See http://ec.europa.eu/information_society/policy/rfid/index_en.htm.

The definite (corporate) reference on EU data protection law. With a hefty price tag of £140, and more details on corporate compliance than you ever cared to know, this book is probably not what you want for your home bookshelf. But if you have a law faculty close-by, their library might hold a copy!

Daniel J. Solove, and Paul M. Schwartz: *Information Privacy Law*, 3rd Edition. Aspen, New York, NY, 2009.
A fascinating read that, despite its dry title, manages to vividly illustrate the many challenges and issues of modern (U.S.) information privacy law. With the help of many examples, the book provides a comprehensive overview that does not require a law degree nor law school attendance to understand.

3.2.4 Interpersonal Privacy

Laws and the Fair Information Principles are important tools in defining the roles of institutionalized data collectors such as companies or government agencies. However, they do not help if privacy issues arise between private parties, that is, between neighbors, friends, or family members. Neither does the law require parents to announce that they are monitoring the time that their children come home, nor do friends exchange privacy policies with each other before allowing continuous access to their instant messenger status. Clearly, the fair information principles are only of limited applicability when it comes to social etiquette and social norms.

Although tort law provides remedies for some privacy violations, such as intrusion or disclosure, these seem to be only last resorts when all friendship, neighborliness, or family bonds have already ended. How can ubicomp applications that offer ubiquitous information exchange between peers be designed in order to avoid having to go to court to ensure one's privacy? What aspects of privacy matter for people when they are not dealing with faceless companies or agencies, but with acquaintances, friends, and family?

In the 1970s, psychologist Irwin Altman looked at how people regulate their environmental privacy, that is, being alone versus joining social interactions (Altman, 1975). Instead of the simple view of privacy as a state of solitude, Altman saw it as a dynamic boundary negotiation process that encompassed the entire spectrum of social interactions, a "selective control of access to the self or to one's group." For Altman, the optimum level of privacy is reached when the *achieved* level of privacy reaches the *desired* level of privacy (i.e., the level of contact with others). Having more privacy than desired leads to the feeling of loneliness, having less than what is desired leads to annoyance and the feeling of being crowded. Privacy regulation, in Altman's sense, is thus the control of one's openness and closedness to others in response to one's desires and one's environment.

As a psychologist, Altman looked at behavioral mechanisms that support such privacy regulation: verbal interactions with others ("inputs and outputs") as well as spatial interactions ("personal space and territory"). These mechanisms are the tools by which one regulates one's privacy, by listening to others (input), talking to others (output), positioning oneself in relationship to others (personal space, i.e., distance, angle, etc.), and choosing one's location (territory).

Although Altman developed his theory for real-world interactions, there is much that can be learned from this theory, even in the context of ubicomp privacy:

- Privacy as a nonmonotonic function. By conceptualizing privacy not simply as one end of the social interaction spectrum (i.e., being alone), but applying it to the entire range of interactions, Altman shows that more privacy is not always better. Both "crowding" and "isolation" are suggested as "examples of privacy regulation gone wrong" (Palen and Dourish, 2003).

- Privacy as a social process. Humans do not use one-off policies and rules to manage their everyday, interpersonal privacy. Instead, they continually adjust their accessibility along a spectrum of "openness" and "closedness" with a variety of mechanisms, in order to match the achieved privacy state to the desired one.

Dourish and Anderson (2006) summarize it succinctly: "Privacy is not simply a way that information is managed but how social relationships are managed." Ubicomp systems that facilitate communication and awareness between peers must thus provide similar tools to allow users to dynamically adjust their inputs and outputs, their levels of openness and closedness, their personal space and territory so that they can achieve their desired level of privacy with respect to other users. In ubicomp systems, Altman's privacy regulation mechanisms of verbal, paraverbal (e.g., personal space or territoriality), and nonverbal communication need to be augmented with (or replicated by) explicit and implicit controls that allow in situ adjustments, rather than a simple configuration panel that allows one to set the desired privacy level to high or low. Privacy is thus not so much a particular state of being, but a tool for social discourse.

3.2.4.1 Conclusions

1. Interpersonal privacy cannot be approached by requiring purpose specifications and recipient lists. Between individuals, privacy is a tool for social discourse that allows one to manage peer groups.

2. In this context, privacy is an ongoing boundary negotiation process and not a monotonic function where more secrecy translates to more privacy. Yes, you can have too much privacy (resulting in isolation).

3. Privacy tools must support in situ adjustments, rather than separating configuration and action (e.g., in a separate control panel).

3.2.4.2 Further Reading

Paul Dourish and Ken Anderson: Collective information practice: Exploring privacy and security as social and cultural phenomena. *Human–Computer Interaction*, Vol. 21, Lawrence Erlbaum, pp. 319–342, 2006.

> Dourish and Anderson argue that the commonly formulated abstract goals of attaining privacy and security should be replaced by thinking about information practices—common understandings of the ways how information should be shared, managed, and withheld. They offer a range of pointers to ethnographic studies that illustrate how privacy is used as a boundary regulation process between social groups.

Battya Friedman, Peter H. Kahn Jr., and Alan Borning: Value sensitive design and information systems. In: P. Zhang and D. Galletta (Eds.), *Human-Computer Interaction in Management Information Systems: Foundations*. M.E. Sharp, New York, NY, pp. 348–372, 2006.

> Value sensitive design is a process that explicitly prompts designers to think about and incorporate human values in a principled and comprehensive manner. In the context of privacy, this process may help to understand the reasons behind and the mechanisms of existing and planned information sharing practices.

Leysia Palen and Paul Dourish: Unpacking "Privacy" for a Networked World. *Proceedings of the SIGCHI Conference on Human Factors in Computing Systems (CHI '03)*, Ft. Lauderdale, FL, April 5–10, 2003. ACM Press, New York, NY, pp. 129–136, 2003.

> Based on Altman's theory, Palen and Dourish define a framework of privacy in the context of modern information systems and point out tensions that govern interpersonal privacy regulation in everyday life.

3.3 TECHNICAL SOLUTIONS FOR UBICOMP PRIVACY

By first understanding what privacy is and what we are trying to solve, we can now set out to design and implement privacy features in ubicomp applications. The previous section illustrated the underlying legal and social aspects of information collection and sharing, thus helping ubicomp application developers to properly formulate their privacy goals. This section attempts to illustrate various technical solutions to common ubicomp privacy challenges. We focus in particular on three examples: smart spaces (Section 3.3.3), RFID (Section 3.3.4), and location-based services (Section 3.3.5). Obviously, these three areas do not fully capture the variety of challenges presented in ubicomp (as discussed in Section 3.3.1), but they hopefully serve as useful examples. Also, since many ubicomp privacy issues apply to information systems in general, additional context can be found from the fields of computer networking, databases, and information retrieval (see Section 3.3.2).

3.3.1 Novel Ubicomp Challenges to Privacy

Privacy and data protection have always been closely related to what is technically feasible. At the end of the nineteenth century, it was the invention of modern photography that prompted Warren and Brandeis to rethink the concept of legal privacy protection. At the beginning of the twentieth century, laws had to be reinterpreted again to take into account the possibilities of modern telecommunication (again, then Supreme Court Judge Brandeis played a large part in that). And in the 1960s and 1970s, it was the implementation of efficient government through the use of modern databases that required yet another update of privacy laws, resulting in the first of today's modern data protection laws with their focus on data self-determination. In each instance, technology changed what was possible in the everyday setting and thus prompted—if sometimes with considerable delay—a realignment of our notion of privacy.

After the rise of Internet e-commerce in the 1990s had initiated the last round of updates, the dawn of ubiquitous computing promises the next revolution of smart things. Even though many ubiquitous computing visions sound like artificial intelligence revisited, applications such as the "intelligent car" or the "smart home" might not face the same fate as the dreams of intelligent machines that some 20 years ago researchers thought of being just around the corner. Ubiquitous computing often solves a much more mundane yet important problem, namely, crossing media boundaries. Using miniature sensors, cheap microchips, and wireless communication, computer technology can penetrate our everyday lives in a completely unobtrusive manner. Similarly, real-world facts and phenomena can be mapped on a computer with an unprecedented reliability and efficiency. The boundary between the real and virtual world seems to disappear—it will soon be possible to comprehensively track real-world interactions in real time on a computer system, making it look like a game simulation (think SimCity®).*

Data protection and privacy are all about these mappings: translating facts of the real world into bits of information that can be stored for later retrieval. Ubiquitous computing is about the digitalization of information about our lives in order to allow computer systems to automatically process it. It comes as no surprise that ubiquitous computing has the potential to yet again change our perception of privacy in a significant manner. This qualitative quantum leap can be traced along five aspects of ubiquitous computing systems: the collection scale, the manner and the motivation, as well as the data types and the data accessibility.

3.3.1.1 Collection Scale
The conscious surveillance of the actions and habits of our fellow men is probably as old as mankind. In the "good old times," when people lived in small villages and close-knit social circles, this type of observation was implicit in our daily interactions: Everybody knew everybody else, and local news and gossip spread fast. Noncompatible people often preferred to move out into the large cities, in which the large number of citizens and their high variance rendered this classical method of direct social monitoring impractical.

With the rise of automated data processing, machines began to take over the role of the curious neighbor. At first only available to governments, automated data processing soon found its way into commerce, in both cases facilitating a much more efficient management by providing detailed population or inventory information. However, although our neighbors would quickly note anything out of the ordinary, machines were now used to actually determine what was ordinary: Not the deviations of the norm were noticed and tracked, but the average citizen and his or her ordinary everyday.†

With ubiquitous computing, real life monitoring—the surveillance of the ordinary—will extend beyond today's credit card transaction, telephone connection records, and Web server logs. Even without assuming a single homogeneous surveillance network such as Orwell's Big Brother, the sheer applicability of ubiquitous computing technology in diverse

---

* See, for example, Sandy Pentland's SenseNetworks for a taste of what is to come: http://www.sensenetworks.com/.
† See, for example, Partridge and Golle (2008) for a description of several large-scale activity studies in the United States.

areas such as hospitals and nurseries, kindergartens, schools, universities, offices, restaurants, public places, homes, cars, shopping malls, and elderly care facilities, will create a comprehensive set of data trails that will cover us anywhere we would go.

The "always on" vision of ubiquitous computing—alleviating us from laboriously switching various devices on and off as everything stands ready to our attention, right when we need it—will drastically extend this coverage over time. Instead of the spotty trails that can be obtained through our Internet logs when we are online, say, after work for an hour or two, smart homes and intelligent environments will not be switched off at night or while we are gone for lunch. In fact, it might not be even possible to turn such devices off, as they would not feature a corresponding on-off switch, but would sleep most of the time to preserve energy and wake up on their own whenever something of interest happens to them. The digital coverage of our lives will be anywhere and anytime, from sunrise to sunset, from cradle to grave, 24 hours a day, seven days a week. As Grudin (2001) points out, the actual selection of data that is captured and stored will at the same time significantly alter the value of that information: "Anything that is recorded instantly achieves a potential pervasiveness and immortality that it did not have before ... Anything that does not 'make the cut'... is invisible to someone inspecting the digital record at a different location or time."

3.3.1.2 Collection Manner
When little children play hide and seek, they often cover their eyes with their hands assuming that if they cannot see, others will not see them in turn. Although they will learn eventually that the principle of reciprocity does not hold in this case, this apparent childish belief is much more difficult to unlearn than we might want to believe. Even years after playing their last game of hide and seek, many will assume that if they cannot see anybody else around, their actions will go unnoticed.

In the old days, this principle of reciprocity was actually a reasonable approximation of the collection manner in which people's actions were observed. Only when one was out in public, were others able to see and draw their inferences. Once we entered the sanctuary of our own homes or those of others, we were shielded from the prying eyes of the public. This dichotomy of public and private was closely associated with the realities of space—the architecture of walls, windows, and doors, or the natural environment of woods and dense thickets: The presence and quality of a physical boundary provided an immediate indicator of the (potential) quality of privacy. With the rise of electronic transactions, day-to-day actions such as talking to a friend (over the phone) or buying groceries (using a credit card) became noticeable beyond such physical boundaries. The presence or absence of others was not a good approximation of privacy anymore, as the digital trace of a transaction could be observed, stored, and retrieved from potentially anywhere in the world.

The deployment of ubiquitous computing technology will make it even more difficult to differentiate between public and private actions. As ubiquitous computing tries to hide the use of technology, to make computers practically invisible, the level of awareness for such electronic transactions will drop drastically from today's implicit awareness through the use of physical tokens such as credit cards or mobile phones. In a fully computerized environment, potentially any item could take fingerprints and wire them halfway around the world, take pictures, measure body temperature, or observe one's gait in order to draw far-reaching conclusions about a person's physical and mental state. Neither data collection nor continuous surveillance activities will have recognizable markers that would indicate the publicity of actions—ultimately requiring us to assume that at any point in time, in any location, any of our actions could potentially be recorded electronically and thus made public.

3.3.1.3 Data Types

With ubiquitous computing, also the type of information that is collected will change. The village gossip was based on the observation of neighbors and fellow citizens and on a person's discussions with others. This information was by definition "soft" information, that is, it was based on an individual's personal reception and more often that not, two different people observing the same fact would retell widely different accounts of it. Although this would often result in rather exaggerated claims, it nevertheless retained some level of deniability.

Modern data processing seems far away from the village gossip of old. It concerns itself with "hard" information—with facts, rather than hearsay. Instead of capturing the individual (and error-prone) human perception, it collects factual information such as names, birth dates, addresses, income levels, or lists of purchases. Using statistical models, this information can subsequently be used to draw inferences on a person's life based on his or her residence and shopping preferences.

Ubiquitous computing will extend this selection of hard facts beyond traditional information types: smart shirts and underwear will be able to record health data such as blood pressure, heart rate, perspiration, or glucose levels in real time; smart supermarket shelves will not only know what items a person bought, but also in what order and how long he or she hesitated before reaching out; mobile phones with global positioning system (GPS) locator already allow friends and family to know one's whereabouts at anytime—in the future this will be standard unless one decides to turn the service off and find a good excuse for doing so.

Data mining technology will allow researchers, politicians, and marketers to make sense of this ever-increasing stream of minute details, by correlating widely disparate information such as chocolate consummation and shower habits (e.g., to infer the beginning of a new relationship) and through comparing information from hundreds of similar people in order to discern broad population patterns. This also has significant implications for the anonymization of such data, as perceived information such as one's location over the course of a day, or the particular way of walking as registered by floor pressure sensors, or one's individual breathing pattern, might turn out to be easily identifiable even if collected in a completely anonymous manner.

With a wide array of new sensors and collection mechanisms, ubiquitous computing technology will potentially allow inferring the "soft" gossip of old, based on the "hard" facts of today, thus not only giving it new credibility (by being based on facts, not hearsay) but also eventually incapacitating our own judgments about personal beliefs and feelings based on computerized self-assessments, for example, inferring our emotional attachment to our partner based on our heart rate and eye blinking rate.

3.3.1.4 Collection Motivations
As we have seen in the previous section on privacy and its motivations, incentive (i.e., the "Why?") plays an important role when it comes to facilitating or preventing data collection. And just as the reasons for wanting privacy have changed over the years, so have the motivations for collecting this data.

Our neighbor's eyes and ears looked for the unusual, the out-of- ordinary events that would make for attractive gossip. Consequently, people who were adept at "blending in," those who hardly attracted attention due to their ordinary lives and average physical features, would get the least scrutiny. With automated data processing, attention shifted from the unusual to the ordinary: Governments tried to make better policies by having better data on whom they governed, and that meant finding out what the average citizen did, liked, or feared. Companies tried to find out what goods consumers wanted (or did not yet know they wanted). Questionnaires were used (and still are) to solicit the preferences of the masses, in order to better understand what products would work and which would not.

With modern data analysis methods, large amounts of statistical information, such as family income, street address, or political preferences, can be statistically correlated in order to segment population groups and predict human behavior (e.g., a family moving into the suburbs might soon decide to buy a lawn mower, as most families living there own one already).

Providing better services and/or better products will still be at the heart of many future ubiquitous computing systems, yet what data are necessary to predict this become less and less clear, as more different types of information can be collected. With better data mining capabilities than ever before, virtually anything can be of importance, if only enough statistical data on it can be collected. Context awareness is one of the main paradigms in ubiquitous computing, as it is thought to enable otherwise "dumb" systems to predict the user's needs and intents without involving any actual intelligence. Not surprisingly, the more such context information is available, the better these systems are expected to perform. Instead of targeted data collections of specific information for a certain purpose, future ubiquitous computing systems could easily attempt to collect any and all information possibly available, thus maximizing their chances for correctly determining the user's context and intent from it.

3.3.1.5 Data Accessibility

Information is only of worth if one can find it: collecting large amount of data without having efficient retrieval mechanisms in place suggests not collecting it in the first place. In the old days, retrieving gossip was typically limited to a particular village or neighborhood. By moving into a different town or even into a larger, anonymous city, the previously assembled body of "knowledge" would typically be rendered inaccessible for the newly acquired neighbors, requiring them to start out anew.

With modern information networks, information can travel quickly around the globe, and modern database management systems allow for the efficient retrieval of minute details out of huge, federated databases from a wide variety of sources. However, even though standardized interface definitions exist, integrating these sources is far from a trivial problem, as the large number of failed data integration projects in both government and industry have shown.

In the vision of ubiquitous computing, such types of information systems would not be primarily designed with humans in mind (and thus will lead to often noninteroperable systems), but directly target machine-to-machine interactions: Smart things would "talk" to other smart things in order to collaboratively determine the current context, and large networks of autonomous sensor nodes would send sensor readings back and forth in order to arrive at a global state based on hundreds of individual sensor readings. Similarly, improved human-computer interfaces (see Chapter 6) would allow easy access to nontraditional data formats such as video and audio streams, for example, for automated diary applications that would document one's everyday in a continuous multimedia format. Living in a world of smart cooperating objects, the freedom of movement for personal information would be greatly increased, both between humans and computers (How well can I search your memory?) and between cooperating artifacts (What is my artifact telling yours?).

3.3.2 The Basics: Privacy Enhancing Technologies

Using (information) technology to protect privacy is as old as the first databases that appeared in the 1960s. Consequently, a number of techniques exist that form the general toolbox of privacy-compliant data processing, often abbreviated as PETs—privacy enhancing technologies. Many of them operate on the networking layer or databases, and as such might be readily applicable to ubicomp applications. Others are more reliant, for example, on traditional Web interfaces and thus require some sort of "ubicompification" to be of use. Today's PETs can be grouped roughly into two sets: transparency tools and opacity tools.

Opacity tools describe the more traditional security approaches, namely, support for authentication and confidentiality. Traditional communication link encryption, for example, secure socket layers or secure shell, that prevent an attacker from learning the contents of a communication, can be complemented by *unobservability* tools that attempt to prevent attackers from even learning that a communication took place. This line of research was instigated in the 1980s by David Chaum's Mix-Net protocol (Chaum, 1981), in which email is not sent directly from sender to recipient, but is routed through a cascade of "mixes"—network nodes that combine (delay) and forward messages in a seemingly random fashion. Using public key cryptography, the sender embeds the message—like a Russian doll—in several layers of encryption. Each layer is readable to only one particular mix node and reveals only where to send the message next. An external observer will not be able to trace the flow of messages through the system, even if a significant fraction of the mix nodes get compromised. *Online* mixes carry this concept over to Internet traffic in general, allowing participants to route any Internet connection through a "cloud" of mixes, making it impossible for an attacker to associate, for example, Web site visits to a particular client.* A similar application is anonymous or pseudonymous payment, again pioneered by David Chaum in the late 1980s, where completely anonymous online payments can be made (Chaum, Fiat, and Naor, 1988).†

In the area of authentication, so-called *identity management tools* attempt to disconnect identity from authority, that is, allowing a user to prove his or her authority to access a resource without revealing his/her identity. Although early commercial systems often equated the term with simple "single-sign-on" solutions where multiple logins would be seamlessly performed by a central login service, today's systems such as Microsoft's CardSpace (which is integrated into Microsoft Vista®) use cryptography-based certificates to prove, for example, that one is older than 18 years, without the need for disclosing any identifying personal information.‡ Figure 3.7 shows the Identity Selection Screen in Windows Vista®.

Maintainers of medical databases have long since identified the need to anonymize their records, without changing important statistical properties that might help researchers, for example, epidemiologists, to identify important patterns in the collected data. Research in so-called *statistical databases* offers guarantees on the viability of several statistical measures while anonymizing individual datasets. A common measure for the degree of anonymity achieved is *k-anonymity*, indicating that a particular record or piece of information could be from *k* possible people (Sweeney, 2002). The higher the value of *k*, the higher the level of anonymity.

Opacity tools are complemented by another set of PETs, so-called *transparency tools*, which in turn attempt to improve the data subject's understanding and control of his or her data profile. *Watermarking* systems allow the visible or invisible marking of information in order to trace the origin of a particular piece of data. This is already in widespread use

---

* For commercial implementations see, for example, JonDo at https://www.jondos.de/en/ or Anonymizer.com at http://www.anonymizer.com/.
† Obviously, the majority of today's online payments are not anonymous, which just goes to show that it takes more than good technology to change a multibillion dollar business (market forces and political preferences clearly dominate such developments).
‡ A Canadian company called Zero Knowledge Systems (now called Radialpoint) pioneered this approach in the late 1990s. Their Freedom Network product allows paying customers to create several digital identities (called "nyms") that they could use to associate their Internet activity with. While the technology was impressive, the company ultimately failed to get consumers to pay for this (the product was discontinued in 2002). Clearly, the idea of "choosing a nym" in advance of each and every Internet session, and the added overhead of "managing" one's set of nyms (i.e., "which one should I choose for this particular action?") ultimately did not appeal to Internet users.

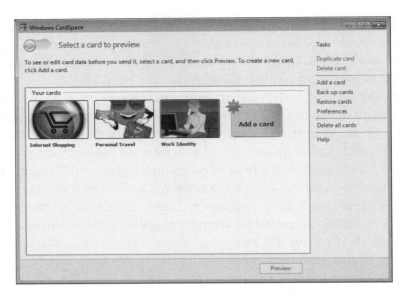

FIGURE 3.7 Windows CardSpace™ (built into Windows Vista®) is an example of a modern identity management system. (Microsoft product screenshot reprinted with permission from Microsoft Corporation.)

in digital rights management system, for example, in digital photography, where the rights holder can use such watermarks to prove authorship/ownership of a particular image. Although such techniques can readily be used to tag personal pictures, the ability to watermark other personal data is limited. People have long since used spelling errors in names and addresses when signing up, for example, for mail-order catalogs, in order to track the origin of (real-world) spam mail. The same approach can, of course, be used with one-time email addresses, but its use with other data items (e.g., preferences) is doubtful.

Policy tools use a similar approach in storage systems in order to allow for the privacy-compliant data processing of personal information. All collected data are stored together with a set of metadata that describes the allowed recipients, uses, and storage duration. Each database operation requires operators to specify the identity of the requestor and the reason for the query (purpose). This can then be used by the policy engine to return only those entries with a compatible privacy policy. If the requester is using a compatible system, the queried data in turn will carry metadata that allows further privacy-compliant processing. This approach is sometimes called the "sticky policy" paradigm (Casassa Mont et al., 2003) or

"Hippocratic databases" (Agrawal et al., 2002). With such a system in the background, data subjects can receive detailed information on the use and storage of their personal information, whereas both companies and data protection agencies can perform automated audits to verify the compliance with data protection law and regulations. Much work has also been undertaken in policy languages, that is, describing both the data processing guarantees by data collectors and the privacy preferences by data subjects. The Platform for Privacy Preferences (P3P) is probably the most prominent one: it uses XML syntax to describe both data processing policies and subject preferences—the latter in an addendum specification called "APPEL—A Privacy Preferences Exchange Language" (Cranor, 2002).

Note that all of the above PETs are centered around traditional data processing and Web use. Although PETs are an important building block to any ubicomp privacy solution, they must be complemented and extended in order to be useful in ubicomp settings, as outlined in Section 3.3.1.

3.3.3 Example: Protecting Smart Spaces

Smart environments are probably the archetypical ubicomp application: a smart room or house that constantly detects the context and activities of its inhabitants and visitors, and seamlessly adapts its services and functions to provide the best possible experience. Initial work on ubicomp privacy has thus focused on both infrastructures that collect these data and on interfaces that allow data subjects to inspect and control the corresponding information flows.

One prominent privacy-aware ubicomp infrastructure is the Confab Toolkit by Hong and Landay (2004). Data in Confab are managed in InfoSpaces, network-addressable logical storage units that store context information about a single entity, that is, a person, a location, a device, or a service. In- and out-filters manage data flows between different InfoSpaces, with *in-filters* only allowing the storage of data from trusted sensors or entities, and *out-filters* enforcing access policies and adding privacy tags to all outgoing data (see Figure 3.8).

Privacy tags are similar to the general "sticky policy" concept presented in Section 3.3.2, as they represent metadata that can be used to enforce[*] privacy-compliant usage and retention. However, Confab's privacy tags are more custom tailored to the exchange of dynamic context data, featuring

---

[*] Enforcement, of course, relies on the use of InfoSpaces throughout the entire data life cycle.

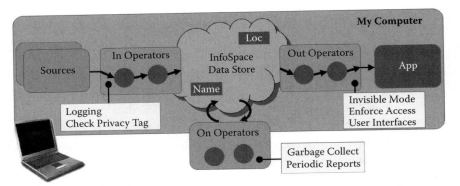

FIGURE 3.8 Confab architecture. Personal data are stored in InfoSpaces that contain attribute tuples. Data flows are governed by in- out-, and on-filters. (From Hong, J., 2004. Reprinted with permission from http://www.cs.cmu.edu/~jason/research.hml)

elements that declare how many "sightings" the other party may amass of a particular attribute (e.g., only retain the last five locations a person was in) and a "garbage collect" declaration that can contain data deletion triggers (e.g., deleting information if the data subject leaves a particular area). To provide plausible deniability, information that is deemed too sensitive to be released will simply be marked by out-filters as "unknown," making it indistinguishable from technical failures or lack of connectivity.

Whereas Confab provides a framework for disseminating context information that data subjects collect themselves, the PawS system (Langheinrich, 2002) focuses on third-party data collection instead. PawS addresses smart environments that can communicate and enforce data collection practices for various optional and mandatory data collections. Based on a Hippocratic Database[*] in the back (privacy DB), it uses *privacy proxies* to control data collection and access for smart devices such as cameras or printers. *Privacy beacons* advertise the URLs of these proxies so that user devices can download information about a smart environment and then directly configure the available services as needed. Figure 3.9 shows an overview of the architecture, using the example of a smart room equipped with a camera and a publicly accessible printer.

As the concept of privacy beacons illustrate, it is difficult to convey the act of a data collection taking place in smart environments, which should

---

[*] Cf. transparency tools in Section 3.2.

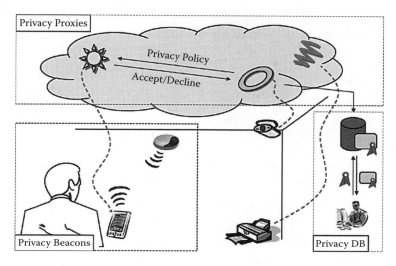

FIGURE 3.9 The PawS System uses privacy beacons to advertise data collections taking place. Data subjects can use privacy proxies to configure and control available devices. A PrivacyDB stores collected data using sticky policies. (From Langheinrich, M. (2002). *Ubiquitous Computing, 4th International Conference Proceedings*, pp. 315–320. New York: Springer. With permission.)

operate unobtrusively in the background. A handheld user device is needed in PawS to report the current "smartness" of a room to a data subject, and to give the data subject control over the respective data flows and services. However, PawS does not address how to best present such information and controls to the user, which is further complicated through the use of small screens on mobile devices. What type of information is important to users in such smart environments? What aspects of such a user interface are the most critical?

One of the earliest projects that tried to answer such questions was undertaken by Bellotti and Sellen (1993) at Xerox's EuroPARC in the early 1990s. As part of an audio-video presence and collaboration environment called RAVE, cameras, monitors, microphones, and speakers were deployed in offices to allow EuroPARC's staff to glance into other offices (i.e., get a few seconds of video-only transmission), make *v-phone calls* using both audio and video, or install a longer lasting office-share (i.e., a semipermanent v-phone call). Hong and Landay's (2004) Confab toolkit also addresses user interface issues. Confab not only attempts to deliver

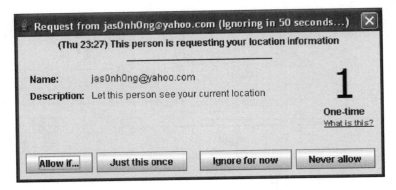

FIGURE 3.10 A location prompt in the Lemming Location-Enhanced Instant Messenger. If the user ignores the request, Confab reports back an "unknown" location, thus facilitating the principle of "plausible deniability." (From Hong, J. Reprinted with permission.)

simple and appropriate controls, as well as clear and timely feedback of the current data collection status, but also incorporates the principle of *plausible deniability* into its design.

A sample application built with Confab was the "Lemming Location-Enhanced Instant Messenger," which allowed users to automatically and semiautomatically publish their current location to their contacts. Figure 3.10 shows an example prompt. The large "1" illustrates a one-off query and contrasts it with queries that seek to continuously track the user's movements. If the user ignores the request, the Confab toolkit returns his or her location as "unknown" to the requestor, which does not allow one to differentiate between network errors, system errors, and user decisions. Users can thus plausibly deny that they ever got the request.

Common to all these examples is trust users need to have in the smart infrastructure surrounding them. Once personal data leave the protection that these spaces offer, no guarantees can be made. Equally important is the challenge of properly configuring these spaces, so that the infrastructure does what data subjects want them to do.

3.3.4 Example: Protecting RFID Tags

RFID tags (Figure 3.11) represent probably the most prominent ubicomp technology, at least when it comes to privacy issues. Their increasing use, especially on consumer products in showcase supermarkets, as implants

FIGURE 3.11   Set of example RFID tags (all passive).

for entering "in"-clubs, and as a security feature of electronic passports, have prompted widespread public concern over RFID privacy issues.[*] The privacy challenges of RFID tags are fourfold:

1. Automation. Reading an RFID tag typically does not require the help of the person carrying the tag, nor any manual intervention on behalf of the reader. Thus, simple reader gates can easily scan large numbers of tags, making data acquisition much easier.

2. Identification. The ability to identify individual items instead of only whole classes of items significantly improves the ability to identify an individual. This would facilitate, for example, the creation of detailed consumer or citizen profiles.

3. Integration. Not only the act of reading a tag can be completely hidden from the tag carrier (especially when operating at larger distances), but also the fact that a tag is present in a particular product will be hard to ascertain for an individual without special detection equipment.

---

[*] There is a bewildering variety of technologies that are often lumped together under the umbrella term "RFID." This section will focus on passive RFID tags, that is, those that do not come with their own power source (battery) but instead receive the energy to operate from the reader's field. For the definite source on RFID technology, see Finkenzeller (2003).

4. Authentication. The above points become especially critical given the increasing amount of sensitive information, for example, health information, payment details, or biometric data that are stored on or linked to tags used in authentication systems.

These four attributes of RFID applications threaten two classes of individual privacy: data privacy and location privacy. The location privacy of a person is threatened if a tag ID that is associated with that person is spotted at a particular reader location. For example, by knowing that a person's car has passed a certain toll station, or that a person's shoes have entered a particular building, others might be able to infer (though not prove) the location and ultimately the activity of that person. These IDs do not need to be unique—certain combinations of non-unique tags might still form unique constellations of items that can be used to identify an individual (Weis, 2003).

Once tags carry more than just an identifier, but also a person's name or account number, data privacy may be violated. This happens if unauthorized readers eavesdrop on a legitimate transaction, or if rogue readers trick a tag into disclosing its personal data. These types of attacks are typically called *skimming*. A special case of data privacy are product IDs that disclose details of the (otherwise not visible) belongings of a person, for example, the types and brands of clothing one is wearing, the items in one's shopping bag, or even the furniture in a house. Note that in the latter case, the actual identity of the victim might very well remain unknown—it might be enough to know that this person carries a certain item.

There are three principal means of violating an individual's data and/or location privacy: clandestine scanning, eavesdropping, and data leakage:

- Clandestine Scanning. The tag data is scanned without the tag carrier's consent. This might disclose personal information (data privacy) either indirectly, for example, by revealing the contents of bags that one cannot see through otherwise, or directly, for example, by revealing personal data such as the name of a user or the date that a particular item was bought. If several clandestine scans are pooled, clandestine tracking can reveal a data subject's movements along a tag reading infrastructure (location privacy).

- Eavesdropping. Instead of reading out a tag directly, one can also eavesdrop on the reader-to-tag channel (or even the tag-to-reader channel) and receive the IDs of the tags being read due to the used anticollision protocol (cf. Section 3.3.4.1).

- Data Leakage. Independent of the actual RFID technology is the threat of having applications read out more information from a tag than necessary, or storing more information than needed. This, of course, is a threat common to all data gathering applications, although the envisaged ubiquity of RFID-based transactions renders it highly relevant in this context.

Security is obviously the primary issue here, that is, ensuring the confidentiality of the information stored on such tags so that only authorized parties are able to detect, read, and potentially write to such tags. Ensuring the confidentiality of information, both through the encryption of transmissions and the use of authentication mechanisms to limit access to it, is an old problem and many solutions exist that make it virtually impossible for an attacker to succeed. However, RFID tags add two novel challenges to the problem that render many existing approaches infeasible:

- Limited resources. Like other small devices (e.g., sensor nodes) in ubicomp, RFID tags are extremely limited in their computational capabilities. Implementing advanced encryption algorithms such as AES* or public key systems has, so far, been only feasible for high-end chips that are specifically engineered for security applications. Cheap mass market tags that are envisioned to be embedded in virtually all products or product packages currently cannot be protected this way.

- Key selection. In order to authenticate an authorized party against an RFID tag, some shared secret must exist between the two.[†] This, in itself, is no different from securing, say, a computer with a password. However, with potentially hundreds or thousands of tags, knowing which particular secret to use becomes a problem. This is difficult because one typically does not know which tag one is interacting with—after all, being able to identify an RFID tag is exactly what one wants to prevent.

The following sections will present a number of approaches that have been developed to enhance the privacy of RFID.

---

* AES stands for Advanced Encryption Standard. It is a symmetric encryption algorithm (a "block cipher" chosen by the U.S. government in 2001 for encrypting all unclassified information).

† A shared key assumes the use of symmetric encryption. Although public key cryptography does not require a shared key, the general principle remains, as one still needs to select the right public or private key.

3.3.4.1 Communication Confidentiality and Anticollision Protocols

Encrypting the wireless communication channel between an RFID tag and a reader is the obvious solution to eavesdropping attacks. A large body of work in the crypto community specifically targets low-power and low-complexity ciphers, and RFID chips used in ePassports or contactless train passes typically use some sort of communication encryption. As pointed out above, however, key management is still critical and thus this approach is only usable for selected applications.

However, even if tag communication is encrypted, the nature of RFID communication might still allow an attacker to track the tag. This is because RFID readers first need to *singularize* a tag before they can read from it; that is, it must determine which tags are present in its range and then address exactly one of them with a command.* To this end, many RFID protocols use a unique ID (UID), similar to the hardware MAC address of a WiFi card or a Bluetooth adapter, to aid tag singularization at a lower protocol level. If this UID is fixed, rogue readers can simply perform a round of singularization and thus be able to track individual tags, even if the tag payload (e.g., the name of a product and its product ID) is encrypted. This process is called an *anticollision* protocol.

Some RFID anticollision protocols offer variants where the UID is randomized upon every reader session; that is, whenever a tag enters the field of a reader, it chooses this UID anew and uses it only until it leaves the field again. This, of course, drives up tag prices, as each tag needs to include a pseudorandom number generator (PRNG) for this. However, if this PRNG is well implemented, tracking tags by their UID is much harder, if not impossible.

A particular attribute of RFID communication is the power asymmetry of the wireless link: although tag replies can only be read from up to a few meters or even centimeters, reader commands can typically be received from up to 100 meters away. This is because the battery-less, passive tags are receiving their energy supply through the reader radio field, which thus needs to be very powerful. If any of the reader commands contain the ID (or UID) of an identified tag, an attacker could obtain this information from a very large distance, even though the defined readout range of the tags is only several centimeters.

Although this is not much of a problem if random UIDs are used, many high-performance RFID systems use fixed UIDs in their anticollision

---

* The need for singularization stems from the fact that RFID tags are unable to detect the presence of other tags. Thus, it is up to the reader to detect collisions between tags.

protocols in order to lower readout times. In these so-called *deterministic* anticollision protocols, readers probe for conflicting UIDs by systematically querying for all possible prefixes, that is, "all tags beginning with a 0," "all tags beginning with 1," "all tags beginning with 00," "all tags beginning with 01," and so on. If the UIDs of two or more tags begin with a queried prefix, all of them will reply at the same time, causing a collision that the reader is able to detect. If this happens, the reader simply increases the length of the prefix by one digit (e.g., by adding a "1" to it) and tries again, until only a single tag replies. After it successfully singularized a single tag this way, it will first instruct this particular tag to be silent in the upcoming rounds, then backtrack and replace the last bit that it added with its inverse (e.g., if it added a "1" before it will now use a "0" instead) and continue. Should more collisions occur, it again increases the length of the prefix until it can singularize a tag, then backtrack. Such protocols are called *binary tree walking* protocols, as this behavior can be seen as traversing a binary tree, in which the individual bit positions are the branches and the tags are the leaves (see Figure 3.12 for an example).

An attacker can overhear the prefixes sent from a reader to its tags and thus infer many partial UIDs of the tags present. A simple fix for this is to have the reader traverse the tree one bit at a time, that is, the reader never explicitly sends any prefixes, but only uses the command "transmit next bit." Tags, in turn, only reply with the *n*th bit of their UID, not with their full UID. As long as the corresponding bit positions of all available tags are identical, no collision occurs and the reader is able to clearly receive the common bit prefix, building it up incrementally. Once two or more tags

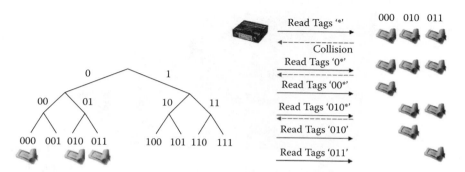

FIGURE 3.12   Binary tree walking protocol with three tags present. Note how the reader commands transmit the tag IDs in the clear.

differ at a particular position, the reader detects a collision and now has to branch into one of the two subtrees. Instead of explicitly naming the subtree that it wants to explore (i.e., sending "0" or "1"), the reader XORs its selection with the previous, error-free bit in the prefix. As the value of this bit was only sent from the tags to the reader, an attacker outside the tag's communication range (but inside the reader's forward channel) will not be able to know the true value of the next selected bit. The tags, on the other hand, know their own ID, and accordingly the bit value at the previously queried position, thus sharing a common secret with the reader that can be exploited for every conflicting bit position.

3.3.4.2 Access Control/Tag Deactivation

A simple solution to access control is to obstruct the reader signal via a metal mesh or foil that encloses the tag. With the inclusion of RFID tags into passports, a number of vendors now offer coated sleeves for protecting the passport while not in use (Figure 3.13). Other options would be aluminum-lined shopping bags for groceries. However, tagged clothing or personal items can often not be protected in this extreme manner.

For tree-based anticollision protocols, researchers have proposed a specially engineered "blocker tag" to jam readers that attempt to traverse the UID tree. Such a blocker tag uses two antennas to simply send out both a "0" and a "1" for each reader query, thus creating a collision at each branch of the tree.

FIGURE 3.13 Low-tech privacy solution—an electromagnetically shielded sleeve for ePassports. (Copyright 2009 Paraben Corporation. Image reprinted with permission from http://www.paraben-forensics.com/.)

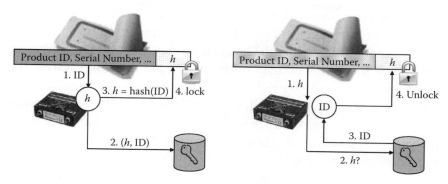

FIGURE 3.14 Example of an ID-based hash lock.

This creates the impression of trillions[*] of tags being present and will in effect fully stall even the fastest reader. The difficulty of this approach lies in controlling the range of tags that are being blocked. Otherwise, a single blocker tag could accidentally disable any (tree-based) RFID system it comes close to.

Alternatively, some form of password could be set on each tag, only allowing readers that use the right password to read out any data from the tag. This could be realized in a very simple manner using so-called *hash locks*. An RFID tag is, by default, open and readable, but it can be locked by sending it a lock command with a password. The tag then creates a hash of this password (which could simply be its true ID number) and hitherto only replies to a query if an unlock command is sent that, when hashed, matches the stored hash value (Figure 3.14).

As elegant and simple as the solution is, it squarely falls into the *key selection* issue mentioned above: how would a reader (or user) know which password to use with which tag? Although individual passwords work well for single accounts or special items such as an RFID-based health card, it seems unlikely that one would be able to remember hundreds of passwords for the individual groceries one buys each day. If users have only a few tagged items, fast readers might simply try out all possible passwords until one of them eventually unlocks the tag. Clearly, such an approach does not scale well.

One option might be to have locked tags always reply with the hashed password, but nothing else. This would still protect the tag's contents without disclosing the real password, while allowing authorized readers to

---

[*] Fully simulating all possibilities of a, say, 64-bit ID would be actually more than just a few trillions. An (implausibly) fast reader able to read 100,000 tags per second would be busy for more than 4 billion years reading all 2^{64} tags.

lookup the right password in a hash-to-password lookup table. However, such an approach might still protect the data privacy of the tag owner but fail miserably at protecting the location privacy, as static identifiers can trivially be tracked. A number of researchers are thus investigating the use of *hash chains*, where a locked tag still replies with its stored hash but continually rehashes it after each reply (Ohkubo et al., 2005). This results in constantly changing hashes, which makes tracking much more difficult (although not impossible). Authorized readers that know the right password can similarly follow such a chain and can thus properly predict the password to use. The challenge of hash-chain schemes lies in synchronizing the two, even in the presence of an attacker who might try to desynchronize them. Another important aspect of such schemes is their *forward security*; that is, if an attacker learns the current key of a tag, she is nonetheless unable to identify previous outputs of the tag (e.g., in log files).

Whether hash chains are used, the problem of managing various passwords for a plethora of tagged items remains the biggest challenge in any access control scheme for RFID. In addition to the sheer number of items that need protection, it is also the number of ownership changes that challenge traditional approaches: An item is shipped from a supplier to a distributor to a retailer, who sells it to a consumer, who might later give it to a friend or even resell it. At each step along the way, the current password must either be passed along or some means for setting a new password must be given. This problem of *ownership transfer* has been the focus of several approaches, that is, allowing either the temporary delegation or full transfer of ownership to a product, without allowing the previous owner to know the new key or the new owner to know the old key (Molnar et al., 2005).

3.3.4.3 Proxies

A common response to many of the problems detailed above has been the proposition of a powerful *proxy device* that would locally manage one's tagged items. By incorporating an RFID reader into a commodity consumer device such as a mobile phone or a wristwatch, this device could act as a proxy for all tag interaction: individually blocking or allowing access to its owner's tagged items (i.e., acting like a smart blocker tag), setting individual passwords on items bought in the supermarket or resetting those of items given to others. Other proposed features are logging and alerting functions that would require readers to identify themselves and offer links to machine-readable privacy policies that could be the basis for automatically regulating access to the owner's tags.

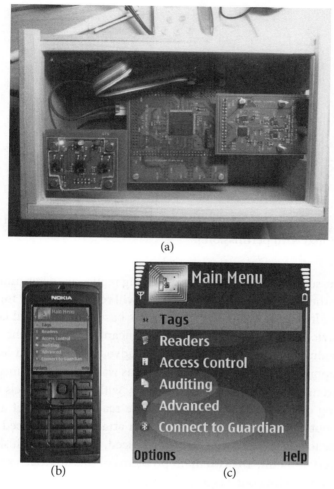

FIGURE 3.15 Privacy Guardian reader hardware and user interface. (From Rieback M. Reprinted with permission.)

Although such devices have been proposed several times in the literature, few actual implementations exist. One such example is the *Privacy Guardian* at the Free University of Amsterdam (Rieback et al., 2006). A portable RFID reader in a 10 × 20 centimeter box is paired via Bluetooth with a control application running on a mobile phone (see Figure 3.15, left to right). Users can scan for tags in the vicinity, control access to them, and log readout attempts (and successes) of other readers. The privacy guardian works similar to a blocker tag, jamming the channel whenever a third-party reader attempts to access a protected tag (see Figure 3.16). This only works, however, for a

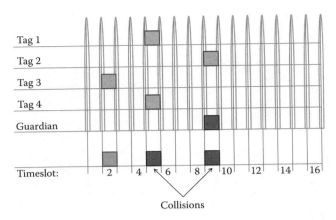

FIGURE 3.16 Privacy Guardian blocking access to tag no. 2. (From Rieback, M. Reprinted with permission.)

particular set of RFID protocols in which tag replies are deterministic, that is, where their UID determines when they will reply to a reader inquiry.

As an alternative to control approaches based on powerful consumer proxy devices (which might imply a significantly management overhead for individuals), several researchers have advocated password-less access control approaches. One of the earliest ideas was to have tags measure the signal strength of a reader signal and reply with different levels of detail, depending on the inferred distance of the reader (Fishkin et al., 2005). To read out the highest level of detail, an attacker would need to come very close to a target and thus become noticed, making stealth skimming attacks very difficult. The two main problems of this approach are the difficulty of controlling one's tags in this manner ("How close is enough for a level 3 disclosure, without reaching the too detailed level 2?") and the costs and reliability of an integrated signal strength module on an RFID tag (an attacker could use a very high powered reader to simulate closeness).

An alternative, password-less solution wraps tag data into several encryption layers requiring continuous read access for significant amounts of time, thus slowing down an attacker (Langheinrich and Marti, 2007). Together with small antennas that restrict tag read ranges to several centimeters, an attacker would not only need to come close but also stay there for, say, a few minutes. Based on Adi Shamir's theory of shared secrets, the tag's real ID is encoded into several pieces ("shares") and these get stored on the tag instead. Due to the properties of Shamir's theory, the original tag ID can only be reconstructed if all of those pieces are known. Although all shares are stored on the same tag, readout is complicated by allowing only a random trickle of bits from the tag. Together with a short read range, this requires an attacker to spend a considerable amount of time in close proximity to the "target," making quick unnoticed readouts difficult. At the same time, however, legitimate owners are able to use simple caching strategies to identify their items instantaneously, as an initial burst of disclosed bits is enough to probabilistically identify a tag from a known set. To prevent the repeated querying of such a larger initial subset, which would give an attacker faster access to the entire key, tags use random temporary IDs for tag singularization, thus making it more difficult for an attacker to correlate two such bit strings across consecutive queries.

3.3.5 Example: Protecting Location Information

Knowing when a particular person was at a particular point in time is at first no different from knowing other facts about a person, for example, her age, place of birth, or favorite hobby. However, location information is typically associated with a particular place (e.g., home or pawn shop), which in turn often implies an activity (e.g., visiting family or selling items) or a certain personal interest (e.g., betting). As such, location information gives rise to a large number of implications, with more or less correlations. Moreover, knowing the *trajectory* of a person's movements often allows an observer to corroborate such implications through causal connections (e.g., "he first went to a gun dealer and then to the victim's home"). Knowing a person's current or favorite location can moreover threaten her personal well-being (e.g., stalker attacks) or make her more susceptible to disturbances (e.g., spam).

Location information is thus in many cases a particularly sensitive piece of personal data, even though people might not realize it (cf. survey results reported in Section 3.2.2). In particular, the problem of location privacy can be framed as the problem of separating three distinct pieces of information: Who? Where? When? Knowing only one or at most two of these pieces typically lowers the value of this data significantly (but not in all cases). For example, knowing that someone entered a supermarket at 2:00 p.m. might not carry much value, but knowing that someone entered a particular bedroom at 3:00 a.m. might tell parents when their son came home on Friday night.

Many technical systems for providing location privacy thus attempt to disassociate such time-identity-location tuples. One of the first location systems, the Active Badge System at Xerox's Palo Alto Research Center (Spreitzer and Theimer, 1993), already separated location tracks from users with the help of two separate architectural entities: *User Agents* and *Location Query Services*. Active Badges emitted a *pseudonymous* ID while being tracked by the location infrastructure. The badge sighting for each place was managed by one Location Query Service, which would register all sightings of these pseudonyms without knowing the user's identity. The true identity behind such an ID was only known to the user's personal User Agent, which would explicitly register an interest in a particular pseudonym with each Location Query Service. To query a user's identity, one would need to contact the user's User Agent, which could then handle the appropriate access control verification.

The location privacy of Active Badge users thus rested on the integrity of the location infrastructure: As long as an attacker does not have access to the full log files, the only way to find out about someone's present and past whereabouts was through his or her personal User Agent. However, once the system is compromised, all static pseudonyms can be trivially associated with a user, instantly providing access to all past and future location traces. Even if an attacker gains only access to the pseudonymous logs, without being able to find the matching User Agents to resolve the pseudonyms, it might be trivial to infer the true identity behind each pseudonym, as movement patterns often allow the inference of one's home or office (see Figure 3.17).

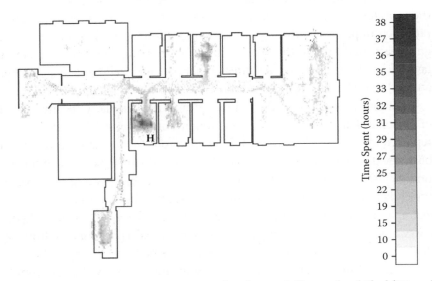

FIGURE 3.17 Static pseudonyms can often be trivially resolved if additional information is available. For example, the above location trace is most likely from the owner of office "H." (From Beresford, A. *Location Privacy in Ubiquitous Computing*. PhD dissertation, University of Cambridge, U.K. 2005. With permission.)

Consequently, in order to truly decouple identity from location-time tuples, pseudonyms must frequently change. This change cannot be simply changing one random number into another, as the two location tracks from these pseudonyms would be trivial to join again later. Instead, switching pseudonyms has to be done in a manner that prevents the location track of the new pseudonym to be associated with the previous one. Beresford and Stajano (2003) propose so-called *mix zones* for this—areas in which no location tracking takes places and which are large enough so that at any point in time, a large enough number of targets are present that can be mixed. Figure 3.18 shows an example of three users entering a mix zone, using pseudonyms *a*, *b*, and *c*. With location tracking disabled inside the mix zone, it is difficult to infer who *q*, *r*, and *s* are. The challenge of this approach is the proper definition of mix zones (would people mind those dead zones where no tracking is available?), as well as adjusting them dynamically to the actual traffic (late night traffic might require much larger zones in order to enclose enough users to mix). Algorithms that attempt to use the crossing of two or more location tracks in order to increase the chances that an attacker confuses the path of different users are also called *path perturbation* algorithms.

One obvious solution to location privacy seems to be the use of *self-positioning systems*, such as GPS, Cricket (Priyantha et al., 2000), or BlueStar (Quigley et al., 2004), where the infrastructure sends out positioning information that clients can use to compute their own location. Although this obviously alleviates concerns that arise with a positioning infrastructure that tracks its clients, it might still disclose this information once a location-based service is used. For example, a

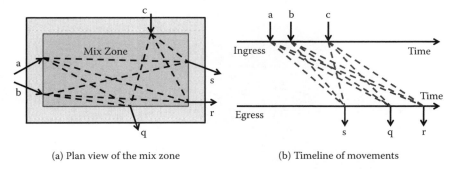

FIGURE 3.18 Three users enter a mix zone; three different ones exit it some time later. Who went where? (From Beresford, A.R. and Stajano, F. *IEEE Pervasive Computing*, 2(1), 46–55. With permission.)

GPS-enabled mobile phone that uses an online mapping tool to visualize its location obviously discloses its location to the online service.

Instead of trying to separate identity from location and time, one can also try to *obfuscate* them, that is, lower their precision or accuracy. For location information, this could mean either to widen the area in which the subject is located in (e.g., from exact coordinates to street level to city level) or to offset the reported position by some (seemingly random) value from the true position (see Figure 3.19). A related technique is to create additional queries from "dummy positions" and to hide the true position in there. Identity information can similarly be obfuscated, meaning that one enlarges the pool of possible identities, for example, making it impossible for an attacker to distinguish between several potential senders of a location-based request. Because even anonymous location information might reveal an identity, given additional information about one's home or office, this process of identity obfuscation thus necessarily implies location obfuscation as well.

(a) Original GPS data

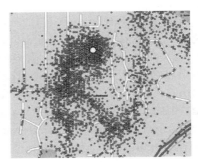
(b) Additive Gaussian noise

(c) Discretized to points on grid

FIGURE 3.19 Obfuscation can simply mean adding some noise to the true position, or to discretize it. (Courtesy of J. Krumm.)

It is important to realize how obfuscated location systems must be used in practice. One obvious option is the use of stored location tracks that are only of statistical or historical significance. This might prevent attackers from using such logs to their advantage, while still allowing, for example, statistical inference over many such tracks. Their online use is more problematic, because degrading the accuracy or precision of position information when using a location-based service might directly affect the quality of the service, or increase the cost of its use (e.g., sending 19 false queries to hide one's true location might incur 20 times as much service charge).

An often used measure for the strength of such an approach is using the above-mentioned concept of k-anonymity within the context of a location system: The location infrastructure pools location requests from at least k users and changes all queries so that the location information given by each user is sufficiently imprecise to make them indistinguishable from the $k - 1$ other users. This pooling can also be done by the clients themselves, by using ad hoc communication and routing location requests among several peers first—similar to a mix network (see Section 3.3.2)—until an "exit" node then actually sends off the request (using a properly adjusted location area that also comprises the original node).

Kriplean et al. (2007) propose an access control system to stored location information that takes the past location information of the inquirer into account: queries are only executed on data that the inquirer could have seen herself at the time; that is, the returned data are centered around a small area of the inquirer's own location trace. This can still offer useful services, for example, for finding lost objects of the owner ("Where did I last see this item?") or to refresh one's own memory ("Who did I meet this morning in the hallway that I wanted to send email to later?"). Obviously, users need to trust the infrastructure to properly enforce this concept of *physical access control*. Thus, this solution is much more about how to give users control over who can access their data—in this case, a default rule would limit access to people who were colocated.

3.3.5.1 Conclusions
1. Ubicomp challenges privacy due to its novel technical capabilities, large coverage (both in time and space), novel types of data collected, vague collection purposes, and envisioned data interchange. Although standard technological fixes exist (PETs), they do not fully cover the many peculiarities of ubicomp applications.

2. Research in privacy for smart environments attempts to uncover and control information flows between data subjects and system operators, and between users of the system. "Sticky policies" allow for the propagation of use policies, but defining and adjusting such policies is a challenge for users. Smart environments require novel user interfaces to allow for the inspection and control of their information flows.

3. RFID applications envision highly automated environments that allow for the unobtrusive detection of tagged artifacts, and implicitly the people carrying those. Due to the high number of items and the lack of user interface, offering users to control who can read what tags is difficult. The low resources available on passive RFID tags additionally challenge the use of traditional security protocols.

4. The data created by location-based services can either be anonymized or obfuscated to protect the identity of the data subject, yet with the help of some simple background information, attackers can circumvent much of this. These techniques also affect service quality, as the fidelity of information is often significantly decreased. Even if the localization service is trusted, there still remains the problem of allowing users to simply control location disclosure to others.

3.3.5.2 Further Reading

Alastair R. Beresford: Location Privacy in Ubiquitous Computing. PhD dissertation, published as technical report UCAM-CL-TR-612, University of Cambridge, Cambridge, UK, January 2005.
    A good overview of the threats and issues in location privacy, and a detailed discussion of the concept of mix zones.

Matt Duckham and Lars Kulik: A formal model of obfuscation and negotiation for location privacy. In: *Pervasive Computing*. Third International Conference, PERVASIVE 2005, Munich, Germany, May 8–13, 2005. LNCS 3468, Springer, Berlin, pp. 152–170, 2005.
    Duckham and Kulik define a formal model for obfuscation and outline methods for using obfuscation in practice.

John Krumm: A survey of computational location privacy. *Personal and Ubiquitous Computing*, 13, 2009. Special Issue on Privacy in Ubiquitous Computing (forthcoming). A prior workshop submission with the same title is available at www.vs.inf.ethz.ch/events/uc07privacy/program.html.
    Provides a concise overview of the literature and serves as an excellent starting point.

3.4 HOW TO ADDRESS PRIVACY IN YOUR UBICOMP WORK

What is there to do if you want to accommodate privacy in your ubicomp system? How would you design your ubicomp application to make sure it respects and supports the privacy of its users? As noted in Section 3.1, there is no simple answer, no set of algorithms or routines that, when applied, will fix the privacy issue once and for all.

What the preceding sections hopefully illustrated is the *scope* of the privacy problem (technical, legal, social), and that simply having a good firewall or implementing strong 128-bit security are not enough. Instead, one needs to carefully analyze each and every ubicomp application: What is it supposed to do, and how does it do it? How are users using it in their daily routines? And what technical and organizational tools are available to support privacy and security under these circumstances? The three steps described below try to illustrate the type of question you should be asking, in order to increase the chances of "getting it right."

3.4.1 Understand Your Application (Consider Users and Use)

If there is no single answer, no single set of "to do's" that will ensure the privacy of our users, then each solution must be more or less unique, depending on the particular ubicomp system and application one wants to build. Consequently, you will first need to understand how this application (or the potential applications that are being built on top of this system) is supposed to work: What problem does it try to solve? How do people currently solve or work around this problem? How are people expected to use your application? And maybe also: How are people *actually* using your system?

These questions are, by no means, specific to privacy, but simply good interaction design practices, as summarized, for example, by Reimann (2001):

> Interaction Design is a design discipline dedicated to defining the behavior of artifacts, environments, and systems (i.e., products) and therefore concerned with:
>
> - Defining the form of products as they relate to their behavior and use
> - Anticipating how the use of products will mediate human relationships and affect human understanding
> - Exploring the dialogue between products, people, and contexts

Understanding users and their use of your application is critical for assessing the privacy implications (Lederer et al., 2004). As discussed in Section 3.2, privacy issues do not happen in a vacuum: privacy is not a monotonically behaving function (i.e., the more, the better), so one will need to understand when someone needs what access and what type of control to what type of data. Simply providing anonymity or access control does not create "privacy."

Smith et al. (2005) illustrate this point nicely in a field study involving a location-aware mobile phone application called "Reno." Participants could use Reno to automatically or manually share their current location with their friends or family. Users who shared their location information were mostly spouses or close friends, so a high level of trust existed and thus the willingness to share one's current location was high. However, users still reported incidents of perceived privacy violations, for example, when an automatic trigger that they configured suddenly notified others of their location, but in a context that they did not anticipate:

> Participant g: "My phone disclosed my location to [participant a] last night when I was out running an errand and was returning to my house ([a] has a trigger for my house). I'm now reconsidering that trigger because I felt weird about that one. I didn't feel weird when he got notified I was "home from work" but did about the late-night errand running."

In addition, Friedman et al. (2006) propose to explicitly incorporate *values* into the design of information system, a process they call *value sensitive design*. Value sensitive design incorporates conceptual, empirical, and technical investigations in order to identify both direct and indirect stakeholders, uncover their values and beliefs, describe the social context in which technology is used, and uncover the hidden values that are inherent in a particular technological solution. An example on how to apply this methodology in the context of ubicomp can be found in the work of Freier et al. (2005) or in the works of Camp and Connelly (2007).

3.4.2 Define the Problem (Think Attacker Model in Security)

Once you understand how your application is supposed to work, how you expect your users to interact with it, and what assumptions and values are coded into the system design, you can start defining what needs protection, that is, what type of privacy you want to provide to your users. If you do not know what you are trying to protect, and what you want to protect it against, how can you know when you are successful? Defining privacy is a fractal problem and if you do not set out clear limits, you will find yourself endlessly chasing "but, what if …?" situations.

Felton (2003) discusses these types of questions as having a long tradition in computer security, running under the term *threat model* or *attacker model*[*]:

> The first rule of security analysis is this: understand your threat model. Experience teaches that if you don't have a clear threat model—a clear idea of what you are trying to prevent and what technical capabilities your adversaries have—then you won't be able to think analytically about how to proceed. The threat model is the starting point of any security analysis.

As we described in Section 3.2, privacy is not only about an active "adversary" that is "threatening" your privacy (although this, of course, also happens). Privacy is also about opportunistic data use (i.e., data that has been given for one purpose is recycled for another) or involuntary disclosures (e.g., someone who is entitled to receive information *in general* should not have gotten this information in a *particular*, unexpected situation). Consequently, we will need to cast a slightly broader net and not only think about *attacks* on our data, but on *opportunities* for unwanted disclosure, based on actual and potential information flows (cf. Figure 3.4) within and between applications. Hong (2004) consequently prefers to talk about *privacy risk models* instead of privacy *threat* models.

Hong and Landay (2004) propose to analyze personal information flows in an application from two viewpoints: the *social and organizational* context, and the *technological* context. The social and organizational context looks at application stakeholders and their use of the system (see Section 3.4.1), but with a particular focus on the actual information flow, for example, "Who are the *data sharers*, who are the *data observers*?" or "What kind of data are shared and what is the value proposition for sharing it?" The technological context then examines the mechanisms collecting

---

[*] There is actually a subtle but important difference between the two terms, but for our purposes this does not really matter. See, for example, McGraw (2006) for a discussion.

information (push vs. pull; one-time vs. continuous; granularity, accuracy, precision), the means of solicitation (opt-in or opt-out; automatic or manual), and how storage and access are handled (retention, encryption, access control). This analysis then forms the basis for privacy risk management, where identified privacy risks are prioritized and translated into design guidelines. Prioritization is based on a cost-benefit analysis that factors the likelihood for an unwanted disclosure, the damage resulting from this, and the cost of protection.

It is important to note that similar types of exercises have become common in industrial and governmental projects involving the collection and processing of personal data. So-called *privacy impact assessments* are recommended by several data protection agencies to ensure legal compliance to national and international privacy laws, for example, by the Australian Privacy Commissioner[*] (Clarke, 2008) or the UK Information Commissioner.[†]

3.4.3 Know Your Tools (Get the Technical Details Right)

Last but not least, you have to make sure that you know the capabilities and limits of current security and access control technology, so that you neither reinvent the wheel nor prematurely assume a problem solved. This is especially true if multiple technologies converge. Ubicomp applications often face severe technological challenges, both in terms of resources (energy, processing power, storage) and interfaces (e.g., small screens, gesture input). Relevant work includes security for wireless sensor networks (Perrig et al., 2004) or RFID (Juels, 2005), as well as security and privacy interfaces (Cranor, 2005).

The data collected in ubicomp applications also pose novel challenges to anonymization methods (Agrawal et al., 2002)[‡] and online subject access (Roussopoulos et al., 2008). As discussed in Section 3.3, even anonymous data collected in a ubicomp application may easily be deanonymized, for example, in the context of location privacy. The problem of subject access[§]

---

[*] See http://www.privacy.gov.au/publications/pia06/.
[†] See http://www.ico.gov.uk/upload/documents/pia_handbook_html/html/foreword.html.
[‡] See http://en.wikipedia.org/wiki/AOL_search_data_scandal and http://www.aolstalker.com/ for an example of how difficult it is to anonymize search records.
[§] Many privacy and data protection laws guarantee the data subject access to the data that are collected about him or her.

is exacerbated by the latent identifiabilty of anonymized human activity, physiognomic and movement data. The rapidly growing area of *reality mining* (Eagle and Pentland, 2006) might not only provide means to make sense of such sensor data, but also offer clues as to their proper anonymization.

3.4.3.1 Conclusions

1. "Solving" the problem of personal privacy in ubicomp applications requires the careful analysis of application use and users, information flows and privacy risks, and technical options.

2. Privacy risk analysis and privacy impact assessments can help to identify privacy issues in actual and planned deployments of ubicomp applications.

3. Traditional technology tools to provide security, anonymity, and privacy need to take the special challenges of ubicomp applications into account, for example, resource-constraint operations and large-scale behavioral data sets.

3.4.3.2 Further Reading

James Waldo, Herbert S. Lin, and Lynette I. Millett, *Engaging Privacy and Information Technology in a Digital Age*, Computer Science and Telecommunications Board, National Academies Press, Washington, DC, 2007
  The book is the result of a multiyear study committee on Privacy in the Information Age, sponsored by the Computer Science and Telecommunications Board of the U.S. National Research Council. It presents a comprehensive and multidisciplinary examination of privacy in the information age, not directly focusing ubicomp applications, but with a wide enough view that looks toward current and future technological developments. The full text is available for free from http://books.nap.edu/catalog.php?record_id=11896#toc.

David Wright, Serge Gutwirth, Michael Friedewald, Elena Vildjiounaite, and Yves Punie (Eds.), *Safeguards in a World of Ambient Intelligence*, Springer, Dordrecht, 2008.
  Based on the results of an EU research project with the same title (SWAMI for short) the book illustrates the threats and vulnerabilities of ubicomp applications via four "dark scenarios." The authors analyze and identify safeguards to counter the foreseen threats and vulnerabilities, and make recommendations to policymakers and other stakeholders on how to protect privacy in future ubicomp scenarios.

BIBLIOGRAPHY

Agrawal, R., Kiernan, J., Srikant, R., & Xu, Y. (2002). Hippocratic databases. *Proceedings of the 28th International Conf. on Very Large Databases (VLDB 2002)*, Hong Kong, China, pp. 143–154.

Altman, I. (1975). *The Environment and Social Behavior: Privacy, Personal Space, Territory and Crowding*. Monterey, CA, USA: Brooks/Cole.

Bellotti, V., & Sellen, A. (1993). Design for privacy in ubiquitous computing environments. *Proceedings of the Third European Conference on Computer Supported Cooperative Work, ECSCW '93*, Milano, 13–17 September 1993, pp. 75–90, Kluwer.

Beresford, A. R. (2005). *Location Privacy in Ubiquitous Computing*. Cambridge, UK: University of Cambridge.

Beresford, A. R., & Stajano, F. (2003). Location privacy in pervasive computing. *IEEE Pervasive Computing*, 2(1), 46–55.

Beresford, A. R., & Stajano, F. (2003). Mix Zones: User privacy in location-aware services. *Proceedings of the First IEEE International Workshop on Pervasive Computing and Communication Security (PerSec) 2004*. Piscataway: IEEE.

Camp, L. J., & Connelly, K. (2007). Beyond consent: Privacy in ubiquitous computing. In: A. Acquisti, S. De Capitani di Vimercati, S. Gritzalis, & C. Lambrinoudakis (Eds.), *Digital Privacy: Theory, Technologies and Practices*, pp. 327–346. Boca Raton, FL, USA: Auerbach Publications.

Casassa Mont, M., Pearson, S., & Bramhall, P. (2003). Towards accountable management of identity and privacy: Sticky policies and enforceable tracing services. *Proceedings of the 14th International Workshop on Database and Expert Systems Applications (DEXA '03)*, September 1–5, 2003, Prague, Czech Republic, pp. 377–382. IEEE Society.

Chaum, D. (1981). Untraceable electronic mail, return addresses, and digital pseudonyms. *Communications of the ACM*, 24(2), 84–90.

Chaum, D., Fiat, A., & Naor, M. (1988). Untraceable electronic cash. *Proceedings of the 8th Annual International Cryptology Conference on Advances in Cryptology (CRYPTO 1988). LNCS 403*, pp. 319–327. New York: Springer.

Clarke, R. (2008). Privacy impact assessment in Australian contexts. (L. Ralph, Ed.) *eLaw Journal*, 15(1), 72–93.

Cohen, J. E. (2000). Examined lives: Informational privacy and the subject as object. *Stanford Law Review*, 52, 1373–1438.

Cranor, L. F., Reagle, J., & Ackerman, M. S. (1999). *Beyond Concern: Understanding Net Users' Attitudes About Online Privacy*. AT&T Labs Research.

Cranor, L. (2005). *Security and Usability*. Sebastopol, CA, USA: O'Reilly.

Cranor, L. (2002). *Web Privacy with P3P*. Sebastopol, CA, USA: O'Reilly.

Danezis, G., Lewis, S., & Anderson, R. (2005). How much is location privacy worth. *Fourth Workshop on the Economics of Information Security*, Harvard University.

Dourish, P., & Anderson, K. (2006). Collective information practice: Exploring privacy and security as social and cultural phenomena. *Human-Computer Interaction*, 21(3), 319–342.

Eagle, N., & Pentland, A. (2006). Reality mining: Sensing complex social systems. *Personal and Ubiquitous Computing, 10*(4), 255–268.

Felten, E. (2003). DRM, and the First Rule of Security Analysis. Retrieved January 30, 2009, from Freedom to Tinker: http://www.freedom-to-tinker.com/blog/felten/drm-and-first-rule-security-analysis

Finkenzeller, K. (2003). *RFID Handbook* (2nd ed.). Chichester, UK: Wiley & Sons.

Fishkin, K., Roy, S., & Jiang, B. (2005). Some methods for privacy in RFID communication. In: C. Castelluccia, H. Hartenstein, C. Paar, & D. Westhoff (Eds.), *European Workshop on Security in Ad-hoc and Sensor Networks (ESAS 2004), LNCS 3313*, pp. 42–53. Berlin: Springer.

Freier, N. G., Consolvo, S., Kahn, P. H., Smith, I., & Friedman, B. (2005, April 3). *A Value Sensitive Design Investigation of Privacy for Location-Enhanced Computing*. (A. Light, P. J. Wild, A. Dearden, & M. Muller, Eds.) Retrieved January 30, 2009, from Workshop on Quality Value(s) and Choice: Exploring Wider Implications of HCI Practice (QVC) at CHI 2005, Portland, OR, USA. http://www.eng.cam.ac.uk/~pw308/workshops/QVC/papers.html.

Friedman, B., Kahn, P. H., & Borning, A. (2006). Value sensitive design and information systems. In: P. Zhang, & D. Galletta (Eds.), *Human-Computer Interaction in Management Information Systems*, pp. 348–372. New York: M. E. Sharpe, Inc.

Grudin, J. (2001). Desituating action: Digital representation of context. *Human-Computer Interaction, 16*(2), 269–286.

Hong, J. I. (2004). Privacy risk models for designing privacy-sensitive ubiquitous computing systems. *Proceedings of the 5th Conference on Designing interactive Systems: Processes, Practices, Methods, and Techniques (DIS '04)*, Cambridge, MA, USA, August 1–4, 2004, pp. 91–100. New York: ACM.

Hong, J. I., & Landay, J. A. (2004). An architecture for privacy-sensitive ubiquitous computing. *Proceedings of the Second International Conference on Mobile Systems, Applications, and Services (MobiSys 2004)*, June 6–9, 2004, Hyatt Harborside, Boston, Massachusetts, USA, pp. 177–189. New York: ACM.

Juels, A. (2005). RFID privacy: A technical primer for the non-technical reader. In: D. S. Raicu, & K. Strandburg (Eds.), *Privacy and Technologies of Identity: A Cross-Disciplinary Conversation*, pp. 57–74. Berlin: Springer.

Kriplean, T., Welbourne, E., Khoussainova, N., Rastogi, V., Balazinska, M., Borriello, G., et al. (2007). Physical access control for captured RFID data. *IEEE Pervasive Computing, 6*(4), 48–55.

Krumm, J. (2008). A survey of computational location privacy. *Personal and Ubiquitous Computing, 13* (Online First).

Kumaraguru, P., & Cranor, L. F. (2005). *Privacy Indexes: A Survey of Westin's Studies*. Institute for Software Research International, School of Computer Science. Pittsburgh, PA, USA: Carnegie Mellon University.

Langheinrich, M. (2002). A privacy awareness system for ubiquitous computing environments. In: G. Borriello, & L.-E. Holmquist (Eds.), *Proceedings, Ubiquitous Computing, 4th International Conference (UBICOMP 2002), LNCS 2498*, Göteborg, Sweden, September 29–October 1, 2002, pp. 237–245. New York: Springer.

Langheinrich, M., & Marti, R. (2007). Practical minimalist cryptography for RFID Privacy. *IEEE Systems Journal, Special Issue on RFID Technology, 1*(2), 115–128.

Lederer, S., Hong, J. I., Dey, A. K., & Landay, J. A. (2004). Personal privacy through understanding and action: Five pitfalls for designers. *Personal and Ubiquitous Computing, 8*(9), 440–454.

Litman, J. (2000). Information privacy/information property. *Stanford Law Review, 52*(5), 1283–1313.

Marx, G. T. (2001). Murky conceptual waters: The public and the private. *Ethics and Information Technology, 3*(3), 157–169.

McGraw, G. (2006). *Software Security: Building Security In.* Amsterdam, The Netherlands: Addison-Wesley Longman.

Molnar, D., Soppera, A., & Wagner, D. (2005). A scalable, delegatable pseudonym protocol enabling ownership transfer of RFID tags. *Selected Areas in Cryptography—SAC 2005, Kingston, Canada, August 2005. LNCS 2897,* pp. 276–290. Heidelberg: Springer.

Ohkubo, M., Suzuki, K., & Kinoshita, S. (2005). Cryptographic approach to "privacy-friendly" tags. In S. Garfinkel, & B. Rosenberg (Eds.), *RFID: Applications, Security, and Privacy.* Upper Saddle, NJ: Addison-Wesley.

Palen, L., & Dourish, P. (2003). Unpacking "privacy" for a networked world. *Proceedings of the 2003 Conference on Human Factors in Computing Systems, CHI 2003,* Ft. Lauderdale, Florida, USA, April 5–10, 2003, pp. 129–136. New York: ACM.

Partridge, K., & Golle, P. (2008). On using existing time-use study data for ubiquitous computing applications. In: J. McCarthy, J. Scott, & W. Woo (Eds.), *UbiComp'08,* September 21–24, 2008, Seoul, Korea, 344, pp. 144–153. New York: ACM.

Perrig, A., Stankovic, J., & Wagner, D. (2004). Security in wireless sensor networks. *Communications of the ACM, 47*(6), 53–57.

Priyantha, N. B., Chakraborty, A., & Balakrishnan, H. (2000). The Cricket location-support system. *Proceedings of the 6th Annual International Conference on Mobile Computing and Networking (Mobicom 2004),* pp. 32–43. New York, NY: ACM.

Quigley, A., Ward, B., Ottrey, C., Cutting, D., & Kummerfeld, R. (2004). BlueStar, a privacy centric location aware system. *Position Location and Navigation Symposium (PLANS 2004),* pp. 684–689. IEEE.

Reimann, R. (2001). So You Want To Be An Interaction Designer. Retrieved January 30, 2009, from Cooper Interaction Design: http://www.cooper.com/journal/2001/06/so_you_want_to_be_an_interacti.html

Rieback, M. R., Gaydadjiev, G., Crispo, B., Hofman, R., & Tanenbaum, A. (2006). A platform for RFID security and privacy administration. *Proceedings of the USENIX/SAGE Large Installation System Administration Conference,* pp. 89–102. New York: ACM.

Rössler, B. (2001). *Der Wert des Privaten.* Frankfurt, Germany: Suhrkamp.

Roussopoulos, M., Beslay, L., Bowden, C., Finocchiaro, G., Hansen, M., Langheinrich, M., et al. (2008). *Technology-Induced Challenges in Privacy and Data Protection in Europe.* European Network and Information Security Agency. Heraklion, Crete, Greece: ENISA.

Smith, I., Consolvo, S., LaMarca, A., Hightower, J., Scott, J., Sohn, T., et al. (2005). Social disclosure of place: From location technology to communication practices. In: H. W. Gellersen, R. Want, & A. Schmidt (Eds.), *Pervasive Computing. Proceedings, Third International Conference, PERVASIVE 2005, LNCS 3468*, Munich, Germany, May 8–13, 2005, pp. 134–151. Berlin: Springer.

Solove, D. J. (2006). A taxonomy of privacy. *University of Pennsylvania Law Review, 154*(3), 477–560.

Solove, D. J. (2007). *The Future of Reputation: Gossip, Rumor, and Privacy on the Internet*. New Haven, CT, USA: Yale University Press.

Solove, D. J. (2008). *Understanding Privacy*. Cambridge, MA, USA: Harvard University Press.

Solove, D. J., Rotenberg, M., & Schwartz, P. M. (2006). *Privacy, Information, and Technology*. New York, NY, USA: Aspen.

Solove, D. J., & Schwartz, P. M. (2009). *Information Privacy Law*, 3rd Edition. New York, NY, USA: Aspen.

Spreitzer, M., & Theimer, M. (1993). Providing location information in a ubiquitous computing environment. *Proceedings of the Fourteenth ACM Symposium on Operating Systems Principles (SOSP '93)*, pp. 270–283. New York, NY: ACM.

Sweeney, L. (2002). k-Anonymity: A model for protecting privacy. *International Journal on Uncertainty, Fuzziness and Knowledge-based Systems, 10*(5), 557–570.

TNS Emnid. (2006). *Kundenkarten etablieren sich als feste Einkaufsbegleiter*. Bielefeld, Germany: TNS Emnid.

Warren, S., & Brandeis, L. (1890). The right to privacy. *Harvard Law Review, 4*(5), 193–220.

Weis, S. A., Sarma, S. E., Rivest, R. L., & Engels, D. W. (2003). Security and privacy aspects of low-cost radio frequency identification systems. In: D. Hutter, G. Muller, W. Stephan, & M. Ullmann (Eds.), *Security in Pervasive Computing*, pp. 50–59. Berlin: Springer.

Weiser, M. (1991). The computer for the 21st century. *Scientific American, 265*(3), 94–104.

Westin, A. F. (1967). *Privacy and Freedom*. New York, NY, USA: Atheneum.

CHAPTER 4

Ubiquitous Computing Field Studies

A. J. Bernheim Brush

CONTENTS

4.1 Introduction	162
4.2 Three Common Types of Field Studies	164
4.2.1 Current Behavior	165
4.2.2 Proof of Concept	166
4.2.3 Experience Using a Prototype	168
4.3 Study Design	169
4.3.1 What Will Participants Do?	169
4.3.1.1 Control Condition	171
4.3.2 What Data Will You Collect?	173
4.3.2.1 Logging	174
4.3.2.2 Surveys	175
4.3.2.3 Experience Sampling Methodology	176
4.3.2.4 Diaries	177
4.3.2.5 Interviews	178
4.3.2.6 Unstructured Observation	180
4.3.3 How Long Is Your Study?	180
4.4 Participants	181
4.4.1 Ethical Treatment of Participants	181
4.4.2 Participant Profile	182
4.4.3 Number of Participants	184
4.4.4 Compensation	185

4.5 Data Analysis	186
4.5.1 Statistics	187
4.5.1.1 Descriptive Statistics	187
4.5.1.2 Inferential Statistics: Significance Tests	189
4.5.2 Unstructured Data	191
4.5.2.1 Simple Coding Techniques	191
4.5.2.2 Deriving Themes and Building Theory	192
4.6 Steps to a Successful Study	194
4.6.1 Study Design Tips	194
4.6.1.1 Have a Clear Research Goal	194
4.6.1.2 Create a Study Design Document	194
4.6.1.3 Make Scripts for Participant Visits	195
4.6.1.4 Pilot Your Study	195
4.6.2 Technology Tips	196
4.6.2.1 Make Your Technology Robust Enough	196
4.6.2.2 Consider Other Evaluation Methods	196
4.6.2.3 Use Existing Technology	196
4.6.2.4 Get Reassuring Feedback	196
4.6.2.5 Negative Results	197
4.6.3 Running the Study	197
4.6.3.1 Have a Research Team	197
4.6.3.2 Make Participants Comfortable	197
4.6.3.3 Safety	197
4.6.3.4 Be Flexible	198
4.6.4 Data Collection and Analysis	198
4.6.4.1 Be Objective	198
4.6.4.2 The Participant Is Always Right	198
4.6.4.3 Do Not Make Inappropriate Claims	198
4.7 Conclusion	199
Acknowledgments	199
References	200

4.1 INTRODUCTION

Ubiquitous computing (ubicomp) weaves computing into our everyday environments and devices. People and their use of technology are at the center of this vision, necessitating an understanding both of people's needs and their reactions to new ubicomp applications and experiences. The field of human-computer interaction, drawing from other fields including psychology and anthropology, has developed numerous approaches to understanding how people interact with technology. These methods include user studies, focus groups, ethnography, and heuristic evaluations. Although using a variety of methods to incorporate user needs and feedback throughout the process of designing technology is critical, this chapter describes how to plan and conduct a *field study*, also referred to as an in situ study. Field studies are a particular type of user study conducted outside a research laboratory or controlled environment (i.e., "in the field"). Field studies offer the opportunity to observe people and their use of technology in the real world, in contrast to a *laboratory* user study, where participants[*] come into your controlled environment and complete tasks you specify.

As other researchers have argued (e.g., Consolvo et al., 2007; Rogers et al., 2007), field studies are often the most appropriate method for studying people's use of ubicomp technologies. Studies conducted in situ allow researchers to collect abundant data about the use of technologies they have developed, observe the unexpected challenges participants may experience, and better understand how their technology impacts participants' lives. The trade-off for increased realism is a loss of control over the participant's experience, so field studies are not appropriate for all evaluations; indeed, for many research questions, a laboratory study where you have complete control over the environment may be more appropriate. When considering a field study, it is critical to think carefully about why you want to conduct a study and what you hope to learn from doing one. Field studies require considerable time and effort and should not be undertaken lightly. You should not undertake a field study because you think it is a requirement to get a paper accepted to a conference or because you would just like to see how people use your ubicomp application, but rather because your research questions requires it. Field study can be very valuable and even necessary in order to understand user needs or technology usage in a particular domain (e.g., emergency response, homes, etc.) or to evaluate use of a novel ubicomp application more realistically than can be done in a laboratory environment.

Approaches to conducting field research in ubicomp draw from many different disciplines and research traditions, each with its own style. This chapter takes a pragmatic approach to field studies and focuses on helping

---

[*] Many researchers use the term participant for people participating in studies because it seems more respectful than users or subjects.

you understand the questions you will need to answer to design a study that meets your needs, and introducing data collection and analysis techniques that are commonly used in field studies. You are highly encouraged to read Chapter 5, which describes an alternative approach to conducting research in the field. Section 4.2 describes three common types of field studies and introduces examples of each type. Section 4.3 focuses on study design choices, including what your participants will do during the study, what data you will collect, and the length of the study. Section 4.4 outlines considerations around choosing participants for your study and treating them in an ethical manner, whereas Section 4.5 discusses data analysis methods. Section 4.6 outlines pragmatic steps that will help ensure the success of your study. Section 4.7 concludes by emphasizing the value of field studies.

4.2 THREE COMMON TYPES OF FIELD STUDIES

The type of study you are conducting and why you are conducting a study will help you determine the *research question* for your study. Examples of research questions include "Is the mobile phone a suitable proxy for the owner's location?" and "Does context-aware power management have the potential to save energy?" When designing your study, you will rely on your research question to inform all of the many decisions you need to make, such as what type of participants to recruit, how long the study should be, and what type of data to collect. To craft your research question, think carefully about what you wish to learn by conducting the field study. What questions do you want to answer? What will your contribution be that will influence the research community or inform others working in this area?

Three common types of ubicomp field studies are

- Studies of current behavior: What are people doing now?

- Proof-of-concept studies: Does my novel technology function in the real world?

- Experience using a prototype: How does using my prototype change people's behavior or allow them to do new things?

Although other types of field studies exist including those exploring playful interaction or ludic engagement with ubicomp technologies (e.g., Gaver et al., 2006, 2007), these three types of studies will be the focus of this chapter. To demonstrate the different choices researchers make when conducting field studies, examples of each type of study were selected from conferences in the area of ubiquitous computing. The examples highlight a range of different approaches and the types of questions used to frame the research. Interested readers are encouraged to consult the papers describing studies in their entirety for studies that are particularly relevant to them. Throughout the rest of the chapter, more details about these examples will be used to illustrate study design, data collection, and data analysis choices.

4.2.1 Current Behavior

Understanding how people are making use of technology in their lives today can provide researchers with insights and inspiration. This type of field study explores how people use existing technology. The contributions of this type of study are an understanding of current behavior and implications for future technology. Research questions for studies of this type typically emphasize how people use technology and may be very open-ended. The two examples below focus on studies of current behavior with specific research questions that make use of interviews and logging, in contrast to more open-ended observations traditionally used in ethnography, described in detail in Chapter 5.

Example 1: Home Technology Sharing and Use (Home Technology)

Brush and Inkpen (2007) conducted an interview study of 15 families in the United States that examined the types of technologies families own, including TV, music players, phones, and computers; where they are situated within the home; and the degree of shared ownership and use. During the visit, the participants were interviewed, sketched the layout of technology in their homes, and gave the researchers a tour of their homes focusing on where computers were located.

Research Question: How do families use and share technology in their homes?

*Example 2: Proximity of Users to Their
Mobile Phones (Phone Proximity)*

Patel et al. (2002) used logging on cell phones and interviews to study the proximity of 16 people to their cell phones over a 3-week period. During the study, participants wore a small Bluetooth beacon, used phones with logging software, and participated in weekly interviews.

Research Question: Is the mobile phone a suitable proxy for the owner's location? What type of information (e.g., cell ID, date, and time) are the best predictors for how close the owner is to the phone?

Other examples of studies on current behavior are Sohn et al.'s (2008) diary study on mobile information needs and Woodruff et al.'s (2007) study on the use of laptops in homes.

4.2.2 Proof of Concept

Ubiquitous computing projects often develop novel technology or seek to validate new algorithms and approaches. For this type of study, technological advance is the primary contribution of research rather than field study. However, it may be important to conduct a field study to validate the feasibility of an approach or prototype in a real-world environment. These field studies may be shorter than the other two types and the research questions generally focus on whether the prototype or algorithm functions appropriately in a real environment.

Example 3: Context-Aware Power Management (CAPM)

Harris and Cahill (2007) conducted a 5-day field study of 18 participants to investigate the potential for using context information to improve power management on personal computers (PCs). The computers of participants were augmented with logging software that asked after a minute of idle time if they were using the PC and sensors that included web cameras, microphones, and ultrasonic object range sensors. Participants also carried a Bluetooth tag on their key chains to provide additional location information.

Research Questions: Does CAPM have the potential to save energy? How accurately can it be inferred that the user is not using the device or about to use it?

Example 4: TeamAwear

The TeamAwear system, developed by Page and Vande Moere (2007), is a novel wearable display system for team sports. Augmented basketball jerseys, shown in Figure 4.1a, are worn by players and display game-related information such as the number of points scored and fouls. The researchers used an iterative user-centered design process that involved participants at several points including initial ethnographic observations of basketball games, a set of discussions with representative users,

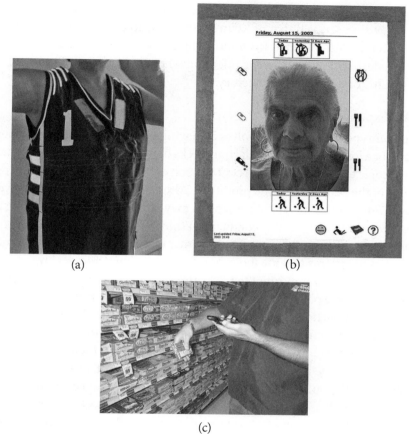

FIGURE 4.1 (a) Augmented jersey from the TeamAwear system. (Courtesy of Mitchell Page.) (b) CareNet display showing information about an elder's activities: (Courtesy of Sunny Consolvo.) (c) Using the AURA system to scan barcodes on items at a grocery store.

and a case study evaluation with 11 participants that included a discussion of game-like scenarios and a 15-minute, half-court, 2-2 basketball game with one team wearing the augmented jerseys.

Research Questions: Are augmented jerseys usable and useful? How wearable are the jerseys? Do people understand the displays? Do augmented jerseys increase the awareness of game-related information for athletes, coaches, referees, and spectators?

Other examples of proof-of-concept studies include Patel et al. (2007), who investigated whether simple plug-in sensors can be used to recognize events in homes such as turning on and off a television set, and that of Krumm and Horvitz's (2006), use of GPS driving data collected from more than 150 people to validate their predestination method for inferring where a driver is going based on his or her driving trajectory.

4.2.3 Experience Using a Prototype

Another type of field study involves deploying ubiquitous computing prototypes for a longer period, often lasting weeks or months, to better understand how people use them. The main contribution of this type of study is the experience of the people using the prototype. Although the technology deployed is typically not commercially available, it may not be a novel contribution. In some cases, researchers may conduct a *Wizard of Oz* study, where aspects of a prototype or system are simulated in order to understand the participants' reactions to systems that are too expensive to fully build and deploy.

It is particularly important to take care in specifying your research question for this type of study. People sometimes frame their research questions as "How will participants use the prototype?" However, it is important to recognize that your prototype is one example of how a novel interaction or experience might be instantiated. Other researchers are typically less interested in raw usage information and are definitely not interested in usability problems with your prototype. So, rather than focusing specifically on how participants will use a prototype, better research questions focus on the concept the prototype embodies or tests, for example, "Does sharing location information lead to privacy concerns?" or "Will peripheral displays enhance family awareness?" Framing research question as a hypothesis is one way to help clarify how you think participants will experience your prototype.

Example 5: CareNet

To explore the value of ambient displays and sensed data for elder care, Consolvo et al. (2004) built and deployed the CareNet display. CareNet, shown in Figure 4.1b, displays data sensed about an elder's activities (e.g., medication taken, activity levels) to members of the elder's care network on an ambient display in a picture frame form factor. Researchers used the *Wizard of Oz* technique to gather data for the CareNet display by phoning the elders several times a day.

Hypothesis: Ambient displays can positively impact the local members of an elder's care network.

Example 6: Advanced User Resource Annotation (AURA)

The AURA system (Brush et al., 2005) allows a person to scan barcoded objects, including CDs, DVDs, and packaged grocery products, using a wireless pocket PC with an attached barcode reader. After scanning a barcode, users can view, store, and share related metadata and annotations about the scanned object. Figure 4.1c shows an AURA field study participant scanning a grocery item. This study also provides an example of reporting both positive and negative results, and includes the authors' reflections of issues they encountered during the field study.

Research Questions: How do people use the system? What do they scan? Do people find the system functional and useful? Does the privacy model meet users' needs? How do people use the sharing features?

Other examples of studies on experiences using a prototype include Ballagas et al.'s (2008) evaluation of REXplorer, a mobile pervasive game for tourists in Regensburg, Germany, and Matthews et al.'s (2006) feasibility study of Scribe4Me, a mobile transcription tool for the deaf.

4.3 STUDY DESIGN

After deciding what type of field study you plan to conduct and your specific research question(s), the next step is to design the field study. It is important to realize that there are very few "right" decisions about how a field study should be run. Instead, there are many decisions that you will need to justify to yourself and your audience (e.g., other researchers, reviewers, funding agencies, etc.) as appropriate and sensible in order to gather the data needed to address your research question. In determining your study design, the three important questions to consider are

- What will your participants do during the study?
- What data will you collect?
- How long will the study be?

4.3.1 What Will Participants Do?

In planning your field study, you need to decide what you will ask participants to do during the study. What participants do clearly depends on the type of field study and your research question. To study current behavior, you might interview participants or log their behavior, whereas in other studies participants typically use a prototype.

Given that field studies are a choice to sacrifice control of the participant's experience for realism, experimental design techniques used in laboratory studies are typically less appropriate for a field study. However, having a high-level understanding of how experimental laboratory studies are designed can help you think about how your field study should be structured. This is particularly relevant when your study involves giving the participants new technology and understanding their experience.

In traditional laboratory studies, researchers specify a hypothesis based on a research question and then gather data through the study to support or reject the hypothesis. To test the hypothesis, researchers identify a variable, called the *independent variable*, that they will vary between different values, called *conditions*, during the experiment in order to understand the effect of variation on the *dependent variables* they are measuring (e.g., task time or user preference). The two main laboratory study designs are within-subjects and between-subjects. In a within-subjects design, also called repeated-measures design, each participant experiences all conditions. So, if your independent variable was versions of an interface and you have two versions (A and B), in a within-subjects study each participant would use both versions. Within-subject designs offer the advantage that your dependent variables can include asking participants to directly compare different conditions. For example, you can ask participants whether they preferred version A or version B of your interface. You can also directly compare any numeric data you collect, such as number of times the participant used the interface, for the same participant across conditions. However, in within-subject designs you need to worry about whether there will be any learning effects. For example, participants might favor version B or be faster in using it just because it was the second version they used. Counterbalancing or varying the order that different participants experience the conditions is used to mitigate any potential learning effects.

In a between-subjects design, you divide your participants into different groups, typically randomly, and each participant experiences only one condition of the independent variable. So half of your participants would use version A of the interface and the other half would use version B. This approach avoids any potential learning effects, but you generally need to have more participants because you cannot directly compare the behavior of a single user across the conditions. Finally, some studies use a mixed design where some independent variables are within-subjects and some are between-subjects. Section 14.4 in the work of Preece et al. (2002) has more information on experimental design.

In choosing to conduct a field study instead of a laboratory study, you are emphasizing studying people's behavior or use of your prototype in the real world. This makes it impossible to control the environment to the extent that you can have confidence that any one independent variable (or set of independent variables) that you choose to vary is the only thing causing changes in the dependent variables. For this reason, using a within-subject design is preferred for studies with conditions so that you can compare the participants' behavior in one condition against their behavior in another condition, since actions by different participants can vary widely (e.g., AURA researchers had three participants who scanned more than 100 items during the study, and six who scanned fewer than 20 items). A within-subjects design also allows you to ask participants for their qualitative comparisons between conditions (e.g., different versions of the same interfaces). One particularly useful type of within-subjects condition to consider having in studies that involve a prototype is a control condition.

4.3.1.1 Control Condition

In a control condition, you measure the dependent variables for a certain period before you introduce the technology (e.g., logging for a week a behavior that you think might change), then introduce your technology and measure the dependent variables again. This within-subjects design allows you to compare a participant's behavior with and without your technology. For example, if you are studying a new location-based application for mobile phones, you might first collect data on how the participants used their mobile phones before introducing your application so that you can compare participants' behavior after they have used your application to their previous behavior. Even if you do not have a full control condition, it is very common in field studies to collect some data before introducing new technology to obtain baseline information about your participants and their expectations. However, collecting control data is not appropriate for all ubicomp studies, because your prototype may afford a behavior that was impossible without it and thus there is no meaningful control condition to compare against. For example, if you wanted to give the location-based mobile application to people that had never used a mobile phone before, you could not compare against previous use of mobile phones, but you might try to collect data about how often the participant communicated using landline phones or other communication methods to compare against.

In addition to deciding on your study method, if you are introducing a new technology in your study, there are a number of pragmatic considerations:

- Will participants use the technology as they wish or to complete specific tasks? For some studies you want participants to behave normally, as researchers sought in the Phone Proximity and CAPM studies. However, other studies are exploring whether a technology works in a specific setting and ask participants to do something specific. For example, part of the TeamAwear study included participants playing a 15-minute, half-court, 2-2 basketball game with one team wearing the augmented jerseys. In the REXplorer study, participants played a location-aware game (Ballangas et al., 2008).

- Will you give the participant technology to use or augment the technology the participant already owns? For example, are you installing a new application on the participant's laptop or cell phone, or switching their SIM card to a phone you provide? Or perhaps putting a novel display in the participant's home? There can be advantages in augmenting devices that participants already have since they are familiar with those devices. However, you can run into challenges supporting a diverse set of platforms and dealing with unique configurations as the AURA researchers did. On the other hand, providing your own technology, as was done in the Phone Proximity study, allows you more control and a consistent setup, but it may be unfamiliar to participants. Readers considering studies involving mobile phones are encouraged to read Section 2.4 of Consolvo et al. (2007) for a more in-depth discussion of the advantages and disadvantages of providing mobile phones or augmenting participants' phones.

- Should you simulate any part of the participant's experience? Using the *Wizard of Oz* technique, where aspects of a prototype or system that are too expensive or time consuming to build and deploy are simulated, can be a valuable approach. Sometimes, the experience you wish to study may not even be possible with current technology. For example, in the CareNet study, the information that was shared on the ambient display about the elder was not automatically sensed. Instead, the researchers spoke with the elder or their caregiver multiple times a day and then manually updated the data shown on the ambient display. In a feasibility study of the Scribe4Me system (Mathews et al., 2006), human transcribers were used to transcribe audio recorded on the mobile phone.

4.3.2 What Data Will You Collect?

During a field study, you can collect *quantitative* and *qualitative* data. Quantitative data can be objectively observed and represented numerically (e.g., timing, errors, usage) and can help you understand what happened during the study. For example, how many times a participant used your prototype or how long they used it. On the other hand, qualitative data include a participant's unstructured feedback and reactions as well as field observations you make. Qualitative data help you understand what people think and hopefully why people behaved in certain ways during the study.

For field studies, it is valuable to collect both quantitative and qualitative data. If you collect only quantitative data you have insight into how people behaved, but may have trouble understanding why. If you collect only qualitative data you will have insight into why participants did certain things, but may have trouble comparing participants or understanding how closely what participants thought they did mapped to what they really did. When considering what specific questions to ask during data collection, first return to your research question. However, it may also be helpful to look at evaluation metrics used by others doing related research. Scholtz and Consolvo (2004) put forth an evaluation framework for ubicomp applications that proposes the evaluation areas of attention, adoption, trust, conceptual models, interaction, invisibility, impact and side effects, appeal, and application robustness. The definitions and metrics they provide in their paper may give you ideas for your study.

It is obviously impossible to include everything you need to know about data collection methods in one short subsection; instead, different data collection techniques will be introduced with a focus on their use in ubicomp field studies. During your study you will likely use more than one of these methods to gather data. In general, you probably want to collect demographic data about your participants, usage data that tell you about their current behavior or how they use your prototype, and reactions to any technology you have introduced.

Finally, no matter what data collection methods you choose for your study, you must pilot them before the study starts to make sure that you are collecting the data you need and that you know how you will analyze the data. For surveys, this means asking colleagues to take them, for logging it means generating logs and checking data generated by yourself or colleagues. If at all possible, conduct a full pilot of your study design with "friendly" participants (e.g., coworkers, friends) before deploying to your participants in order to find as many problems as possible with the study design and prototypes in advance.

The rest of this section introduces different methods for collecting data.

4.3.2.1 Logging

In field studies, logging is often the main method for collecting quantitative data about usage, either of existing technology or your novel technology. When logging data, your prototype typically writes information to a data file when things occur that you want to know about. For example, both the Phone Proximity and CAPM studies logged data from many different sensors to collect a data set on which researchers could do machine learning. The AURA study used logging to determine what objects participants scanned and when they uploaded them to the server to share.

Considerations for logging:

- How will you use the logged data? Although logging is an incredibly useful tool, you must have a plan about how you will use the data you are logging. Too often people are tempted to try to log everything with the notion that they will figure out after the study how to analyze it. This is a recipe for an analysis nightmare as you struggle to abstract out the meaningful information in the log from the noise.

- Have you forgotten to log something important? Make a list of specific questions that you expect to answer from the log data. For each question, identify the events you are logging that you will use to answer the question and then walk through the analysis using data from your own use or a pilot study. Make sure that the data you are logging will answer the questions you expect it to.

- Will your logging help you know if the study is going smoothly? Another valuable use of logging data is for reassurance that the technology is operating as expected. If logging data are collected on a central server (e.g., instead of the device), the presence of data can reassure you that nothing disastrous has happened or warn you that you may need to intervene to avoid the devastating discovery that your field study needs to be rerun because of a technical problem. This feedback is so valuable that you should consider having your technology "phone home" with an "I'm ok" message at regular intervals even if logging data are stored locally to a device.

4.3.2.2 Surveys

Surveys are often used to gather data before a field study begins (presurvey), after any changes of condition in a between-subjects study (postcondition), and at the end of the study (postsurvey). In a longer field study, they may also be used at regular intervals (e.g., a weekly survey) to measure how a participant's reactions might be changing. The AURA study had pre- and postsurveys, the CareNet study had a midpoint survey and postsurvey, and participants in the TeamAwear study were surveyed after they played basketball.

Common types of survey questions are open-ended questions, multiple choice questions, and Likert questions. Open-ended questions ask participants to enter free text answers. In a multiple choice question, participants are given several options to choose from, for example, "Did you like A, B, or C better?" or participants might be asked to select all options that apply. Likert questions are statements that participants are asked to agree or disagree with. For example, "I like chocolate ice cream" with the option to select Strongly Agree, Agree, Neutral, Disagree, or Strongly Disagree. Likert questions with five possible answers are very common; although sometimes a larger scale (e.g., seven or nine answers) is used. The AURA study includes examples of several Likert scale questions participants were asked. See Section 13.3 of Preece et al. (2002) for more information about designing surveys.

Below are several factors to consider when creating your survey questions.

- Are the questions stated to allow both positive and negative responses? For example, "How much do you love the prototype: a little, a lot, an immense amount" is a leading question with bad options. Instead, you might ask a Likert question: "The prototype was easy to use" with options ranging from Strongly Agree to Strongly Disagree.

- Are the questions clear to others? Pilot your survey with colleagues to make sure the questions are clear to them. Watch out for "and" questions. If you ask "The prototype was fun and easy to use," you will not know if participants were answering about the prototype being fun or easy to use.

- Will the questions obtain the information needed from participants? Review your research questions and make sure the survey contains all the questions you need. Collecting pilot data and formally writing them up to present to others can help ensure that you are collecting all the data you need to answer your research questions.

- Is the survey appropriate in length? Be careful not to make your survey too long; there can be a tendency to include every possible question you can think of on a survey. This can cause participants to get tired and frustrated with your survey. Make sure your survey contains only questions that really matter.

- Have you looked for questions that others have used? Although each field study is unique, it is worth looking at papers for projects related to yours to see if there are any survey questions that you should adopt for your study or if standard surveys such as the Questionnaire for User Interaction Satisfaction (http//lap.umd.edu/quis) or scales from psychology such as the National Aeronautics Space Administration Task Load Index [NASA TLX] may be appropriate. Sadly, the text of survey questions is not often included in research papers; however, other researchers are often willing to share their survey questions if you contact them. This may allow for some comparison between field studies if appropriate.

4.3.2.3 Experience Sampling Methodology

Ubicomp researchers have long recognized the value of using experience sampling methodology (ESM), a technique borrowed from psychology, for field studies (e.g., Consolvo and Walker, 2003). In ESM, participants are asked to fill out short questionnaires at various points throughout their day, asking about their experience at that time. ESM allows the researcher to collect qualitative data throughout the study, which has advantages over asking participants later to try to recall what they were thinking or feeling, or why they took some action. Participants can be asked to complete a survey either randomly throughout the day, at scheduled times, or based on an event. For example, in the AURA field study, researchers used event-based ESM and asked participants a few questions about the object they had scanned after every fifth item they scanned. In CAPM, the computer queried the user after 60 seconds of idle time if they were using the device. Although it is most often used to gather qualitative data, you can also use ESM to gather quantitative data based on events, for example, recording the location of a participant every time he or she answers a call on their mobile phone.

MyExperience (http://my experience.sourceforge.net), which was developed by Froehlich et al. (2007), is a popular open-source toolkit for collecting ESM data using cell phones. Screenshots of using MyExperience to collect data are shown in Figure 4.2.

FIGURE 4.2   Two examples of ESM data collection using the MyExperience toolkit. (Courtesy of Jon Froehlich.)

Considerations when using ESM:

- How often should you ask participants to answer questions? Should it be random, event-triggered, or on a regular schedule? Asking participants for feedback too often will be annoying, but you need to ask frequently enough to collect the appropriate data. Make sure it is easy for participants to ignore the survey if they need to. For example, the dialog in the CAPM study asking participants whether they were using the computer was dismissed if the user moved the mouse.

- How many questions will you ask on each survey? Try to keep the number of questions very small and quick to answer. One way to compensate for each survey being short is to ask different sets of questions at different times.

- Do you want to collect sensor data using ESM techniques? Are there quantitative data that you want to collect randomly throughout the day, at a scheduled time, or based on a particular event?

4.3.2.4 Diaries

Similar in spirit to ESM, some studies gather data by asking participants to record information about what they do, typically referred to as a "diary." This method is frequently used when participants are making diary entries about something that would not be possible to sense using an ESM tool, thereby rendering event-based ESM inappropriate. For example, asking a participant to track when he wants to do something or feels a certain emotion. In some studies, participants are given small paper notebooks to carry around, whereas other researchers have used mobile technology for recording diary entries. In Sohn et al.'s study of mobile information needs, participants used short message service (SMS) to make entries and then added details at a Web site (Sohn et al., 2008). Many of the considerations for diary studies are similar to those for ESM, such as what you will ask your participant to record in each diary entry. However, for diary studies there are typically greater concerns about participation, because participants are typically not carrying a device that interrupts them as in an ESM study. Another option is asking participant to retrospectively construct a diary, as the Phone Proximity study had participants do at the weekly interview for the previous day using the Day Reconstruction Method (Kahneman et al., 2004).

Considerations:

- How will you remind participants to complete their diary entries? Will you email them daily? Call them occasionally or send SMSs?

- How will you incent participants to complete their diary entries? Will you reward them per entry, which can cause people to generate extra entries to earn money or perhaps with a set amount of money for each day of participation?

4.3.2.5 Interviews

Interviewing your participants can be an excellent way to gather qualitative data. During field studies, researchers frequently conduct "semistructured interviews." In a semistructured interview you bring a list of specific questions, but ask follow-up questions about interesting things that participants say in addition to the predefined questions. Interviews were the main method used in the Home Technology study, and also in Phone Proximity, TeamAwear, AURA, and CareNet studies. Researchers also sometimes bring data collected earlier in the study, perhaps through logging or ESM, to show to participants in order to help them remember particular events that the researchers want to ask questions about. For example, in the Phone Proximity interviews researchers compared the diary data generated by the participant with visualizations of the logging data. In the TeamAwear interviews, participants viewed a video recording of the basketball game with the augmented jerseys and were retrospectively interviewed as a group. Retrospective interviews using video can be a valuable method to use for asking participants about situations in which they cannot be interrupted (e.g., playing basketball, performing surgery).

During interviews, it is highly recommended to record the interview and take photos, although make sure to ask for participant's consent to do so (see Section 4.4.1). Having an audio recording can be very helpful either to have transcribed or to refer back to later (e.g., for exact quotes). However, you need to recognize that transcribing audio is either a time-consuming process for you to do or expensive if you have it done professionally. So, it is usually still valuable to take notes during the interview if only to identify specific parts of a recording that are interesting. It is also very valuable to take photos that can help you remember the context and be used in presentations of your work. In the CareNet study, researchers took pictures of where people placed the prototype in their home. Consider taking some photos that do not have people or at least their faces in them; these can be easier to use since they better protect participants' privacy. See Section 13.2 of Preece et al. (2002) for more information about conducting interviews.

Considerations for interviews:

- Are your questions phrased as neutrally as possible? Be careful in how you to phrase your questions so that participants feel comfortable telling you about negative experiences as well as positive ones. If participants say something interesting, prompt them for more information with neutral language such "Ummm" (do not underestimate the power of a well-timed ummm), or "Can you tell me more about that?" The word "Why" can sometimes be interpreted as accusatory, so avoid it if possible.

- Are you prepared to take negative feedback without becoming defensive? Your job during an interview is to ask questions, record answers, and follow up on interesting information. The participants are telling you about their experience and they are always right about what they experienced. For example, if they tell you the logging software you installed slowed their computer down, tell them you are sorry and that you will look into to it. Do not argue with them about how it is impossible, even if you believe that to be true. During an interview, you must leave your own opinions at home and collect feedback without judging the participant. Always remember that they are helping you by participating.

4.3.2.6 Unstructured Observation

Although interviews tend to last 2–3 hours and researchers ask specific questions, when collecting data through observation participants are observed as they engage in their normal lives, possibly while using a novel prototype. In the AURA study, eight participants were observed on shopping trips while using the system. Chapter 5 discusses in detail the history and use of ethnography in ubicomp field research.

4.3.3 How Long Is Your Study?

People often agonize over how long their study should be. Like most aspects of a field study, there is no definitive answer, only questions to consider based on your research goals and how long the study needs to be to help answer your research question. The ubicomp research literature has examples of field studies that last hours, weeks, months, and even a year.

Considerations:

- What type of study is it? Proof-of-concept studies may be on the shorter side if less time is needed to prove the feasibility of the prototype. For example, the CAPM study gathered the data the researchers needed in 5 days. Studies of experience using a prototype are usually longer (e.g., CareNet study lasted 3 weeks and the AURA study was 5 weeks long) because the study is the contribution, whereas studies of current behavior vary widely. The Home Technology study used 2-hour interviews to gather data, whereas the Phone Proximity study collected data for 3 weeks.

- Do you expect novelty effects to be an issue? Often, when using new technology, people start out very enthusiastically using it and then decrease their usage. Unfortunately, there is no guarantee about how long novelty effects last. If you are worried about novelty effects, try to make your study as long as possible and be wary of basing too many of your findings on usage from the beginning of the study. A within-subjects design can help provide metrics for comparing between usage in different conditions. Choosing a within-subjects design would have helped the AURA researchers better understand the novelty effect they observed.

- How much work do participants have to do? The more effort your study requires from participants, the shorter you may need to make it. Researchers in the CareNet study chose 3 weeks because of the effort involved for the elders to provide data about their activities each day.

- How frequently will participants use your technology or engage in the behavior you are trying to study? If participants use your technology frequently, say multiple times a day, then your study length can typically be shorter than if participants use your technology less frequently, because you will be able to gather more data in a shorter period.

- How many times during the study will you interact with participants? How noticeable is your technology or logging to the participant? Consider how intrusive your study will be for your participants. If you need to meet with them every day or week, or your data collection includes ESM that frequently asks participants for feedback, you will probably need to have a shorter study than if you are studying current behavior using logging that is essentially invisible to the participant. The intrusiveness of the CAPM study, which queried the participant after 60 seconds of idle time, probably contributed to the choice of a 5-day-long study period.

4.4 PARTICIPANTS

A key part of any field study is the participants. This section outlines important aspects of dealing with participants including treating them in an ethical manner, selecting the type of participants, determining how many are appropriate for your study, and how to compensate participants.

4.4.1 Ethical Treatment of Participants

Researchers must be very careful to treat participants in an ethical manner. It is critical that participants understand what will be required of them if they participate in the study and how you will report on what they did so they can make an informed decision whether they wish to participate.

Participants should also be given the option of discontinuing participation at any time if they choose.

Participants should receive a consent form at the beginning of the study to review and sign to signify that they have consented to participate. The consent form should tell participants what data you are collecting and have a privacy statement describing how the data will be used and how long they will be kept. Ubicomp studies often collect data that can be considered sensitive (e.g., location of participants, activities) and thus researchers must take particular care in making clear to participants what data are being collected and who will have access to it. Participant data should be kept as secure as possible with access limited only to the people who need to see it. Whenever possible, store the data using identifiers (e.g., participant 1) rather than the participant's name. When you report your data, you should never use participants' real names; instead, use participant numbers or pseudonyms.

The exact wording of your consent form will depend on your organization and the country of your study, as different countries have different laws. Your organization should have some review process to ensure that your study is treating participants in an ethical manner. For example, in the United States many universities have an institutional review board (IRB), which reviews research involving humans (U.S. Institutional Review Guidebook, http://hhs.gov.ohrp/irb/irb_chapter3.htm). To conduct studies at these institutions, researchers must submit an application to the IRB and obtain its approval. Companies may have legal teams that can advise on consent forms and privacy statements.

4.4.2 Participant Profile

Identifying the participants you would like to recruit for your field study depends on the research goals of your study. In general, you want to select participants who match the realistic usage of your prototype or the demographic that you are most interested in. For example, having computer science graduate students pretend to be elders and use the CareNet system for elder care would not have been appropriate. It is also best to have participants who are not involved in any way with your research. This reduces the chances that they are biased by knowledge they might have of your study or goals. Finally, recruiting different types of participants and comparing between them is a common type of independent variable. However, you need to be careful that you have enough participants of each type to be able to make reasonable comparisons. In the Home Technology studies, researchers recruited five families that had one computer, five families that had multiple computers, and five families that had one or more computers per household member.

To come up with your *participant profile*, the description of participants to recruit, consider the following:

- Age. Does your research question suggest a particular age range for participants (e.g., teenagers or elders)?

- Gender. Often, equal numbers of men and women are recruited so that you can use gender as an independent variable and compare the experiences of male and female participants. However, some studies may benefit from a different balance; in particular, if you need certain types of technology experience or job roles, you may have trouble recruiting and decide to have all women or all men participants to hold this variable constant.

- Technology use and experience. Your research question may require recruiting people who use certain types of technology (e.g., have a laptop, use a smart phone) with a particular frequency of use (e.g., send text messages every day). One thing to carefully consider is whether it is most appropriate to recruit people already using cutting-edge technology (sometimes called early adopters) who are already familiar with a device you want them to use (e.g., a smart phone) or people whose usage patterns are more consistent with the general population.

- Other characteristics. You may need to recruit people or groups with special characteristics that go beyond their experience with technology. For example, the Home Technology study recruited families with different numbers of computers in their homes, whereas the TeamAwear study recruited athletes, spectators, a referee, and a coach.

Last, recognize that your research question will help you decide how to rank the importance of different aspects of your participant profile. Depending on what is important for your study, you may be forced to make trade-offs in other criteria. For example, in the AURA study, the researchers recruited participants who owned a Pocket PC with access to the Internet in 2004. However, this likely explains why the gender balance of their 20 participants was 16 men and 4 women instead of a more equal division.

4.4.3 Number of Participants

Determining the exact number of participants you need for your study can be a difficult question. Factors to consider when identifying the number appropriate for your study are

- Are there any conditions in your study (e.g., between or within subjects)? Having conditions typically mean you should have more participants, particularly for a between-subjects design, so that the groups are large enough to justify making comparisons between them.
- What claims are you trying to make? Is this a proof-of-concept study? If so, you can typically have fewer people and may want to have many different types of people and/or conditions to show different people can use your prototype in a wide range of environments. For example, the Phone Proximity study recruited a diverse set of 16 mobile phone users ranging in age from 21 to 66 years with very wide variety of income levels, professions, and phone usage.
- The length of your study and amount of data you will collect. A field study needs to collect enough data to convince people its findings are valid. If the study will collect a considerable amount of data per participant, then you can typically have fewer participants than a study that collects less data per participant. For example, the CareNet and AURA field studies, each of which lasted multiple weeks, had 13 and 20 participants, respectively, whereas the Home Technology study, where data collection consisted of one interview, had 50 participants in 15 homes.
- What is feasible? Although one might ideally like to have a large number of participants, each participant will require a considerable time investment for you and your team. In trying to decide what is feasible, make sure to estimate the time per participant. This time should include the length of any visits, time for travel to the participant's locations as necessary, preparation time for each visit, and support for participants using prototypes or logging software.
- Plan for participants to drop out. Because field studies typically require considerable effort on the part of the participant over a long time, it is practically guaranteed that some participants will drop out of your study. Replacing participants that drop out midway through

a study can be quite difficult and cause you problems if you wanted the study to finish by a certain time. Thus, if possible, recruit one or two more participants than you think you need, so that if participants drop out of the study you still have a reasonable number (and maybe you will get lucky and no one will drop out and you will have more participants).

- Time to recruit participants. Do not underestimate the time it may take to recruit people to participate in your field study, especially if you have a specific participant profile. In the best case, you work for a company that has a team dedicated to finding people for user studies or you have funding to outsource finding participants to a company that specializes in recruiting, but this is very rare. Most people recruit their own participants using a variety of methods such as advertising on websites (e.g., Craigslist in the United States) or newspapers, or passing out flyers as the TeamAwear researchers did in sports halls and after basketball games.

In the end, there is not one right answer about the number of people you need in your study. Consider the factors discussed above and then try to include the largest number of participants that is feasible and seems appropriate for your research question.

4.4.4 Compensation

Study participants are typically given a gratuity or compensated in some way for their time and effort. Sometimes if the technology is interesting, it can be possible to entice people to participate for no compensation as participants did in the TeamAwear field study. However, as researchers in that study noted, when you do not compensate your participants you need to consider bias. Your participants may have chosen to participate because they were very excited about your technology, which could bias their feedback. In general, it is best to compensate people in some way, and compensating people fairly can help with recruiting. You should always make clear when presenting your study how participants were compensated.

Considerations:

- How much effort is required? Study compensation typically varies based on how much effort the study requires and can be anything from food to money to software. To get an idea about what is fair, you can ask other people in your organization or look at recent papers in your field. The longer the study, the more compensation is needed. In the Phone Proximity study, participants were compensated with $200 for 3 weeks.

- Will the compensation method affect the data collected? If you would like to have people use a prototype, it might seem like a good idea to compensate people based on their amount of interaction. But if you are going to make claims about usage, you have to be very careful that your gratuity method does not distort any effects. For example, if you paid participants a small amount every time they used your prototype, this could lead to distorted usage. One approach is to reward people for any day or week they use your prototype at least once and then base any analysis on additional usage above the requirement for the gratuity. However, if you are collecting feedback using ESM or other methods, you would want to incent participants to provide as much feedback as possible. In this case, giving participants a small incentive (e.g., entries into a raffle) can be a useful strategy to increase the amount of feedback you receive. For example, in the CareNet study, elders and other people who provided data received between $75 and $300 for a 3-week-long deployment, depending on how often they provided the researchers with updates, whereas participants with the displays received $150. In the AURA study, participants were compensated with coupons to the company café, and each week they scanned at least one item during the final 3 weeks of the study they were eligible for a $50 Amazon gift certificate that was raffled each week.

- Are there other incentives besides money that would also appeal to participants? When determining your compensation strategy, do not underestimate the power of food. In the Home Technology study, researchers brought a pizza dinner to the families to help break the ice and entice families to participate, in addition to giving families their choice of two pieces of Microsoft software.

4.5 DATA ANALYSIS

Using appropriate data analysis techniques will help convince people of the validity of your findings. The type of field study and the data you have collected will determine the depth of the analysis that is appropriate and choice of techniques. For example, in a proof-of-concept study, the analysis may be a very straightforward account of whether the technology worked in the field and participants' reactions collected through surveys or interviews. Other types of studies may involve a more in-depth analysis of the data you have gathered.

Take care that your analysis section does not become a very dense description of results. It can be helpful to use your research question to frame this section and then tell a story about how the data address the research question(s) and any themes or surprises that emerged. When presenting your results, also try to support your findings using more than one type of data. For example, having both logging data and participants' qualitative feedback makes the findings of the Phone Proximity study more convincing and helped the authors explain logging anomalies. Quotes from participants may also help explain logged data or actions of the participants. This section introduces the most common analysis methods and references where you can learn more as needed, first discussing statistical analysis methods and then ways of analyzing unstructured qualitative data.

4.5.1 Statistics

To analyze numeric data, there are two main types of statistics: *descriptive statistics*, which describe the data you have collected, and *inferential statistics*, which are used to draw conclusions from the data.

4.5.1.1 Descriptive Statistics

Common descriptive statistics reported include the frequency of occurrence or count, mean (averages), and median. The statistics that are appropriate to use depend on how a variable was measured, referred to as its *level of measurement* (for more details, see Level of Measurement, http:/ en.wikipedia.org/wiki/Level_of_measurement and Chapter 6 in de Vaus, 2002). The three common levels of measurement for field study variables are described below.

4.5.1.1.1 Nominal   Variables where the possible answers represent unordered categories are referred to as nominal, or sometimes categorical. For nominal variables, you can only report the frequency that each category occurred. For example, gender is a nominal variable where the count of responses can be reported (e.g., Phone Proximity study had 10 male and 10 female participants), but there is no concept of ordering between the response categories.

4.5.1.1.2 Ordinal   Variables measured on an ordinal scale represent a rank order preference without a precise numeric difference between different categories. For example, a survey question with five possible responses of daily, weekly, monthly, and almost never, is measured on an ordinal scale, because the response options can be ordered from more to less frequent, but not added or subtracted. For ordinal variables, both the frequency that each category occurred and the median value can be reported.

Answers to Likert scale questions on a survey are the most common example of ordinal variables collected during a field study. You can compare whether different participant's answers are more positive or less positive than another, but they cannot be added or subtracted. For example, trying to subtract between an answer of Strongly Agree and Strongly Disagree does not yield meaningful information. Note that you may sometimes see means (averages) reported for Likert scale data because some people believe Likert variables can be treated as being measured on an interval scale.

4.5.1.1.3 Interval   For variables measured on an interval scale, the difference between any two values is numerically meaningful. Interval variables can be added and subtracted—for example, a person's age in years, the number of times someone performed a particular action, how long an action took, or the number of ESM surveys a participant answered. Descriptive statistics valid for interval data include sum, mean, and median.

It is important to examine interval data for outliers. Outliers affect the mean, so always report the standard deviation if you report the mean value for a variable. The median can sometimes be more appropriate to report. As an illustration, imagine you have five participants in your study and collect how often each participant performs an action resulting in five data values of {1, 2, 3, 5, 40}. If you report the mean you would say on average participants performed the action 10.2 times (standard deviation = 16.7), whereas if you report the median you would report 3 times. However, neither of these values may accurately describe the data, so you might instead discuss how one participant performed the action much more frequently than the other participants. For some variables, it may be best to report the raw values as well as the descriptive statistics if the variations are large and important.

Considerations:

- Do the descriptive statistics you report give an accurate picture of your data? Have you checked to see if your data contain outliers?
- Are the descriptive statistics reported appropriate for the level of measurement of the variable? Are standard deviations reported with the means?

4.5.1.2 Inferential Statistics: Significance Tests

Once you have computed descriptive statistics for a variable, one type of inferential statistics, significance tests, allow you to determine whether the results found in your sample of participants are statistically significant or might be due to sampling errors. The use of inferential statistics in analyzing field study data is rare since the small number of participants typically feasible to have in a field study makes it difficult to collect enough data for many statistical tests to be appropriate. However, it is sometimes useful to compare descriptive statistics calculated for a variable (e.g., average number of items scanned in the AURA study or average times participants performed an action with your prototype) between different conditions or groups of participants.

This section introduces some common significance tests. Readers interested in more detail about statistical analysis are encouraged to refer to two books: *Analyzing Social Science Data* by de Vaus (2002) and *Using SPSS for Windows and Macintosh, Analyzing and Understanding Data* by Green and Salkind (2002). The first addresses and clearly explains 50 common problems in data analysis from preparing data for analysis to determining which statistical tests are appropriate for your data, whereas the second book provides an excellent practical introduction to using SPSS, a common statistical package, using many examples.

To conduct a significance test comparing descriptive statistics, you first determine the variable you wish to compare and the appropriate groups of participants or different conditions to compare between. In a field study, the most common groups to compare between are either different types of participants (e.g., men vs. women or people who scanned many objects with AURA and those who did not) or answers from the same participants in different conditions in a within-subjects design (e.g., when participants were asked the same questions on a prestudy and poststudy survey, and you wish to compare the answers). It is worth noting that even if you did not initially design your study to explicitly have different types of participants, you may observe—as researchers in the AURA study did—that your participants fall into different groups that you want to compare. Researchers in the AURA study grouped their participants into high-, mid-, and low-volume scanners based on the number of items the participant scanned during the field study. They then used the significance tests to verify if the average number of items uploaded by each group were significantly different and compared the number of days participants in each group used the system.

Once you have identified what variable you want to compare across a set of groups or conditions, which statistical test is appropriate is determined by the level of measurement of the variable, number of groups you are comparing, and whether the variable has a normal distribution (when plotted, it follows the normal curve). For example, an independent samples t-test is appropriate to use when comparing the mean of a variable with a normal distribution between two groups, whereas analysis of variance (ANOVA) tests are used for comparing across more than two groups. Chapter 39 in de Vaus (2002) and the detailed description of each test given by Green and Salkind (2005) can help you choose the most appropriate test. You may be most familiar with the independent samples t-test and ANOVA statistical tests, which are most appropriate for interval-level data with normal distributions. However, many of the nonparametric equivalents (e.g., Mann-Whitney U, Kruskal-Wallis) that do not assume a variable has a normal distribution, may be more appropriate for field study data since they make fewer assumptions about the data that have been collected.

Regardless of what statistical test you use, significance tests start with the assumption that there is no difference between the groups for the variable being examined (referred to as the null hypothesis). If a difference is observed (e.g., the means or medians are different), there are two possibilities: there is a difference between the groups or that there is sampling error in the data. The p value indicates how likely it is that the data might be wrong. Researchers often use a cutoff of either $p < 0.01$ or $p < 0.05$ to determine if the test results are statistically significant. If $p < 0.01$, there is a 99% chance that the data collected represent a real difference between the groups rather than a sampling error (or a 95% chance for $p < 0.05$). Chapters 23–26 of de Vaus (2002) are an excellent introduction to significance testing, factors that affect significance levels, choice of sample size, and statistical power analysis.

Considerations:

- Are there variables that I want to compare? Do I have a large enough sample size to run this test? For example, 30 pairs of scores are considered a moderate sample size for a paired-samples t-test.

- Do the data meet the assumptions of the statistical test I am using? All statistical tests make a set of assumptions about your data (e.g., data values are independent, scores are normally distributed) and the validity of the test depends on how well your data matches the assumptions for the test you choose. Green and Salkind (2005) clearly describe the assumption for each test in their book.

- Are you using statistical tests to explore theories that you have rather than running every test you can think of? de Vaus (2002, p. 174) strongly cautions against "data dredging or running every test you can think of in hopes something will turn up as significant." When conducting multiple related tests, it is often necessary to use a method, such as the Bonferroni technique, to adjust the significance cutoff (the p value) used based on the number of tests conducted. For more detail, see Appendix B in Green and Salkind (2005).

4.5.2 Unstructured Data

Most qualitative data, with the exception of some survey data, are unstructured. This type of data includes free response questions on surveys, answers to interview questions, and any field notes you take down while observing participants. Although trying to understand and derive themes from a large amount of unstructured qualitative data may seem daunting, there are methods that you can apply. This section introduces simple coding techniques and methods for deriving themes from data. Readers should also refer to Chapter 5 for additional insights on analyzing qualitative data.

4.5.2.1 Simple Coding Techniques

Strauss and Corbin (1998, p. 3) broadly define coding as "the analytic processes through which data are fractured, conceptualized and turned into theory." However, the simplest coding techniques consist of closely examining your data and counting the number of times a concept or theme reoccurs, essentially turning qualitative data into quantitative data. For example, if you asked the open-ended question "Did you have any problems using the prototype during the study? If so, what were they?" on a postsurvey, you might want to report common problems. To do this, you would first read over the responses from all the surveys to get a sense of the types of answers. After identifying commonly mentioned problems (e.g., "it crashed," "was slow," "too loud"), you would then count the number of times each problem was mentioned. In a write-up or presentation you could report the number of occurrences and might also include some quotes from the survey responses.

Depending on your study, it may be appropriate to have one person code the data. However, multiple coders, sometimes referred to as raters, are often used if there is a large amount of data to code. When multiple raters code, it is necessary to check for *interrater reliability*, agreement between the raters, to make sure different people are coding the data consistently. If there is a relatively small amount of data and all raters code all the data you can identify any places where there is disagreement, and then discuss and come to an agreement between the raters. More typically, multiple raters each code the same subset of data (in addition to mutually exclusive subsets), and then a test such as Cohen's kappa is used to report interrater reliability on the overlapping set of data and show that the raters are coding consistently. In the Phone Proximity study, researchers present both the exact agreement between the raters and Cohen's kappa as measures of interrater reliability.

4.5.2.2 Deriving Themes and Building Theory

Although simple coding techniques may be appropriate for some qualitative data, organizing field observations and understanding interview data require other methods. Two common methods used are affinity diagramming and grounded theory.

Based on the affinity process introduced by Kawakita, Beyer and Holtzblatt (1998) created an affinity diagramming process, as part of their Contextual Design process to develop user-centered systems. Their affinity diagram process is designed to organize a large number of notes captured during observations or interviews into a hierarchy to understand common issues and themes present in the data.

To construct an affinity diagram, researchers start by putting each note captured on its own small slip of paper. Each note is then placed on a table or wall near other notes that are similar. As groups of notes emerge they are labeled and then these subgroups are grouped and labeled to identify higher-level themes. This approach works well for teams of researchers who can discuss where notes belong and how they should be grouped together to come to a common understanding, which can be especially helpful if all members of a research team were not present at every interview or data collection opportunity. Beyer and Holtzblatt (1998) encourage building the affinity in one day if possible, and it is also helpful to have a large amount of space (e.g., a conference room with a large table or wall space) to spread out the notes and move them around as you recognize themes. Ballagas et al. (2008) used affinity diagramming on more than 1000 quotes collected during the REXplorer study, whereas in the Home Technology study researchers used the process to analyze 650 notes including observations and quotes. In the Phone Proximity study, researchers grouped self-reported reasons for the phone's proximity into 15 themes using affinity diagramming. Researchers considering using affinity diagramming are encouraged to read more about the process in Beyer and Holzblatt (1998) or Preece et al. (2002).

The philosophy of affinity diagrams where issues and themes are derived from the data using a bottom-up approach was influenced by the *grounded theory* method, developed by Glaser and Strauss (1967). Grounded theory emphasizes building theories from the observed data rather than starting from preconceived hypothesis or theories. Researchers begin by conducting a microscopic examination of a subset of their data to identify concepts and categories in the data and relationships between them. These categories are then used and adjusted as needed while coding the rest of the data. Researchers interested in learning about and using the grounded theory approach are encouraged to read *Basics of Qualitative Research* by Strauss and Corbin (1998).

Considerations:

- Make sure you leave enough time to do the qualitative analysis. Carefully reading through and immersing yourself in qualitative data takes time. It is also best to schedule a longer period of uninterrupted time (e.g., eight contiguous hours rather than eight 1-hour blocks of time) because starting and stopping the analysis process can be disruptive to your thought process.

- Think about flagging interesting points as you are collecting the data because this can help you during the analysis. For example, during interviews and when rereading or listening to interview data, make notes of particularly interesting points or comments made by participants that you want to return to.

- Think carefully about whether you need a complete transcription of recorded data (e.g., interviews) and whether you will hire someone to do it. For example, if you plan to use affinity diagramming, you may not need to transcribe interviews completely; you may be able to take notes during the process and use the transcriptions where needed or to get the exact text of quotes that you want to use.

- Makes sure when you report the data you do not become sloppy. Too often when people report qualitative results they use vague words such as "Many participants" or "some." Whenever possible, be as specific as possible in reporting the number of participants that expressed a particular concern; this gives readers more confidence in your results.

4.6 STEPS TO A SUCCESSFUL STUDY

Now that you have been introduced to study design, data collection and analysis methods, and considerations for choosing and managing participants, this section offers practical advice on ways to make your study more successful. At a high level, preparation is the key to having your study run as smoothly as possible. Given the time and effort involved, the last thing you want to realize at the end of the study is that you forgot to do or ask something important.

4.6.1 Study Design Tips

4.6.1.1 Have a Clear Research Goal

Knowing why you want to conduct a field study and what you hope to learn from the study is necessary so that your research questions can inform the numerous choices you need to make when designing the study. Without a clear research question, it may be hard to explain your findings and why they are a contribution to other researchers working in the field. You can also slip into focusing on usability problems, which are typically not interesting to other researchers, if you are studying a prototype.

4.6.1.2 Create a Study Design Document

A study design document should capture the decisions you make when planning your study. Ask your colleagues to review this document and help you identify any problems. A well-written study design document can also be an excellent start on the methodology section of any presentation you need to make. The Study Design Document should contain:

- Your research question
- Participant profile
- Compensation plan
- Your methodology including any conditions you might have in the study
- Timeline: how often you will visit participants, what will happen at each visit
- Types of data you will collect
- How you will analyze the data

4.6.1.3 Make Scripts for Participant Visits
If you are interacting with participants, create a script document for each visit. The script should roughly outline what you will say to the participants and anything you need to do. For example, a script for the first visit would include telling participants about the study and giving them the consent form to sign. In addition to making sure you remember to do everything during a visit, a script also helps ensure you are giving the participants similar instructions and information. Note that it is typically not necessary to read from your script word for word, but it will keep you on track during the visit. Depending on the complexity of your participant visit, some of your scripts may be simple checklists of things to do (e.g., collect logs, give compensation).

4.6.1.4 Pilot Your Study
In a pilot study, you run a group of people through the entire study from the beginning to end as if they were real participants. This is a dress rehearsal for the real study. The pilot will help you identify many potential problems from technical challenges (e.g., you did not take the leap year into account) to issues with surveys and interviews (e.g., your initial visit takes 4 hours instead of the 2 hours you expected).

4.6.2 Technology Tips

Below are some specific tips for studies that involve deploying technology.

4.6.2.1 Make Your Technology Robust Enough

The "enough" part of "robust enough" is very important in managing the effort involved in the study. Based on your research question, determine which aspects of the prototype absolutely have to work and where you can scrimp or cut corners. For example, perhaps it is difficult and confusing to set up your prototype, but since participants will receive the prototype already configured you do not need to spend time on a nice setup experience.

4.6.2.2 Consider Other Evaluation Methods

Before taking the large step of deploying your technology in the wild, consider other evaluation methods to identifying as many usability problems as possible. In heuristic evaluation, developed by Nielsen and Molich (1990), a set of evaluators (which could be you and your colleagues) uses a small set of heuristics to critique your technology and identify problems. Section 13.4 in Preece et al. (2002) further describes the process. Mankoff et al. (2003) have developed additional heuristics for ambient displays that may be relevant to your project. Laboratory studies before your field study can also be very valuable to ensure that your technology is usable.

4.6.2.3 Use Existing Technology

Leverage existing toolkits (e.g., MyExperience) or commercially available prototyping hardware (e.g., Phidgets) when appropriate to make developing your technology easier.

4.6.2.4 Get Reassuring Feedback

Once your technology has gone into the field, look for means to reassure yourself it is working as you expect. Do not count on your users to always tell you if things are going wrong. You do not want to get to the end of a deployment and then discover that, for some reason, the data you expected were not being collected. As mentioned previously, if your technology is not logging data to a central server, consider having it send you periodic "everything's fine" messages so you can detect problems as soon as possible.

4.6.2.5 Negative Results

Deploying a technology that people do not use can be a very painful experience. Decide before the study what you will do if people are not using your technology as much as you expected. Is naturalistic usage a variable you care about? Or will you intervene during the study to encourage people to continuing using your prototype or find out what challenges they are having? Within-subject designs that compare versions of an application can be helpful so that you can get qualitative feedback from participants and also gauge if they used one version more than another, since it can sometimes be hard to figure out how much usage constitutes adoption. Do not plan a study that relies on adoption and usage as the only dependent variable, because you will be in trouble if people do not adopt your technology. If your study results in a negative outcome, as the AURA study did to a certain extent, it can still be very worthwhile to present and publish your results although it is sometimes more difficult, because it is easy for people to dismiss negative results as due to usability issues. Use pilot studies and laboratory evaluations to detect and fix usability problems before your field study. Also make sure your research question focuses on understanding the concepts your technology embodies rather than "Do participants like my prototype?"

4.6.3 Running the Study

4.6.3.1 Have a Research Team

Recognize that each phase of the study from recruiting, to installation, visiting participants, providing technical support, and analyzing the data will take more time than you expect. If you are not already working with a research team, you will want to enlist other people to help you with the field study to make it more manageable.

4.6.3.2 Make Participants Comfortable

Participating in a study can be an awkward experience. Participants have invited people, often strangers, to study them and may be self-conscious. Make clear to participants that you respect and value their feedback and participation. On your first visit, make time for small talk; chat about the weather or other general topics to establish some rapport with participants.

4.6.3.3 Safety

Field studies take place in a variety of environments, and both participant and researcher safety should be taken into consideration. Consolvo et al. (2007)

suggest mixed-gender research teams when visiting people in homes, particularly women participants who may be alone. Use common sense when meeting with participants. For example, consider sending more than one person to meet with a participant, rather than a single researcher.

4.6.3.4 Be Flexible
Field studies are always exciting and things often happen that you were least expecting. For example, participants might suddenly decide to go on vacation in the middle of the study or may take apart your prototype even though you asked them not to. Be flexible and be prepared to make adjustments.

4.6.4 Data Collection and Analysis
4.6.4.1 Be Objective
If you have a hypothesis you strongly believe in or perhaps have spent many years developing a technology, it can be very easy to see what you want to see or ask participants leading questions. If you are studying something you built, try to downplay or avoid mentioning your investment in the technology. For example, if you tell participants "This is something I've been building for the last 5 years and my graduate degree depends on this field study," they may be less likely to share their honest feedback with you. There can also be a tendency for participants to want to please you and do what you want, so watch carefully for this. For example, did a participant tell you they love the prototype but only used it twice during the study?

4.6.4.2 The Participant Is Always Right
In a field study, the participant is always right. No matter what they say, assuming it does not threaten your safety in some way, you need to record the feedback and thank them for it. Do not argue with the participant or get defensive if the participants describes technical problems they are having or reasons why they might not like the technology they are trying. During the analysis phase, you can interpret what participants have said in conjunction with your observations, but it is critical not to argue or disagree with the participant during the study.

4.6.4.3 Do Not Make Inappropriate Claims
Making inappropriate claims based on your findings is one of the most common mistakes people make. Recognize that your prototype represents

only one instantiation of an idea and given that you probably studied a very small number of people, it would be inappropriate to claim that you know that a prototype works for everyone or that you have conclusively answered a general research question. A limitations section in a paper or presentation that acknowledges potential limitations (e.g., a small number of participants from a limited geographic region) of the study helps make clear to the audience that you are not making inappropriate claims. Also, pay attention to the language you use in a written presentation; watch out for words like "prove" and instead use terms such as "suggest" or "support," which are more appropriate to use.

4.7 CONCLUSION

Field studies are a crucial tool for ubicomp research. As part of this research field, we must continue to build and extend tools (e.g., MyExperience (Froehlich et al., 2007), Momento (Carter et al., 2006), tools built by Intille et al. (2003)) that ease the process of conducting field studies. We must also build on and continue the work that has been started through papers (e.g., Scholtz and Consolvo, 2004) and workshops (e.g., USE 2007, 2008) to develop best practices and evaluation strategies for ubicomp systems. By working together, we can reduce the effort required to conduct studies and facilitate comparison between different approaches and applications.

Although the amount of effort involved in conducting a field study may seem a bit daunting, there is really no substitute for the inspiration and understanding you will gain from interacting with participants in the field. Regardless of whether your field study involves observing people's current behavior, conducting a proof-of-concept study, or deploying your technology to participants for a long period, you will learn something that surprises you and helps you to move your research forward. The understanding and insights you gain from a field study can often spark new ideas and directions for future research.

ACKNOWLEDGMENTS

The author thanks Ed Cutrell, Sunny Consolvo, and Beverly Harrison for sharing their wisdom and experience from many studies. Authors of the examples are also acknowledged for allowing the inclusion of their research in this chapter. Finally, Mike Brush is thanked for his support.

REFERENCES

Affinity Diagrams, http://en.wikipedia.org/wiki/Affinity_diagram

Ballagas, R., Kuntze, A., and Walz, S. P., Gaming tourism: Lessons from evaluating REXplorer, a pervasive game for tourists. Pervasive '08.

Beyer, H., and Holtzblatt, K., *Contextual Design: Defining Customer-Centered Systems*. Morgan Kaufmann, San Francisco, 1998.

Brush, A. J., and Inkpen, K., Yours, mine and ours? Sharing and use of technology in domestic environments, in *Proceedings of the Ubicomp 2007*, September 2007.

Brush, A. J., Turner, T., Smith, M., and Gupta, N., Scanning objects in the wild: Assessing an object triggered information system, in *Proceedings of UbiComp 2005*, 2005.

Carter, S., Mankoff, J., and Heer, J., Momento: Support for situated ubicomp experimentation, *CHI 2007*, 2006, pp. 125–134.

Consolvo, S., and Walker, M., Using the experience sampling method to evaluate ubicomp applications, *IEEE Pervasive Computing Magazine: The Human Experience* 2(2), 24–31, 2003.

Consolvo, S., Harrison, B., Smith, I., Chen, M. Y., Everitt, K., Froehlich, J., and Landay, J. A., Conducting in situ evaluations for and with ubiquitous technologies, *International Journal of Human-Computer Interaction* 22(1–2), 103–118, 2007.

Consolvo, S., Roessler, P., and Shelton, B. E., The CareNet display: Lessons learned from an in home evaluation of an ambient display, in *Proceedings of the 6th International Conference on Ubiquitous Computing: UbiComp '04*, Nottingham, England, Sep. 2004, pp. 1–17.

de Vaus, D., *Analyzing Social Science Data 50 Key Problems in Data Analysis*. Sage Publications, London, 2002.

Froehlich, J., Chen, M., Consolvo, S., Harrison, B., and Landay, J., MyExperience: A system for in situ tracing and capturing of user feedback on mobile phones, in *Proceedings of MobiSys 2007*, San Juan, Puerto Rico, June 11–14, 2007.

Gaver, W., Bowers, J., Boucher, A., Law, A., Pennington, S., and Villar, N., The History Tablecloth: Illuminating domestic activity, in *Proceedings of DIS '06: Designing Interactive Systems: Processes, Practices, Methods, and Techniques*, 2006, pp. 199–208.

Gaver, W., Sengers, P., Kerridge, T., Kaye, J., and Bowers, J., Enhancing ubiquitous computing with user interpretation: Field testing the Home Health Horoscope, in *Proceedings of the SIGCHI Conference on Human Factors in Computing Systems, CHI '07*, San Jose, CA, April 28–May 3, ACM, New York, NY, 2007.

Glaser, B. G. and Strauss, A. L., *The Discovery of Grounded Theory. Strategies for Qualitative Research*. Aldine De Gruyter, Hawthorne, NY, 1967.

Green, S. B., and Salkind, N. J., *Using SPSS for Windows and Macintosh: Analyzing and Understanding Data*, 5th Edition. Prentice Hall, Upper Saddle River, NJ, 2005.

Harris, C., and Cahill, V., An empirical study of the potential for context-aware power management, *Ubicomp 2007: Ubiquitous Computing*, 16–19 September, Innsbruck, Austria, 2007, pp. 235–252.

Intille, S., Tapia, E., Rondoni, J., Beaudin, J., Kukla, C., Agarwal, S., Bao, L., and Larson, K., Tools for studying behavior and technology in natural settings, in *Proceedings of the Ubicomp 2003: Ubiquitous Computing*, 157–174, 2003.

Kahneman, D., Krueger, A. B., Schkade, D. A., Schwarz, H., and Stone, A. A., A survey method for characterizing daily life experience: The day reconstruction method, *Science* 306, 1776–1780, 2004.

Krumm, J., and Horvitz, E., Predestination: Inferring destinations from partial trajectories, in *Proceedings of Ubicomp 2006*, 2006, pp. 243–260.

Level of Measurement, http://en.wikipedia.org/wiki/Level_of_measurement.

Mankoff, J., Dey, A., Hsieh, G., Kientz, J., Ames, M., and Lederer, S., Heuristic evaluation of ambient displays, in *Proceedings of CHI 2003*, 2003, pp. 169–176.

Matthews, T., Carter, S., Pai, C., Fong, J., and Mankoff, J., Scribe4Me: Evaluating a mobile sound transcription tool for the deaf, in *Proceedings of the International Conference on Ubiquitous Computing (Ubicomp 2006)*, Newport Beach, CA, 2006, pp. 159–176.

MyExperience ESM Toolkit, http://myexperience.sourceforge.net.

NASA TLX: Task Load Index, http://humansystems.arc.nasa.gov/groups/TLX/.

Nielsen, J., and Molich, R., Heuristic evaluation of user interfaces, in *Proceedings of the ACM CHI'90 Conference*, Seattle, WA, 1–5 April, 1990, pp. 249–256.

Page, M., and Vande Moere, A., Evaluating a wearable display jersey for augmenting team sports awareness, International Conference on Pervasive Computing (Pervasive 2007), 2007, pp. 91–108.

Patel, S. N., Kientz, J. A., Hayes, G., Bhat, S., and and Abowd, G., Farther than you may think: An empirical investigation of the proximity of users to their mobile, in *Proceedings of the Ubicomp 2006: 8th International Conference on Ubiquitous Computing*, Orange County, CA, September 17–21, 2006, pp. 123–140.

Patel, S. N., Robertson, T., Kientz, J. A., Reynolds, M. S., and Abowd, G. D., At the flick of a switch: Detecting and classifying unique electrical events on the residential power line, in *Proceedings of Ubicomp 2007*, 2007, pp. 271–288.

Preece, J., Rogers, Y., and Sharp, H., *Interaction Design: Beyond Human-Computer Interaction*. John Wiley & Sons, Inc., New York, NY, 2002.

Questionnaire for User Interaction Satisfaction, http://lap.umd.edu/quis/.

Rogers, Y., Connelly, K., Tedesco, L., Hazlewood, W., Kurtz, A., Hall, R., Hursey, J., and Toscas, T., Why it's worth the hassle: The value of in-situ studies when designing ubicomp, *Proceedings of UbiComp 2007: Ubiquitous Computing*, 2007, pp. 336–353.

Scholtz, J., and Consolvo, S., Toward a framework for evaluating ubiquitous computing applications, *IEEE Pervasive Computing Magazine* 3(2), 82–88, 2004.

Sohn, T., Li, K., Griswold, W., and Hollan, J., A diary study of mobile information needs, in *Proceedings of the ACM CHI 2008, Conference on Human Factors in Computing Systems*, Florence, Italy, April 5–10, 2008, pp. 433–442.

Strauss, A., and Corbin, J., *Basics of Qualitative Research, Techniques and Procedures for Developing Grounded Theory*. Sage Publications, London, 1998.

U.S. Institutional Review guidebook, http://www.hhs.gov/ohrp/irb/irb_chapter3.htm, describes basic IRB guidelines.

Woodruff, A., Anderson, K., Mainwaring, S., and Aipperspach, R., Portable, but not mobile: A study of wireless laptops in the home, *Pervasive Computing, 5th International Conference, Pervasive 2007*, Toronto, Canada, May 13–16, 2007, pp. 216–233.

CHAPTER 5

Ethnography in Ubiquitous Computing

Alex S. Taylor

CONTENTS

5.1	Introduction	204
5.2	From Ethnography to Design	205
	5.2.1 Ethnography	205
	5.2.1.1 Nuer Time-Reckoning	206
	5.2.2 Design-Oriented Ethnography	210
	5.2.3 Ubicomp	212
5.3	Design-Oriented Ethnography in Practice	214
	5.3.1 Planning	214
	5.3.1.1 Hypotheses	215
	5.3.1.2 Sampling and Generalization	216
	5.3.1.3 Access	218
	5.3.2 In the Field	219
	5.3.2.1 Reflexivity and Indifference	220
	5.3.3 Analysis	223
	5.3.3.1 Data and Its Influence on Analysis	224
	5.3.3.2 Analytic Sensibilities in Ubicomp	225
5.4	What Is It Good For?	228
	5.4.1 Design Implications	228
	5.4.2 Future Directions	230
References		232

> I don't believe in teaching. One learns by looking. That's what you must do, look.
>
> (FRANCIS BACON [1909–1992], A FEW WEEKS BEFORE HE DIED)

5.1 INTRODUCTION

Those introduced to ethnography often struggle to understand what it is, what it has to offer, and—most importantly—how to do it. Many find it hard to define in terms of the commonly used research nomenclature; is it a methodology, method, orientation, technique, or something altogether different? How, they also ask, does it contribute to informing scientific investigation and what steps does an ethnographer follow to successfully do an ethnography?

Broadly speaking, ethnography can be thought of as a sensibility or "way of seeing" one adopts in collecting and interpreting field study materials (see Wolcott, 1999). This sensibility can be influenced by a wide range of ideas and theories. The methods and techniques used to collect data are also varied and can often be shaped by the settings an ethnographer finds himself or herself in. So, to add to the newcomer's confusion, ethnography is motivated by an assortment of intellectual traditions and is only loosely defined by its methods. At the same time, ethnography's traditions—as well as its perspective on how empirical materials are gathered and interpreted—cast doubt on some of the common underpinnings of empirical, scientific research. For example, questions are implied, if not explicitly raised, about notions of validity and generalizabilty. All this is unlikely to be of much comfort to those embarking on ethnography. At this stage, however, it should be enough to recognize that it is not all that surprising that ethnography is the source of trouble for those fresh to it (as well as, it should be said, those who regularly ply its trade).

This chapter initially provides some background to ethnography in the hope of unraveling at least some of this apparent confusion. Much has been written with similar motivations, especially in the social sciences and humanities, and those interested in pursuing ethnography are encouraged to review this work (e.g., Geertz, 1973; Clifford and Marcus, 1986; Bryman, 2001; Wolcott, 1999). Here, though, the emphasis is on how ethnography has made its presence felt in ubiquitous computing (ubicomp) and the associated areas of human–computer interaction (HCI) and computer-supported collaborative work (CSCW). The differences between ethnography and the more general types of technology-oriented field studies are reviewed in Chapter 4. Ethnography's contribution in the areas related to ubicomp is also considered to better understand how it has been applied.

In the chapter's last two sections, closer attention is paid to doing ethnography. Again, there have been efforts to produce how-to guides for ethnography in the social sciences (Hammersley and Atkinson, 1995; O'Reilly, 2005; Wolcott, 1995). In addition, there are several notable commentaries of ethnography when applied to technology design (Anderson, 1994, 1997; Button, 1993; Button et al., 2009; Dourish, 2006; Grudin and Grinter, 1995; Heath, 2000; Macaulay et al., 2000; Randall et al., 2007). Building on these past works, this chapter aims to focus on some particular issues raised by ethnography in ubicomp and the implications these issues have for real-world ethnographic practice. This leads into a discussion of what such a practice is good for in ubicomp. Finally, thought is given to how ethnography in ubicomp is beginning to transform so as to accommodate a number of distinctive aspects of design and technology.

Overall, the reader should be aware that this chapter is not a how-to or a recipe list for undertaking an ethnography in ubicomp research. At a practical level, one would be hard pushed to produce anything useful of this type that captures all the possible contingencies that can arise. More importantly, however, the ethnographic sensibility is not something that can be definitively expressed or prescriptively laid out. Rather, an ethnography is something that one must go out and do, and the ethnographic sensibility is something that only really comes about through experience in the field. The best preparation is to read past ethnographies, prodigiously, in the hope of learning how others have grappled with sensitizing themselves to the settings and peoples they are studying.

5.2 FROM ETHNOGRAPHY TO DESIGN

5.2.1 Ethnography

Although there are exceptions,* ethnography is usually characterized by an ethnographer spending time in a place among a distinct group of people. Thus, the ethnographer may spend time with a Samoan tribe (Mead, 1928), Chicago's hobos (Anderson, 1923), cigar smokers in Kentucky (DeSantis,

---

* One exception is virtual ethnography where the researcher participates in online or virtual communities (see Hine, 2000).

2003), or mobile phone users in Tokyo (Ito et al., 2006); the types of people and places that might be studied using ethnography are endless. Common between all ethnographies, however, is the effort the ethnographer spends in analyzing field materials and writing. Ethnography is a deeply literary practice that places great emphasis on the ethnographer crafting his or her evolving, descriptive analysis. Indeed, the iterative composition of themes and arguments produced in writing is as much a part of ethnography as being in the field and collecting data.

To develop this point, this section presents some past work starting with a classic ethnography from social anthropology, and then moving to more recent cases of greater relevance to ubicomp. In both examples, the emphasis is on helping the reader see how ethnographers start to piece together their field materials and apply certain sensibilities when interpreting them. Importantly, the reader should recognize that even with the older, seemingly less relevant work, there are salient themes that ubicomp research might draw on. The ethnographic sensibility is one that continually casts back to past ideas and theories to discover the world anew. Sometimes, exception is taken to past work, but almost inevitably new arguments are threaded from old.

5.2.1.1 Nuer Time-Reckoning

Influenced by the early pioneers of ethnography in anthropology (e.g., Malinowski, Radcliffe-Brown, and Seligman), Edward Evans-Pritchard played a key role in developing both anthropological theory and ethnographic practice (Beidelman, 1974). Particularly through his studies of the Nuer and Azande in the 1920s and 1930s, he made significant contributions to what were then the burgeoning theories of *structuralism* and *functionalism* in anthropology. His time in the Southern Sudan and Ethiopia with the Nuer, for example, spanning 10½ months, led to his classic monograph detailing how the Nuer's social structures were tightly coupled with their pastoral life (Evans-Pritchard, 1940). For example, Evans-Pritchard described how the personal names used by Nuer men, at least at the time, related to the coloring and form of their favorite oxen and, for women's names, which cattle they milked. This demonstrated a curious tie between the Nuer and their work with cattle. As Evans-Pritchard wrote:

> Sometimes the name of a man which is handed down to posterity is his ox-name and not his birth name. Hence a Neur Geneology

may sound like an inventory of a kraal. The linguistic identification of a man and his favorite ox cannot fail to affect his attitude to the beast, and to Europeans the custom is the most striking evidence of the pastoral mentality of the Nuer.

(EVANS-PRITCHARD, 1940, P. 18)

Similarly, Evans-Pritchard discovered Nuer time-reckoning to be guided by the activities associated with caring for and feeding cattle (Evans-Pritchard, 1939). The Nuer, for instance, have two main seasons: one, *tot*, associated with when they live in their villages and the other, *mei*, when they must set up and move from camp to camp to graze their cattle. These seasonal names are not just signifiers of the time of year. They can also be used to refer to the activities they are associated with, so a Nuer might be said to be going to *mei* (camp) in such a place: "ba wa mei." *Tot* and *mei* are thus not used as abstract points of reference in calendar time, but function more as practical terms loosely influencing social organization.

In fact, during his fieldwork Evans-Pritchard found the Nuer to have no abstract concept of time (as we do). They had no equivalent word for "time," and did not talk of "time" as something "which passes, can be wasted, can be saved, and so forth" (Evans-Pritchard, 1939, p. 208). On a daily basis, the passing of time was reckoned with respect to herding cattle to pasture, milking, churning, drying of dung fuel, and so on. Moreover, the Nuer's division of the day appeared greater during periods when the cattle-related activities were relatively more intensive. Likewise, longer periods between months or years were usually described using events as points of reference; droughts, bouts of cattle disease, weddings, etc., might be used to make reference to the past. As Evans-Pritchard writes:

> Certainly they [the Nuer] never experience the same feeling of fighting against time, of having to coordinate activities with an abstract passage of time, since their points of reference are mainly activities themselves, which are generally of a leisurely and routine character. There are no autonomous points of reference to which activities have to conform with precision.

(EVANS-PRITCHARD, 1939, P. 208).

For the purposes of this chapter, Evans-Pritchard's studies of the Nuer help to illustrate three general points concerning ethnography. First and foremost, they demonstrate how an ethnographer approaches studying a group of people in situ (see Chapter 4). Over the course of the fieldwork, Evans-Pritchard observed, interviewed, took field notes, and generally got to grips with the Nuer's way of life. A central facet of this field research was to build, as far as possible, an intimate understanding of the Nuer and how they organized themselves socially. For this, Evans-Pritchard relied heavily on individual informants with whom he developed close relationships. His informants helped in practical matters such as translation and detailing social practices. Interviews with others were also used, but the informants were usually the ones trusted and relied on. As well as this, Evans-Pritchard also purchased his own herd of cattle and tended to them as a Nuer might (or tried to). Collectively, these various strategies fall under ethnography's broad and, some would argue, defining method for collecting fieldwork materials, *participant observation*. Notable, however, is that no one fieldwork strategy above is essential, nor must a critical balance be struck between interviews, observations, and participation in some activity. The approaches an ethnographer adopts tend to be driven by good (or bad) fortune and opportunism, of which more will be discussed later.

A second point to draw from Evans-Pritchard's work is the importance of interpretation in ethnography, specifically through textual description. In reading the many articles and monographs Evans-Pritchard produced from his studies of the Nuer and Azande, it becomes clear that his work has a distinctive discursive style. In his writings, he often begins by detailing a social practice. He then gradually builds up his descriptions, thickening them with various theoretical claims. In his article on the Nuer and their reckonings of time, for instance, he begins by describing the various systems the Nuer use to account for time. In the later stages of the article, these descriptions interleave with an argument claiming that the Nuer's sense of time functions to coordinate and structure their social relations. More will be discussed of this claim below. At this stage, it suffices to say that Evans-Pritchard provides us with an example of how the ethnographer makes choices to foreground certain aspects and themes in the ethnographic material—that is, there is a strong interpretive character to an ethnographer's written descriptions. In Evans-Pritchard's case, he not only details the Nuer and their social practices. He instructs the reader to see as he does—to interpret the Nuer as he has.

The third and final point to be taken from this example is closely related to this idea of interpretation. As noted, Evans-Pritchard pursued much of his field research, particularly in his early years, with an inclination toward structuralism (or more particularly, structural functionalism), that is, he was concerned with how particular practices functioned to structure societies. For example, how the predominant activities of cattle rearing among the Nuer functioned to organize their patterns of movement, divisions of labor, notions of time, social relations, and so on. Evans-Pritchard's writings were thus inclined toward a very particular analytical standpoint or orientation. Of course, Evans-Pritchard's depictions of the Nuer could well have been wrong (and indeed, debates continue over the veracity of his work). What is important here, however, is that it demonstrates how ethnography is not merely description; it is not a litany of how many times x spoke to y, or a mapping of who is related to whom—it is not the mere survey of a scene. The ethnographer draws on his or her own analytical traditions in the interpretative process to express something more of the setting studied, its peoples, or possibly society *writ large*.

To summarize, Evans-Pritchard's studies of the Nuer offer a reminder of three important, if not defining, features of ethnography. For the purposes of clarity, these can be listed as follows.

Participant Observation. Empirical materials are collected in the field using participant observation. Importantly, the interviews, observations, and participation in everyday life that make up participant observation, do not adhere to a strict, formal method. Rather, they make up a loosely assembled collection of strategies used to investigate a setting and its peoples, in situ. Participant observation is thus driven by the motivation to gain a deep familiarity with a people and their practices, and not by a strict idea of and adherence to method, per se.

Interpretation. In producing his or her texts, the ethnographer unavoidably interprets the people and setting under observation. Clifford Geertz, a central figure in anthropology who wrote cogently on this topic, famously sought to openly reveal this by referring to ethnography as *thick description*. "What we call our data," he elucidates, "are really our constructions of other people's constructions of what they and their compatriots are up to…" (Geertz, 1973, p. 9). The processes of interpretation, then, the layering of construction upon construction, is something to be recognized and worked with in the ethnographic enterprise.

Analytical Orientation. Whether it is made explicit of not, ethnographers invariably adopt one or sometimes several analytical orientations in their interpretive process. In the field, an orientation trains the eye, so to speak, providing the ethnographer with themes and topics to engage with. In writing, the analytical orientation is used to develop the empirical materials, teasing out specific arguments and sometimes contributing to broader theories of social practice and organization. In writing of the value of ethnography in systems design, Anderson (1997) nicely captures the analytical mindset:

> It is the patterns and patterning the ethnographer is looking for and not simply a realistic, behavioralized description or natural history. What an ethnographer is most interested in, and thus what makes an ethnography of particular interest, is not lots of everyday detail about some local scene. Neither is it some hitherto unsuspected, beneficial or deleterious aspects of an activity. Rather the ordinariness is somehow rendered extraordinary and yet, recognizable. The deeper patterns being played out, in and through the detail, come to the surface.
>
> (ANDERSON, 1997, P. 158)

5.2.2 Design-Oriented Ethnography

The *in situ* field studies undertaken in HCI, CSCW, and ubicomp research have been somewhat removed from the types of ethnography found in anthropology. With their emphasis on informing design, the studies have tended to be far briefer and focused on the use of technologies and other material artifacts rather than the broader concerns of social life (Hughes et al., 1994). Nevertheless, in keeping with the ethnographic tradition, there has remained research directed toward the detail of people's interactions with technology and "toward the production of a 'rich' and 'concrete' portrayal of the situation" (Hughes et al., 1997).

To understand how, exactly, ethnography is both understood and used in ubicomp, some background to its uptake in systems design offers a useful starting point. The basis for what might be thought of as a *design-oriented ethnography* began, initially, in HCI and CSCW, and later fed into ubicomp research. In the late 1980s, Lucy Suchman's now much-cited book *Plans and Situated Actions* (1987) was perhaps the most influential factor in introducing a sociological and loosely ethnographic sensibility to HCI. It also shaped what was then nascent research into CSCW. In her book, Suchman presented a study of photocopy use she undertook with her organization at the time. Using her findings as a basis, she made a convincing argument proposing that people's interactions with technology could be seen to be influenced by the particular features of a setting. Specifically, she highlighted the ways in which people's HCIs were situated in social practices and that this *situated action* could be made "visible" through detailed in situ studies. Such insights, she revealed, could provide a significant contribution to the understanding of how people use technologies and thus how future technological solutions might be designed.

Suchman's work was, in part, a reaction to the predominant thinking in HCI at the time. It contrasted with efforts that aimed to model the task-based and cognitive aspects of HCIs and, in doing so, abstract away from the particularities of the setting. Suchman, alongside others (Anderson, 1997; Button, 1993; Harper, 2000; etc.), provided an impetus for a "turn to the social" in HCI and CSCW, where greater emphasis was placed on revealing those moment-by-moment actions particular to a setting and the often coordinated interactions between people.

An early example of applying this type of perspective to informing design consisted of the studies of air traffic control (Harper et al., 1991; Hughes et al., 1992; Hutchins and Klausen, 1996; Mackay, 1999). This ethnographic research drew attention to the collaborative work of air traffic controllers involved in organizing airplanes in airspace. Specific focus was given to flight strips—the paper strips containing the details of planes in the air—and their role in the control room. The various publications produced from this work elaborated on how flight strips aided in planning, helping with the management of plane trajectories, and the coordination between controllers (Mackay, 1999). For example—and to oversimplify the details—the orderly arrangement of the flight strips was found to operate as a proxy for the orderliness of the skies; glances from controllers to the strips arranged on racks provided lightweight means of assessing, reacting to, and anticipating the moment-by-moment conditions of the skies. In one attempt to develop design proposals from these observations, an alternative to flight strips was put forward (Hughes et al., 1992). Here, different visualizations were used to demonstrate how a computer system might support the display of volumes of air traffic and the potential for collisions, and to do so using the same at-a-glance qualities of the physically arranged flight strips. Further proposals were made for enabling controllers to "test out" different flight path solutions in advance, augmenting their established methods for judging and anticipating the busyness of the skies.

This example provides a useful illustration of some of the differences and commonalities between ethnography as it originated in anthropology and how it first took shape in HCI and CSCW. In common between the two were the use of careful *in situ* investigations of a setting, the collection of copious field materials, and interpretive analyses represented through thick textual descriptions. Studies of the workplace differed, however, in that they were less prolonged and focused on specific mediated interactions—for example, the ways controllers interacted with and through things such as flight strips. Moreover, they took on an analytical orientation rarely given much attention in anthropology or sociology known as *ethnomethodology* (the orientation Suchman had also used to inform her ideas of situated action). This use of the ethnomethodological orientation (or so it was argued by its proponents) allowed for far greater emphasis to be placed on the ways work was practically accomplished and specifically the commonsense methods (ethno-methods) people use to get on with the business at hand.[*]

In its initial uptake in HCI and CSCW, then, ethnography took on a distinctive character in four primary ways:

1. The research was undertaken under greater time constraints.

2. The interpretive character of the research fell largely under the analytical auspices of ethnomethodology (even if only loosely).

3. Attention was given to specific interactional features of a setting (rather than the orderings of a society or culture) and a fine-grained level of analyses was used.

4. The general outcome was oriented toward how technology could be designed to support the types of social practices and accomplishments the field studies revealed.

5.2.3 Ubicomp

Since these beginnings, design-oriented ethnography has come in some shape or form to have an established role in HCI and CSCW. As ubicomp

---

[*] Much has been written on ethnomethodology and its use in systems design. For an accessible introduction to ethnomethodology, see Livingston (1997), and for its role in design, see Button (2000) and Randall et al. (2007).

gained momentum as an area of research in its own right, it, too, incorporated the practice. Broadly speaking, research in these areas came to use ethnography as a means of (1) studying new settings to inform design and (2) evaluating the use of newly designed systems in the real world. Ubicomp also played its part in broadening the types of places design-oriented ethnographies were brought into play. Looking at technologies that were intentionally designed to pervade everyday life, ethnographers found themselves studying not only the workplace, but also leisure and domestic environments (Brown et al., 2007; Crabtree et al., 2006), and those spaces in between (Brewer, 2008).

With the diffusion of a design-oriented ethnography into different research programs and its application in different settings, it has also come to incorporate varying analytical orientations. When compared to other disciplines that use ethnography, ethnomethodology continues to have a disproportionate presence as an orientation, arguably because of the early and substantial impact of Suchman's ideas on situated action. Yet, there have been recent trends, especially in ubicomp, to adopt different frames of reference. A continued focus on materiality and the interactions people have in physical space has, for example, helped foster a now reasonably developed position that incorporates theories from phenomenology (Dourish, 2001). Moreover, a growing interest in mobility has led to investigations of space as a topic and the use of relevant theories originating in the social sciences (Harrison and Dourish, 1996; Ito et al., 2006). Cultural practices have also begun to be addressed. Some in ubicomp have examined computing in very distinct cultural groupings ranging from those in African townships (Marsden, 2008), to Filipino and Ghanian transnationals (Williams et al., 2008), to Orthodox Jewish households in North America (Woodruff et al., 2007).

This increasing openness to theory and analytical perspectives, however, has not gone without controversy. A general and reoccurring commentary has emerged around whether much of the field research presented as "ethnographic" in HCI, CSCW, and ubicomp deserves such a title (Anderson, 1997; Button, 2000; Harper et al., 2005). The debates broadly center around three issues. The first concerns the amount of time spent in the field collecting data. The weeks or even days spent doing fieldwork with the users of a technology have been criticized for being far too brief. With respect to ethnography, it is argued that they do not allow sufficient time for a setting to be adequately studied. A second issue concerns the use of an analytical orientation. Here, it is questioned whether a field study that simply reports the details of a scene can legitimately be called an ethnography. So-called scenic fieldwork has been contrasted with ethnography, where an analytical sensibility or orientation is seen as a defining feature (Button, 2000). Third, a debated issue arises over the question of who can make claims to be an ethnographer and thus practice ethnography. As researchers and commercial practitioners in ubicomp, HCI, and CSCW wander out into the field and take on the strategies of participant observation, questions are raised over how well qualified they are to do ethnographies. It is questioned whether anthropologists have an authoritative position from which to examine the in situ use of technology because of their training (Harper et al., 2005).

None of these issues can be easily addressed. In many respects, the debates and the different ways they are resolved come to make up the substance of design-oriented ethnography as it is practiced and as it evolves in areas such as ubicomp. The remaining portion of this chapter aims to offer a greater insight into ethnography as a practice and the various pitfalls one can come across. However, the origins of ethnography and the ensuing debates in ubicomp, HCI, and CSCW should be kept in mind. Not only will they allow a critical engagement with the following materials, they will also shape how one develops one's own ethnographic sensibility.

5.3 DESIGN-ORIENTED ETHNOGRAPHY IN PRACTICE

So far, this chapter has detailed something of the character of ethnography and how it has been taken up in design-related areas such as CSCW, HCI, and ubicomp. The aim has been to express a feel for design-oriented ethnography—to capture something of it origins, how it has developed, and the points of tension in this development. In the following, a few of the more pragmatic details of undertaking a design-oriented ethnography will be discussed. Earlier, it was explained how ethnography is not something to be easily proceduralized. In this vein, three general topics will be discussed: *planning fieldwork*, *being in the field*, and *analysis*. The intention is to give the reader an idea of what to look out for and what to keep in mind when doing ethnographic fieldwork to inform design. Many of the points will be of a practical nature, but a few should be seen as returning to some of the central themes that underlie the ethnographic enterprise.

5.3.1 Planning

One of the first things to do when embarking on an ethnography is to scope and plan for the field research. In design-oriented ethnography, the scope will often be dictated by the types of technology one is interested in. Is the project concerned with investigating sites where ubiquitous computing might be used such as the home or office, or when mobile? What types of people are envisaged to inhabit these settings? Knowledge workers, the young or aged, families, etc.? Alternatively, is the project centered on an evaluation of a technology? If this is the case, where will the technology be deployed and what kinds of people will have access to it? Moreover, how might the different types of interactions be recorded: through observation, interview, video recording, or audio recording, etc.?

Much of the decision making associated with planning empirical research was covered in the earlier chapter on field studies (Chapter 4). However, there are several issues to keep in mind that have immediate relevance for ethnography. Three are considered in some detail below: the role of hypotheses, sampling and generalization, and access to fieldwork sites.

5.3.1.1 Hypotheses
Critically, unlike many other forms of scientific research, ethnographies will not usually be designed around an initial hypothesis or hypotheses. An ethnography is used when the aim is to openly investigate a topic. For example, if a project was to use an ethnography to study teenagers and their use of mobile phones, the research would be framed in an open manner, possibly with a broad question such as: "What is the role of the cell phone in teenagers' everyday lives?"

This investigative nature of ethnographic fieldwork makes it particularly hard to apply any strict structure or schedule when planning an ethnography, especially one that lasts over several weeks or months. Often, circumstances will change or various, sometimes unexpected, themes will emerge that will alter the focus and trajectory of the research. The study of teenagers and their cell phones might, for instance, evolve over time as the fieldwork reveals that a far more salient issue is how cell phones are used to maintain friendships or provoke rivalries. The critical point here is that an ethnography should not be seen as a means to find resolution (to prove or disprove hypotheses), but instead as an exercise in opening up new avenues or possibilities.

As it happens, there are several examples of ethnographies of teenage cell phone use that illustrate this. Weilenmann (2003), for instance, used a study of cell phone talk between teenagers to show how questions such as "Where are you?" prompt responses associated with activity and availability, and not just location. So, an answer such as "I'm in the fitting room" says a great deal about all three. In another example, Taylor (2005) considers the material properties of the phone and its role in teenagers' conversations. He suggests that the taken-for-granted presence of the cell phone in conversations between teenagers provides them with a means of managing the topic of talk and, at times, subverting topical talk; the phone provides a legitimate reason to break off from a conversation and talk about something else, or someone else. Stepping back from the details of talk, Ito (2005) examines how space is configured as teenagers coordinate their activities using their phones. She reflects on the role the phone plays in power relations in Japan, between the teenagers themselves and with the institutions of authority including the teenagers' parents. In each of these examples, what is evident is that the ideas are developed in and through the textual analysis and worked up in a discursive fashion. None of the three studies are used to definitively answer some *a priori* hypothesis of phone use by teenagers. What they accomplish, though, is a starting point for thinking about how phones are used and opening up a set of design possibilities.

The explorative nature of ethnographic research, however, does not eliminate the need for scoping or planning fieldwork. The point here is that the planning and scoping should be done bearing in mind the openness discussed. Room must be left for new and unexpected empirical themes to arise, and for the results that pose questions (and not just answers) around design. A project, then, is best approached with a flexile plan that can adjust with the unfolding research. The project's members should allow for a bit of opportunism in their work rather than sticking doggedly to a method or fixed sequence of stages. The scoping of the research is something that might benefit from more restraint at first. Too often, ethnographies take on topics that are far too broad and, for much of the research, effort is spent managing the quantity, detail, and complexity of the field data. In a similar vein, effort is put into figuring out what to focus on, the scope of the research, and how to justify attending to one thing over another. By limiting the scope from the outset, the research is given space to expand and follow different trajectories. In many ways, an ethnography should be treated as a continuous scoping exercise, where decisions on method, analysis, and interpretation need to be made on an ongoing basis.

5.3.1.2 Sampling and Generalization
The issues of choosing how many participants to study and how long to spend in the field are difficult (and perhaps contentious) ones in

ethnographic research. Empirical research, whether done in the laboratory or in situ, tends to be concerned with collecting and analyzing data that can be generalized to a "population." In studying teenagers and their uses of cell phones, for example, a study would be designed so that claims could be made about cell phone use by teenagers in general and not just the participants in the study. Thus, the procedure is usually to find a representative sample of subjects or participants and to use the appropriate empirical methods to generate generalizable data.

With the small number of participants usually included in a field study and its less-than-structured empirical methods, ethnography has been thwarted by claims that it offers no means for generalization. For the most part, however, ethnographic research has come to operate outside of this empirical framework; the issue of generalization is not entirely resolved, but rather seen in quite a different light. As an early protagonist of ethnography in sociology, Howard Becker, put it:

> If we haven't settled (these epistemological issues) definitively in two thousand years, more or less, we probably aren't ever going to settle them. These are simply the commonplaces, in the rhetorical sense, of scientific talk in the social sciences, the framework in which the debates go on.
>
> (BECKER, 1993, P. 219)

In an ethnography, then, there is no overriding concern for choosing a representative sample from a population. Far more important is how a study's participants (or a setting) will make visible their own commonsense reasonings or social patterns and rituals. The ethnographer is not trying to explain social behavior in terms of whether an entire population does or does not do something—of whether *all* teenagers use their cell phones to maintain friendships. Instead, he or she is interested in the *how*. How is it, for instance, that the cell phone is routinely used by young people to make plans and coordinate with one another? From this perspective, the issue is not so much with the representativeness of the study's participants, as it is with the ways the ethnographer might start to see the established patterns of phone use. The question the ethnographer must ask is who might he or she need to observe or ask to get to grips with teenagers' phone use patterns and where might they look?

Similarly, the number of people participating in a study and the field study's length are not driven by issues of generalizability. Again, it is the need to see how things are socially arranged and accomplished in routine ways that dictates the number of participants and the time spent in the field. Ethnographies in ubicomp often limit their participant numbers to roughly 5 to 15 and may have studies that run for weeks or, at most, a few months. It would be wrong, however, to assume exact numbers can be decided in advance. A common rule of thumb for both the number of participants and length of time in the field is whether the ethnographer starts seeing the same patterns or themes reoccurring in his or her observations and interviews. Once this happens and the ethnographer feels he or she has a grasp of what is being observed, it may be time to either develop another line of investigation or put more time into analysis and writing.

A possibly obvious point to note is that ethnographic fieldwork does not necessarily have to include participants, in any formal sense. The work may involve, instead, the careful observation of a setting or the ethnographer taking on a role. For instance, Livingston (1987) provides a compelling example of what can be learned by both participating in and observing pedestrians crossing the road at a busy intersection. By being on the street and crossing the road oneself, at eye level, he demonstrates how road crossing is something pedestrians accomplish through gaze direction and body orientation with respect to one another. This mutual coordination unfolds moment by moment so that the road is crossed successfully, as it were, through the continuous microcoordination between fellow pedestrians.

Beyond observing a specific setting, an ethnographer may choose to apply an ethnographic sensibility to his or her own practices—undertaking what is known as *autoethnography*. There are, for instance, examples of ethnomethodologists producing accounts of piano playing (Livingston, 1987) or playing with their pets (Goode, 2006). Although these types of ethnography are rare in ubicomp research (for an exception, see, e.g., Aoki, 2007), the possibility should be seen as a serious option. This is especially the case as ubicomp continues to extend its interests to include, for example, urban computing, sports, health monitoring, etc.

5.3.1.3 Access
A third more practical issue to consider with regard to planning an ethnography is gaining access to participants (if they are to be used at all). The time needed to plan and arrange access can be easily underestimated. Perhaps the hardest aspect is finding people who are willing to give up

their time and, as they see it, to have their behaviors (and sometimes private lives) scrutinized. Generally speaking, it often works well to find two or three people willing to participate in a study and then ask whether they are able to make introductions to friends or colleagues. Having personal introductions seems to ease people's discomfort. This, rather bizarrely, has been referred to as the *snowballing method*, as it involves accruing participants on a rolling basis. It is also helpful to explain the motivations and broad focus of the investigation to potential participants. The overall aim should be to help people feel at ease with the research and emphasize that it is the ethnographer who is the newcomer or novice to a situation or setting.

Something else to keep in mind is that participants do not have to be recruited and fully signed up to a study at its outset. Because the broad aim is to better understand the workings of a setting or particular activity, people can be sought out and invited to participate if and when it is felt they are needed.

To briefly review the points made above about planning an ethnography:

1. Plan for the research to be investigative and exploratory, not driven by hypotheses

2. Be open to the study following new trajectories and evolving as new areas of interest and themes develop over the course of the research

3. Scope the fieldwork tightly at first, leaving room for the scope to alter, broaden, and deepen

4. Select the type and number of participants with the aim of observing and detailing *how* a setting is socially organized not with the aim of generalizing to large populations

5. Consider alternatives to recruiting participants, such as simply observing a setting/activity or undertaking an autoethnography

6. Leave plenty of time for getting access to an empirical site and recruiting participants as the effort involved in both can be easily underestimated

5.3.2 In the Field

The prospect of going into the field to interview someone, observe a scene, participate in some activity, or take on some other data collection technique can be daunting. Unfortunately, things do not get any easier once in the field. One can feel awkward, clumsily getting in the way of the very thing being studied. The best that one can do to deal with the discomfort is to recognize that any awkwardness is an ordinary consequence of being somewhere new, with new people, and taking on the role of observer. Indeed, a very real and practical aspect of doing fieldwork is learning to deal with the sense of unease. If one should be prepared for anything, it should be the possibility of asking stupid questions or doing something foolish. Thus, rather than trying to detach oneself from the setting and playing the proverbial fly on the wall, a much more realistic approach to starting off in the field is to simply start trying different ways to engage with a setting. The concern should not be with getting it right so much as getting one's hands dirty, so to speak.

Another point worth remembering is that the collection of data and the analyses of collected field materials go hand in hand. The fieldwork may play a larger role at the beginning of a study and the analysis increases toward the end, but the two should interleave with one another in an iterative fashion. The fieldwork, of course, provides the raw materials. The analysis, though, helps the ethnographer discover a way of seeing and subsequently a perspective from which to revisit the field. Consequently, it is both one's practical and sometimes clumsy efforts to collect data as well as the analytical perspective that guide the unfolding direction of the research. Indeed, the initial fieldwork may help shape the analytic sensibility and the subsequent use of empirical methods that define the ethnography.

5.3.2.1 Reflexivity and Indifference

The flexibility of ethnographic work may appear to confound the objectivity usually thought of as the basis for scientific research. How, one might ask, does an ethnographer remain objective if their methods and analytic orientations are able to change in response to the object of study? Does this really promote sound scientific investigation? There is, of course, a long and complicated response to these questions, a response that can quickly turn to questions regarding the nature of science and how ethnography corresponds to scientific principles. A number of books have been written relating to these concerns (e.g., Clifford and Marcus, 1986). For the purposes of introducing ethnographic research and helping to convey its distinctive character, there are though two important concepts that should be considered. Both are complex, but deserve at least some explanation so they might be kept in mind when embarking on an ethnography.

One concept has its origins in anthropology and has to do with how a researcher reflects on his or her position in ethnographic research, specifically vis-à-vis the study of an established group of people—be it a community, workplace, family, etc. This *reflexivity* has come to be a fundamental feature of modern ethnography (and has also played a part in qualitative research more generally).* The ethnographer, in a manner of speaking, builds a reflective stance into the ongoing fieldwork and analysis, recognizing the inevitable subjectivity of the accounts he or she produces. Reflexivity thus motivates the ethnographer to continually shape and reshape his or her fieldwork and analysis. Effort might be made, for example, to collect and present multiple "voices" from a setting or to adopt a textual style that juxtaposes conflicting perspectives. Whatever the specifics, the reflection and ongoing adjustments aim to acknowledge the perspectives and prejudices that come with being present in the field and taking up particular empirical methods. At best, reflexivity is also something to be incorporated into the writing up and analyses of the field materials so that the analytic sensibilities take into account the processes of reproducing and representing a setting through the written word.

In ubicomp research, there is very little sign of reflexivity reported in the published literature. There is a small trend for ethnographers to situate themselves vis-à-vis their fieldwork, perhaps detailing their personal histories with respect to the studied setting (e.g., Wyche et al., 2006) or the analytical lens adopted (Swan et al., 2008). There is, however, scant reflection on and critical engagement with the research presented in ethnographic works. The criticism of conventional scientific practice implied through such reflexivity is probably seen as beyond ubicomp's scope. Nevertheless, whether reported or not, an ethnography of any type should be seen as lacking without at least some reflection on the ethnographer's part. Thought should be put into how the research is situated with respect to the fieldwork, the participants, and the chosen analytical orientation. Such reflection can only help to understand what types of things are being gleaned from the research and what sorts of implications the results can have for design.

A second concept, the ethnomethodological *policy of indifference*, relates to this notion of reflexivity. In producing an ethnographic account—that is, going into the field, meeting with participants and applying some type

---

* For a review of different perspectives on reflexivity, see Macbeth (2001) and for examples of particular positions on the topic, Lynch (2000) and Woolgar (1991).

of theoretical orientation in producing the analysis—the field worker unavoidably takes on an authoritative or privileged status. The ethnographer is, after all, writing on behalf of one or more people. Even when incorporating a reflexive position, there is the implication that the ethnographer's claims hold a certain importance over and above those he or she is studying; whatever the approach taken, choices are made over what to include and exclude. It is as if the ethnographer is peering into a world from the outside and explaining what he or she sees from that perspective.

It is this problem that the ethnomethodological *policy of indifference* aims to take on. It offers not a solution so much as an alternative perspective from which the ethnographer might approach their fieldwork. The *policy of indifference* is one that prioritizes a setting's members' ways of doing and seeing over and above the themes, theories, and methods of social science. In adopting the policy, the objective is to reveal just how people, as a matter of course, achieve a recognizable order to their everyday doings. The emphasis is consequently on how members of a setting observably produce their own order rather than how abstract theories might help to explain and represent social order. In the case of ethnography, the policy of indifference provides a way for a setting's members to be heard over the authoritative voice of the ethnographer (Taylor et al., 2007).

An instructive example is found, again, in Livingston's (1987) examination of pedestrian traffic flow. Livingston contrasts two ways of getting to grips with pedestrian's crossing a busy intersection. On the one hand, he recounts a sociologist's use of a film camera to record the intersection from above. On the other, he describes the experience of crossing a road, as a pedestrian. The camera's view, he suggests, lends itself to seeing who goes where and what the arrangements of people are. From a manufactured vantage point, as it were, the sociologist thus explains the pedestrian flows in terms of opposing "wedges" that move in "fronts," and are led by "point people." In contrast, the view from eye level, as a pedestrian, gives access to how road-crossing is accomplished on the ground. As described earlier, from the eye-level perspective, one gleans the moment-by-moment glances and shifts in orientation performed to follow, shift, dodge, and eventually get to the other side of the intersection.

What should be evident in this example is that the theory of crossings proposed by the sociologist is not available to the ordinary pedestrian. Pedestrians accomplish the business of road crossing using just those methods they have to hand; they bring to bear their own "lay" methods and theories for crossing roads. The policy of indifference, then, gives priority to the methods and theories of those people on the ground, so to speak. It claims an indifference to theories like the sociologist's because they do not reveal how, exactly, roads are crossed.

The legacy of ethnomethodology in ethnographies undertaken in HCI, CSCW, and ubicomp has seen at least an implied recognition of the policy of indifference. Works from Tolmie et al. (2002) and Crabtree and Rodden (2004), both based on field studies of domestic life, examine how the orderliness of homes is unavoidably occasioned by the doing of domestic routines. That is, they detail the ways in which the social order of home life is locally accomplished in and through ordinary household routines. The policy of indifference, however, is not commonplace. In ubicomp, in particular, a growing number of field studies have seen a closer alignment with the theorizing commonly associated with contemporary anthropology and the social sciences. Recently published work from Bell and Dourish (2007) offers an example. It frames the presence and use of garden sheds in terms of authority, power, and gender. The shed, as the authors make clear, is used as a lens to detail how domestic boundaries are drawn between male and female, outside from in, and the migrations of technology between these categories. Thus, attention is not directed at how the borders of home are routinely and unremarkably produced by its members. Departing from indifference, the work instead aligns itself with contemporary social and cultural theorizing and situates the home/shed in such discourses.

The point here is not to set reflexivity up against the policy of indifference. What the reader should be aware of is that rather than objectivity, ethnographers use such concepts to grapple with producing written texts that resonate or ring true for their participants. They may not capture all that goes on in a setting and the perspectives of all participants, but at the very least the hope is that they reveal something of the peoples studied and of being in the places those peoples inhabit. Reflexivity and the policy of indifference are thus, in some respects, efforts to overcome the biases of the ethnographer. However, they are based on very different theoretical foundations and produce very different outcomes. Those new to ethnography should have these differences in mind when they review past examples of fieldwork and make choices in their own research.

5.3.3 Analysis

The classical image of ethnography in anthropology portrays the lone ethnographer writing up field notes in his or her tent after a day's observations and interviews with the natives. The serious business of writing articles and monographs then happens once back at the office—the office probably located in some ivy-leafed bastion of academia. How often this happens today or, indeed, whether it happened all that often in the past is debatable.

Whatever the case, in applied areas such as ubicomp, an ethnography is far more likely to be part of a larger project in which there are multiple team members, made up probably of social and computer scientists, designers, and other interested parties. The collection of data in the field may well involve more than one team member and these members may not be limited to the social scientists. Similarly, the analysis, as already mentioned, is usually done in parallel with the fieldwork and will also often include researchers from a range of backgrounds. The implications of this multiparty analysis for ubicomp will be considered later. Here, the focus will be on two aspects of the analysis: (1) data and its influence on analysis and (2) analytic sensibilities in ubicomp.

5.3.3.1 Data and Its Influence on Analysis
For many, it is probably obvious that the type of fieldwork data collected has an impact on the types of analysis that can be conducted. Interview transcripts, for instance, are essentially the accounts participants produce of some past occurrence or possibly thoughts they are willing to express on a particular matter. The analysis of interviews can thus focus on the forms of talk used by the interviewees or how in a post hoc fashion, interviewees verbally account for themselves and their actions. Analysis that treats interviews as accurate descriptions of occurrences or what an interviewee actually thinks is common. However, this treatment will always be seen as more suspect and thus it is wise to be clear about the assumptions being made.

For reasonably comprehensive overviews of the different data collection methods used in ethnography and the bearing they have on analysis, it is worth reviewing popular textbooks on ethnography such as Hammersly and Atkinson's (1995) *Ethnography, Principles in Practice* or Wolcott's (1995) *The Art of Fieldwork*. One data collection technique worthy of particular attention, here, however is the use of video cameras to record interviews or events in the field. The use of video has been common in ubicomp research because it offers a powerful means of capturing field materials. It allows recordings to be made from multiple perspectives and/or when the researcher is absent. When analyzing data, the recordings can also be watched repeatedly to observe, in detail, some aspect of talk or interaction. Another benefit is that the analysis can more easily involve others who may not have been in the field but who may, nevertheless, have insights to contribute. Indeed, data sessions where video recordings from fieldwork are viewed by groups of researchers are becoming increasingly popular.

However, there are some issues to keep in mind when using video. It should not be forgotten that video recordings provide a certain perspective on a setting. There is the obvious point on perspective to be made here that a video camera's framing of a setting draws attention to some features and misses others. The camera provides a constrained perspective on the world, so to speak. It is also important to remember though that the qualities of video are very likely to influence the analytical perspective in specific ways. For example, because video can be easily replayed and watched in slow motion, it lends itself to detailed analysis and repeated rewatching. Thus, the emphasis is often placed on fine-grained analysis such as studying patterns of speech, gestures, or nuanced interactions between people and with things. Moreover, because informants' conversations can be transcribed word for word and the video can be played back to an audience who were not present, video materials are often used to prove an empirical point. The visual and audio information captured and then replayed using video appears to be seen as constituting better or more valid evidence. These issues are not necessarily weaknesses or criticisms of video and its use in ethnographies. The point is that video encourages a particular way of seeing the world and making sense of it. It is not, then, to be seen as the panacea for collecting data, but just one of the techniques ethnographers should be willing to use to investigate a setting.

5.3.3.2 Analytic Sensibilities in Ubicomp

A second aspect to analysis that deserves some consideration has to do with the analytic sensibility and how it is applied in design-oriented ethnographic field studies. If there is any type of tradition in applying an analytical sensibility in design-oriented ethnographies, it is to offer countertheses to some of the established topics in ubicomp, topics such as context, privacy, and location (aspects of which are discussed in Chapters 3, 7, and 8). The trend is usually to use fieldwork materials to unpack topics and to illustrate how they cannot be easily abstracted from real-world situations.

Although predating ubicomp and targeted at expert systems design, Suchman's (1987) work is an early but, again, compelling illustration of this perspective. By investigating people's interactions with a photocopier that had been designed to guide users through copying tasks (using predictive models), Suchman was able to critically reflect on the notion of *plans* that was, at the time, fundamental to artificial intelligence (AI). Working within the analytical auspices of ethnomethodology, she revealed how people's interactions with the copier were shaped by the situation at hand. Any initial plans on a user's part could change on a moment-by-moment basis depending on what exactly was happening. In short, real-world planning was found to provide a loose structure to an activity, but not to operate in a step-by-step fashion. This contrasted with plans as they were thought of in expert system design and AI. Here, it was assumed human behavior could be broken down into discrete, sequential actions that could be defined a priori. This mismatch, Suchman suggested, was at the heart of many of the problems users had in operating the photocopiers. Seeming to establish a tradition, Suchman's empirical work was thus the impetus for a rethinking of a taken-for-granted aspect of AI, namely, planning.

A similar critical sensibility is evident in a more recent example of ubicomp research, one involving the deployment and evaluation of a location system called the *Whereabouts Clock* (Brown et al., 2007). The Whereabouts Clock was designed for domestic use, providing those in a home with a lightweight means of seeing the location of other household members (Figure 5.1). A key motivation in designing the Clock was to build on some of the features of a domestic appliance—the functionality

FIGURE 5.1 Whereabouts Clock.

Ethnography in Ubiquitous Computing ■ 227

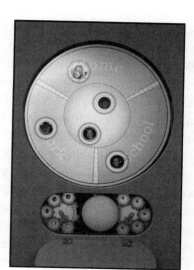

FIGURE 5.2 Whereabouts Clock interface with three specified regions representing work, home, and school, and middle region for locations other than these.

was intentionally kept simple, it was physically constructed to be situated in one place in the home, and the display was designed so that it could be seen at a glance (like a clock). Crucially, the location of household members, tracked using their mobile cell phones, was displayed in a coarse-grained way. The display showed householders at either work, home, or school, and had a region to represent when they were somewhere other than these three locations (Figure 5.2).

There are many aspects of the use of the Whereabouts Clock that could be discussed including, for example, those related to appliance design, privacy, context, home life, and so on. However, one aspect of its use in the field study raised some particularly relevant questions for ubicomp research. Specifically, the work provoked questions around conventional ideas of location and the way it is commonly thought of as something detected and represented using geographical coordinates. Brown and his colleagues were able to use the fieldwork materials they collected by deploying the Clock to demonstrate how location is reckoned in terms of what one imagines others to be doing in a place. In practice, location was treated more like *location-in-action* and not merely in terms of physical geography. The participants in their study described how places such as work, home, and school became meaningful in terms of their interactions, their accountabilities, and their obligations in those places. When dealing with location, family members appear to construct a mental geography, as it were, that expresses a great deal more than just numerical coordinates.

Examples such as these hopefully demonstrate how ethnographers have sought to develop and think critically about some of the major themes in ubicomp. Overall, it should be apparent that through the application of certain analytic sensibilities, efforts have been made to detail, and in some cases defamiliarize, some of the ideas that underlie a good deal of technological development in the area. An important aspect to this, and one that should be considered in undertaking an ethnography, is the role that analytic sensibility plays. In both of the examples discussed above, it is the sensibility that forms the basis for the critique and enables the research to be formulated as a coherent argument. The art of seeing action as situated, seeing it as bound up in the ordinary affairs of everyday life, informs the sensibility and offers a basis to rethink conventions in ubicomp. They hopefully demonstrate the value of drawing on and applying an analytic sensibility in ubicomp research.

5.4 WHAT IS IT GOOD FOR?

So far, the presented materials have offered some background to ethnography and its uptake in design-related areas such as HCI, CSCW, and ubicomp. A number of relevant aspects of ethnography as it is practiced in these areas have also been covered. There remain questions, however, about what ethnography is good for and when it should be used over and above the other methods used to inform design. This concluding section will aim to address some of these questions and also discuss how ubicomp research has begun to have its own influence on design-oriented field studies.

5.4.1 Design Implications

Design-oriented ethnography provides the methods and techniques used to investigate people's real-world (inter)actions with technology and also offers a theoretical basis from which to reexamine some of the commonly held assumptions in technical research. The methods and techniques used enable detailed studies of people's in situ activities and, as the examples above illustrate, the ethnographic sensibility lends weight to some of the more critical positions taken up and investigated in ubicomp. Are there more immediate results, though, that can be had from an ethnographic field study, offering, perhaps, more explicit implications for design?

This has been a long-standing question for practitioners and researchers in HCI and, in recent years, of immediate relevance to those in ubicomp.

Various attempts have been made to reconcile ethnography (in its various guises) with the processes and objectives of design. For example, during ethnography's initial uptake, attempts were made to integrate the results from ethnographic fieldwork with the then established (mostly cognitive-based) user modeling schemes (e.g., *Cognitive Work Analysis*, Vicente, 1999). Efforts were also made to find common ground between the descriptive character of ethnography and the prescriptive aims of design (e.g., work-oriented design, Bloomberg, 1995; and patterns of cooperative interaction, Martin et al., 2001) and, more ambitiously perhaps, alter the practice of systems design itself to be more amenable to the ethnographic enterprise (e.g., technomethodology, Button and Dourish, 1996). The use of more structured methods for assembling the results of field studies and configuring them to be applied to solving problems has also been proposed. These methods have usually entailed the use of diagrams and schematics of fieldwork materials to be used by teams of practitioners (e.g., Beyer and Holtzblatt, 1998). All in all, the attempts have sought to experiment with the intersections between ethnography and design to provide some repeatable means of translating ethnographic results into useful design implications.

Despite these numerous attempts to find a systematic and concrete role for ethnography in systems design, not one proposal has been particularly successful or succeeded in sustaining any long-term interest. Certainly, some have seen uptake in different quarters. For example, Beyer and Holtzblatz's *Contextual Inquiry* has found favor in commercial settings where value may be had in gaining a broad picture of organizational patterns rather than revealing detailed features of social interaction and informing system design. However, there appears to be no tried-and-tested means of getting ethnographic materials to yield concrete design requirements. Indeed, it is a running joke in some circles that the design sections of design-oriented ethnographic publications tend to be notoriously weak, succeeding to do little more than suggesting the blindingly obvious when it comes to design.

Somewhat perversely, the practical value of ethnography in areas such as ubicomp has been recognized through a growing number of exemplary, design-oriented ethnographic case studies rather than the proposal of any specific approach or method. The main lesson has been that an ethnography, at its best, succeeds in opening up a set of possibilities for design by providing a rich and detailed characterization of some setting and/or people. So, rather than being seen as a means of narrowing in on a design, ethnography should be thought of as a way to discover the design spaces and how technological ideas might be subsequently investigated in more detail.

The roots of this idea were proposed early on in HCI by Anderson (1997), an ethnomethodologist who was keen to clarify the role ethnography should have in design. Emphasizing the analytic aspects of ethnographic inquiry, Anderson demonstrated how ethnography "opens up the play of possibilities for design" through its analytic strategies and literary modes of representation. Building on this early position, Dourish (2006) recently criticized how ethnographies have been judged in HCI. He points out that ethnographic studies are commonly measured by their ability to produce fieldwork data as reliable facts, a criteria operating at the empirical level. He suggests this misses out on a key feature of ethnography: that much of the work in an ethnography operates at the analytic level, where the "data are theorized, understood, organized, juxtaposed, interpreted, and presented in order to make an argument that reveals something about the setting under investigation" (p. 548).

What these arguments from Anderson and Dourish illustrate, nicely, is how the analytic sensibility of ethnography has a sound role to play in design and specifically ubicomp. Bringing us back to the beginning of this chapter, both positions reveal that the analytic mindset applied in ethnographic inquiries is not a distraction from the aims of design, but something to be valued and fostered in the process of achieving some vision for design. The specific point worth noting here is that ethnography should not be treated as one more tool for eliciting design requirements or indeed design implications. By understanding that its value is in opening up the possibilities, ethnography can retain its integrity as an analytic and interpretive enterprise but still have relevance to design.

5.4.2 Future Directions

Something hopefully evident in this chapter has been the pliable nature of design-oriented ethnography. In many ways, it can be seen as something that has evolved as its related areas of study have matured. Tracing its trajectory, it contributed to design's turn toward the social setting and was applied first in the workplace but more recently in domestic and leisure settings, as well as to study those on the go. During these shifts, tensions have arisen over what constitutes ethnography and the types of analytical positions that can be incorporated. Ethnomethodology has remained a central influence as an analytical orientation, but a greater impact has been felt in recent years from the social sciences, especially anthropology.

This progress continues unabated. Interestingly, though, ethnography has not just played a contributing role in HCI, CSCW, and ubicomp. It has also begun to feel the effects of some of the more progressive developments in these areas. For example, design-oriented ethnographies have begun to use probes to engage with their participants and provoke discussions specifically around issues relating to design (see Boehner et al., 2007). Postcards, diaries, and cameras might be packaged (often creatively) to give to participants as probes, alongside conducting interviews and observations (Gaver et al. 1999). Similarly, technologies might be deployed in the form of probes not to evaluate a design, but, instead, to learn more about the setting under study (e.g., Sellen et al., 2006).

On one hand, this seemingly inconsequential addition to ethnography in design could be seen as simply adding to the various methods ethnographers have to collect data. However, there are more fundamental implications resulting from the inclusion of probes in an ethnographic study. Whether it is intentional, they immediately transform the role of participants. No longer are they the passive object of investigation. They take on the role of commentator or observer in their own practices; the probes become sources of disquiet, provoking one to question the commonsensical and routine in daily life. From the ethnographer's perspective, the probes and this change in role of the participants leads to a reframing of the empirical exercise. The probe takes on an active role, giving shape to, if not completely rearranging, the participant's practices. The use of probes thus necessitates a degree of reflexivity in the analytic perspective by introducing a new type of dynamism to the studied scene and shifting the authoritative voice of the ethnographer.

Another example of the influence developments in systems design have had on ethnography come from the idea of *critical technical practice*, first proposed in response to the introspective research programs within AI (Agre, 1997), and relatively recently introduced to HCI and ubicomp by various members of the Culturally Embedded Computing Group at Cornell University. Critical technical practice brings to systems design recognition of the role technologies play in propounding a set of social values, of interjecting and enforcing particular cultural mores through the ways technologies are designed and used (and theorized). Although these ideas have been long discussed in anthropology and sociology (e.g., MacKenzie and Wajcman, 1999), critical technical practice involves design in the interplay between the social and technical. It proposes a practice where technology designers reflect on their participation in society by propagating some values and counteracting others. This, consequently, attaches a moral dimension to design, forcing designers to be accountable for their choices and, hopefully, taking seriously their own practice.

Design-oriented ethnography has, in some respects, been forced to play catch up with this initiative (although not without controversy; see Button et al., 2009). Although, as noted, ethnography seeks a degree of reflexivity in its practice, there has been a lack of introspection around the role design-oriented ethnographies play in interjecting theory or values into design practice. The uptake of a so-called "critical practice" in ubicomp and other areas is encouraging ethnography to reinspect its position in systems design and, in some cases, rearticulate the types of contributions it can offer to design (e.g., Bell et al., 2005; Dourish, 2006).

Thus, ethnography, it seems, has an established place in ubicomp research. Its past and current practice has done much to help contribute to the design and evaluation of technological systems. Broadly, its successes have been in opening up the spaces for design and giving a language through which practitioners can imagine new possibilities. It has also played a large part in promoting a deeper engagement with the sites for technology and the impact technological innovations might have on such sites. In practical and theoretical terms, though, ethnography has come to be a practice growing alongside design. Its fluidity, in this sense, is at one and the same time something to be struggled with and celebrated. The ethnographer is continually reminded of the fragility of his or her place in the assemblies of people and things, and how the ways of looking are never done and always to be discovered anew.

REFERENCES

Agre, P. *Computation and Human Experience.* Cambridge University Press, Cambridge, 1997.

Aoki, P. M. Back stage on the front lines: Perspectives and performance in the combat information center. In: *Proceedings of the SIGCHI Conference on Human Factors in Computing Systems, CHI '07.* ACM Press, New York, 2007, pp. 717–726.

Anderson, N. *The Hobo: The Sociology of the Homeless Man.* University of Chicago Press, Chicago, IL, 1923.

Anderson, R. J. Representations and requirements: The value of ethnography in system design. *Human-Computer Interaction* 9(1), 151–182, 1994.

Anderson, R. J. Work, ethnography and system design. In: Kent, A., and Williams, J. G. (Eds.), *The Encyclopedia of Microcomputers (20).* Marcel Dekker, New York, 1997, pp. 159–183.

Becker, H. S. Theory: The necessary evil. In: Flinders, D. J., and Mills, G. E. (Eds.), *Theory and Concepts in Qualitative Research.* Teachers College Press, London, 1993, pp. 218–229.

Beidelman, T. O. Sir Edward Evans-Pritchard: An appreciation. *Anthropos* 69, 553–567, 1974.

Bell, G., Blythe, M., and Sengers, P. Making by making strange: Defamiliarization and the design of domestic technologies. *ACM Transactions on Computer-Human Interaction* 12(2), 149–173, 2005.

Bell, G., and Dourish, P. Back to the shed: Gendered visions of technology and domesticity. *Personal and Ubiquitous Computing* 11(5), 373–381, 2007.

Beyer, H., and Holtzblatt, K. *Contextual Design: Defining Customer-Centered Systems.* Morgan Kaufmann Publishers, San Francisco, CA, 1998.

Bloomberg, J. L. Ethnography: Aligning field studies of work and design. In: Monk, A. F., and Gilbert, N. (Eds.), *Perspectives on HCI: Diverse Approaches.* Academic Press, London, 1995, pp. 175–197.

Boehner, K., Vertesi, J., Sengers, P., and Dourish, P. How HCI interprets the probes. In: *Proceedings of SIGCHI Conference on Human Factors in Computing Systems, CHI '07.* ACM Press, New York, 2007, pp. 1077–1086.

Brewer, J., Mainwaring, S., and Dourish, P. Aesthetic journeys. In: *Proceedings of 7th ACM Conference on Designing interactive Systems, DIS '08.* ACM Press, New York, 2008, pp. 333–341.

Brown, B., Taylor, A. S., Izadi, S., Sellen, A., and Kaye, J. Locating family values: A field trial of the Whereabouts Clock. In: *Proceedings of 9th International Conference, UbiComp '07,* Springer-Verlag, 2007, pp. 354–371.

Bryman, A. E. *Ethnography* (4 volumes). Sage Publications, London, 2001.

Button, G. (Ed.). *Technology in Working Order: Studies of Work, Interaction, and Technology.* Routledge, New York, 1993.

Button, G. The ethnographic tradition and design. *Design Studies* 21, 319–332, 2000.

Button, G., and Dourish, P. Technomethodology: Paradoxes and possibilities. In: *Proceedings of Conference on Human Factors and Computing Systems, CHI '96.* ACM Press, New York, 1996, pp. 19–26.

Clifford, J., and Marcus, G. E. *Writing Culture: The Poetics and Politics of Ethnography.* University of California Press, London, 1986.

Crabtree, A., and Rodden, T. Domestic routines and design for the home. *Journal of Computer-Supported Collaborative work, CSCW.* 13(2), 191–220, 2004.

Crabtree, A., Benford, S., Greenhalgh, C., Tennent, P., Chalmers, M., and Brown, B. Supporting ethnographic studies of ubiquitous computing in the wild. In: *Proceedings of the 6th Conference on Designing Interactive Systems, DIS '06.* ACM Press, New York, pp. 2006, 60–69.

Crabtree, A., Rodden, T., Tolmie, P., and Button, G. Ethnography considered harmful systems design. In: *Proceedings of Conference on Human Factors in Computing Systems, CHI '09.* ACM Press, New York, 2009, 879–888.

DeSantis, A. D. A couple of white guys sitting around and talking: A collective rationalization of cigar smokers. *Journal of Contemporary Ethnography* 32(4), 432–466, 2003.

Dourish, P. *Where the Action Is: The Foundations of Embodied Interaction.* MIT Press, Cambridge, MA, 2001.

Dourish, P. Design implications. In: *Proceedings of the Conference on Human Factors in Computing Systems, CHI '06.* ACM Press, New York, 2006, pp. 541–550.

Evans-Pritchard, E. E. Nuer time-reckoning. *Africa: Journal of the International African Institute* 12(2), 189–216, 1939.

Evans-Pritchard, E. E. *The Nuer: A Description of the Modes of Livelihood and Political Institutions of a Nilotic People.* Clarendon Press, Oxford, 1940.

Gaver, B., Dunne, T., and Pacenti, E. Design: Cultural probes. *Interactions* 6(1), 21–29, 1999.

Geertz, C. *The Interpretation of Cultures: Selected Essays.* Basic Books, New York, 1973.

Goode, D. *Playing with My Dog, Katie: An Ethnomethodological Study of Canine-Human Interaction.* Purdue University Press, Purdue, IN, 2006.

Grudin, J., and Grinter, R. E. Ethnography and design. *Computer Supported Cooperative Work* 3, 55–59, 1995.

Hammersley, M., and Atkinson, P. *Ethnography, Principles in Practice*, 2nd edition. Routledge, London, 1995.

Harper, R., Hughes, J., and Shapiro, D. Harmonious working and CSCW: An examination of computer technology and air traffic control. In: Bowers, J. M., and Benford, S. D. (Eds.), *Studies in Computer Supported Cooperative Work: Theory, Practice and Design.* North-Holland, Amsterdam, The Netherlands, 1991, pp. 225–234.

Harper, R., Randall, D., and Rouncefield, M. Fieldwork and ethnography in design: The state of play from the CSCW Perspective. In: *Proceedings of EPIC 2005*, American Anthropology Association, 2005, pp. 81–100.

Harper, R. H. R. The organisation in ethnography—A discussion of ethnographic fieldwork programs in CSCW. *Computer Supported Cooperative Work (CSCW)* 9(2), 239–264, 2000.

Harrison, S., and Dourish, P. Re-place-ing space: The roles of place and space in collaborative systems. In: *Proceedings of the 1996 ACM Conference on Computer Supported Cooperative Work, CSCW '96.* New York, ACM Press, 1996.

Heath, C., Knoblauch, H., and Luff, P. Technology and social interaction: The emergence of 'workplace studies.' *British Journal of Sociology* 51(2), 299–320, 2000.

Hine, C. M. *Virtual Ethnography*, Sage Publications, London, 2000.

Hughes, J. A., King, V., Roden, T., and Andersen, H. Moving out from the control room: Ethnography in design. In: *Proceedings of the Conference on Computer Supported Cooperative Work, CSCW '94.* ACM Press, New York, 1994, pp. 429–439.

Hughes, J. A., O'Brien, J., Rodden, T., Rouncefield, M., and Blythin, S. Designing with ethnography: A presentation framework. In: *Proceedings of the Conference on Designing Interactive System, DIS '97.* ACM Press, New York, 1997, pp. 147–159.

Hughes, J. A., Randall, D., and Shapiro, D. Faltering from ethnography to design. In: *Proceedimgs of the Conference on Computer Supported Cooperative Work, CSCW '92.* ACM Press, New York, 1992, pp. 115–122.

Hutchins, E., and Klausen, T. Distributed cognition in an airline cockpit. In: Engeström, Y., and Middleton, D. (Eds.), *Cognition and Communication at Work.* Cambridge University Press, Cambridge, 1996, pp. 15–34.

Ito, M. Mobile phones, Japanese youth, and the replacement of social contact. In: Ling, R., and Pedersen, P. (Eds.), *Mobile Communications: Renegotiation of the Social Sphere*. Springer-Verlag, London, 2005, pp. 131–148.

Ito, M., Okabe, D., and Matsuda, M. *Personal, Portable, Pedestrian: Mobile Phones in Japanese Life*. MIT Press, Cambridge, MA, 2006.

Livingston, E. *Making Sense of Ethnomethodology*. Routledge & Kegan Paul, London, 1987.

Lynch, M. Against reflexivity as an academic virtue and source of privileged knowledge. *Theory Culture and Society* 17(3), 26–54, 2000.

Macaulay, C., Benyon, D., and Crerar, A. Ethnography, theory and systems design: From intuition to insight. *International Journal of Human-Computer Studies* 53(1), 35–60, 2000.

Macbeth, D. On "reflexivity" in qualitative research: Two readings, and a third. *Qualitative Inquiry*. 7(1), 35–68, 2001.

Mackay, W. E. Is paper safer? The role of paper flight strips in air traffic control. *ACM Transactions on Computer-Human Interaction* 6(4), 311–340, 1999.

MacKenzie, D. A., and Wajcman, J. *The Social Shaping of Technology*, 2nd edition. Open University Press, Buckingham, 1999.

Marsden, G. Under Development: New users, new paradigms, new challenges. *Interactions* 15(1), 59–60, 2008.

Martin, D., Roden, T., Rouncefield, M., Sommerville, I., and Viller, S. Finding patterns in the fieldwork. In: *Proceedings of the Seventh Conference on Computer Supported Cooperative Work, ECSCW 2001*, Kluwer Academic Publishers, 2001, pp. 39–58.

Mead, M. *Coming of Age in Samoa: A Study of Adolescence and Sex in Primitive Societies*, HarperCollins, 1943.

O'Reilly, K. *Ethnographic Methods*. Routledge, New York, 2005.

Randall, D., Harper, R., and Rouncefield, M. *Fieldwork for Design: Theory and Practice (Computer Supported Cooperative Work)*. Springer-Verlag, London, 2007.

Sellen, A., Harper, R., Eardley, R., Izadi, S., Regan, T., Taylor, A. S. et al. HomeNote: Supporting situated messaging in the home. In: *Proceedings of the Conference on Computer Supported Collaborative Work, CSCW '06*, 2006, pp. 383–392.

Suchman, L. A. *Plans and Situated Actions: The Problem of Human-Machine Communication*. Cambridge University Press, Cambridge, 1987.

Swan, L., Taylor, A. S., and Harper, R. Making place for clutter and other ideas of home. *ACM Transactions on Computer-Human Interaction* 15(2), 1–24, 2008.

Taylor, A. S. Phone talk. In: Ling, R., and Pedersen, P. (Eds.), *Mobile Communications: Re-negotiation of the Social Sphere*. Springer-Verlag, Godalming, 2005, pp. 149–166.

Taylor, A. S., Swan, L., and Randall, D. Listening with indifference. In: *Proceedings of the 3rd Ethnographic Practice in Industry Conference, EPIC*, American Anthropological Association, 2007, pp. 239–250.

Tolmie, P., Pycock, J., Diggins, T., MacLean, A., and Karsenty, A. Unremarkable computing. In: *Proc. Conference on Human Factors in Computing Systems, CHI 2002*. ACM Press, New York, 399–406, 2002.

Vicente, K. J. *Cognitive Work Analysis: Toward Safe, Productive, and Healthy Computer-Based Work*. Lawrence Erlbaum Associates, Mahwah, NJ, 1999.

Weilenmann, A. "I can't talk now, I'm in a fitting room": Availability and location in mobile phone conversations. *Environment and Planning A* 35(9), 1589–1605, 2003.

Williams, A., Anderson, K., and Dourish, P. Anchored mobilities: Mobile technology and transnational migration. In: *Proceedings of the 7th ACM Conference on Designing Interactive Systems, DIS '08*. ACM Press, New York, 2008, pp. 323–332.

Wolcott, H. F. *The Art of Fieldwork*. Sage Publications Ltd, London, 1995.

Wolcott, H. F. *Ethnography: A Way of Seeing*. Rowman Altamira, London, 1999.

Woodruff, A., Augustin, S., and Foucault, B. Sabbath day home automation: "It's like mixing technology and religion." In: *Proceedings of the Conference on Human Factors in Computing Systems, CHI '07*, ACM Press, 2007, pp. 527–536.

Woolgar, S. *Knowledge and Reflexivity: New Frontiers in the Sociology of Knowledge*. Sage Publications, London, 1991.

Wyche, S. P., Hayes, G. R., Harvel, L. D., and Grinter, R. E. Technology in spiritual formation: An exploratory study of computer mediated religious communications. Paper presented at the Conference on Computer Supported Cooperative Work, CSCW '06, New York, NY, 2006.

CHAPTER **6**

From GUI to UUI: Interfaces for Ubiquitous Computing

Aaron Quigley

CONTENTS

6.1 Introduction	238
6.1.1 From Graphical User Interfaces to Context Data	242
6.1.2 Inventing the Future	243
6.1.3 Chapter Overview	246
6.2 Interaction Design	246
6.2.1 User-Centered Design	247
6.2.2 Systems Design	248
6.2.3 Genius Design	249
6.2.4 Design Patterns	250
6.2.5 Discussion	250
6.3 Classes of User Interface	251
6.3.1 Tangible User Interface	253
6.3.2. Surface User Interface	257
6.3.2.1 Large SUIs	261
6.3.3 Ambient User Interfaces	264
6.4 Input Technologies	269
6.4.1 Sensor Input	270
6.4.2 Gesture Input	271
6.4.3 Speech Input	274

6.5 Interface Usability Metrics	277
6.6 Conclusions	278
Acknowledgments	278
References	279

6.1 INTRODUCTION

The user interface represents the point of contact between a computer system and a human, both in terms of input to the system and output from the system. There are many facets of a "ubiquitous computing" (ubicomp) system from low-level sensor technologies in the environment, through the collection, management, and processing of context data through to the middleware required to enable the dynamic composition of devices and services envisaged. These hardware, software, systems, and services act as the computational edifice around which we need to build our ubicomp user interface (UUI). The ability to provide natural inputs and outputs from a system that allows it to remain in the periphery is hence the central challenge in UUI design.

For our purposes, ubicomp is a model of computing in which computation is everywhere and computer functions are integrated into everything. It will be built into the basic objects, environments, and the activities of our everyday lives in such a way that no one will notice its presence [Weiser, 1999]. Such a model of computation will "weave itself into the fabric of our lives, until it is indistinguishable from it" (Weiser, 1999). Everyday objects will be places for sensing, input, processing along with user output (Greenfield, 2006). Take, for example, the multiperson interactive surface in Figure 6.1. Here, dozens of people can interact simultaneously with a large historical record using gesture alone. Coupled with directional microphones, personal displays, and other forms of novel interface, one can imagine this as part of a larger system in the future. This future experience might bring schoolchildren back to past events not just in one room but throughout a city, country, or continent. Ubicomp aims to make information, applications, and services available anywhere and at anytime in the human environment, where they are useful. In keeping with Weiser's original vision of keeping technologies unnoticed (Weiser, 1999), a further aim is to have all this delivered in a fluid manner appropriate to our current context (Coutaz et al., 2005).

Around the world, research and development groups are exploring mobile and embedded devices in almost every type of physical artifact including cars, toys, tools, homes, appliances, clothing, and work surfaces.

From GUI to UUI: Interfaces for Ubiquitous Computing ■ 239

FIGURE 6.1 Shared public display at the Cabinet War Rooms and Churchill Museum, London.

Indeed, anywhere computation will aid the user in solving a problem or performing a task in situ, ubicomp can be viewed as the model of computation. Ubicomp represents an evolution from the notion of a computer as a single device, to the notion of a computing space comprising personal and peripheral computing elements and services all connected and communicating as required; in effect, "processing power so distributed throughout the environment that computers per se effectively disappear" (Greenfield, 2006) or the so-called *calm computing*. It is important to note that the advent of ubicomp does not mean the demise of the desktop computer in the near future. The ubiquity of this technology took decades to advance and we can expect the same gradual evolution in ubicomp technology and scenarios of use.

Many ubiquitous computing scenarios suggest introducing new affordances, features, services, and modes of interaction into the simplest and most basic operations of our daily lives, from turning a door handle while getting haptic feedback to convey a status update through to the augmented cooking experiences in kitchens with radio frequency ID (RFID), computer vision, speech recognition, and projected and embedded displays. For example, Figure 6.2 shows a multidisplay gaming environment suitable for multiperson game playing, coordination tasks, or planning activities. Topics covered elsewhere in this text, such as communication, systems design, context awareness, and privacy, are all crucially important

FIGURE 6.2 Coupled personal (iPhone) and public display (MS Surface) gaming environment.

to realizing such an infrastructure; however, so, too, are the interfaces with which we will interact with such systems.

It is clear that future ubicomp applications and deployments will rely on new devices, services, and augmented objects, each supporting well-understood affordances, but also simple computational adaptations to existing well-known and well-understood devices, services, and objects. As more and more computation is woven into the fabric of our lives, our interaction with such ubicomp systems cannot be the focus of our attention. As noted previously, if every physical object in our home demanded attention in terms of alerts, updates, and configurations in the way our current personal computers do, we would become quickly overwhelmed. Indeed, it has been noted that "… the benefits of inexpensive Ubiquitous Computing may be overwhelmed by its personal cost" (Heiner et al., 1999). Instead, the computational and activity support offered by ubicomp systems must reside in the periphery of our attention and should remain unnoticed until required.

It is important to note that myriad processing elements, sensors, displays, and outputs are not elements of a statically defined user interface akin to the keyboard, mouse, and screen of a desktop computer. Instead, these elements and their related technologies afford the user an experience and a place to interact. Keeping in mind that the goal is to develop and deploy systems and technologies that are calm and invisible, how do we provide UUI cues and feedback to users leveraging the full range of

FIGURE 6.3 Three mute options, four volume controls, five web access methods, and 168 buttons for one television with six inputs.

their human senses? By contrast, take the standard appliance model one might find in a home. Figure 6.3 presents eight control devices that can be used to interact with one television that has six devices connected—DVD, AppleTV, Mac Mini (using VNC on iPhone, and Apple remote), PS3, Wii, and Cable Box. Aside from the 168 buttons available, there are in addition, a myriad of ways to access overlapping features. Figure 6.3 demonstrates there are three ways to mute the television, five ways to access a web browser, and four ways to alter the volume. Although not typically dealing with such a set of controllers, many people feel frustrated with just one or two. Such devices are often unclear and have too many functions available at the same level, which gives rise to the "controller hell" so many people feel. If we can learn anything from this, it should be that, a ubicomp system is made up of subsystems and we must design for the experience not the individual subsystem.

Although this chapter surveys the current state of the art to the user beyond the classical keyboard, screen, and mouse, it is important to also acknowledge that UUIs represent a paradigm shift in human-computer interaction (HCI) with input and output technologies not yet envisaged. UUIs are built around a next-generation technological paradigm that, in essence, reshapes our relationship with our personal information, environment, artifacts, and even our friends, family, and colleagues. The challenge is not about providing the next-generation mouse and keyboard, but instead making the collection of inputs and outputs operate in a fluid and seamless manner.

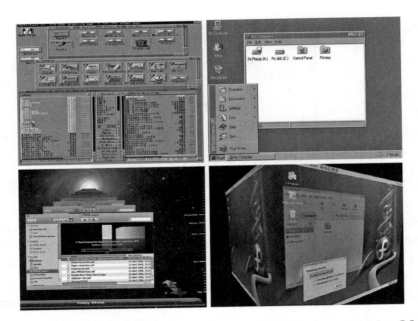

FIGURE 6.4 GUI interface evolution, Amiga UI, Windows 95, Mac OS X 10.5, and Compiz.

6.1.1 From Graphical User Interfaces to Context Data

The current graphical user interface (GUI) of the personal computer is built around keyboard, screen, and mouse devices. GUIs offer the Windows, Icons, Menus, and Pointer (WIMP) metaphor as a more intuitive view of the computer as compared with the more classical command line (textual) interface. GUIs have evolved considerably since they were first introduced, due to a better understanding of the user through HCI research along with advances in hardware and software. From rudimentary graphics through to the exploration of three-dimensional (3-D) elements as shown in Figure 6.4, GUI has grown to offer users better features.

Other technological advancements of the past 15 years have also seen the widespread adoption of mobile phones with keypads, personal digital assistants (PDAs) with styli, tablets with styli, and touchscreens with fingers. More recent developments have seen the adoption of game controller inputs or gesture-driven controls for game platforms. In spite of these advancements, many elements of the WIMP metaphor for HCI live on, be it in gesture, mouse, stylus, or finger form. The UUI draws on the GUI tradition but moves beyond the special-purpose device (PDA, phone, laptop, desktop) and into the support of activities of our lives, some of which are not currently supported by computation.

Drawing on technologies both in the research stage and also in widespread adoption, this chapter describes many of the foundational input and output methods, systems, and technologies that will help us realize future UUIs. Indeed, we must consider the full range of interaction styles available for input beyond the mouse, keyboard, stylus, or controller. Although not yet widely available, technologies and methods exist to both sense or infer a much wider range of human motion, activity, preference, and actions desired. In general a UUI must consider a broader range of inputs than current desktop, mobile or gaming devices, and applications. Examples of data or knowledge a UUI can rely on include spatial information (location, speed of movement), identity (users and others in vicinity), user model (profile, preferences, intentions), temporal data (time of day or year), environmental (noise, light), social (meeting, party), resource availability (printers, fax, wireless access), computing (network bandwidth, memory), physiological (hearing, heart rate), activity (supervision, interview), schedules, and agenda along with data that can be mined and inferred.

Broadly speaking, the range of implicit inputs to a ubicomp system via its UUI can be called context data whether sensed or inferred. Context includes information from the person (physiological state), the sensed environment (environmental state), and computational environment (computational state) that can be provided to alter an applications behavior. Explicit inputs from the user can include speech, gaze, or human movement of any type. Other chapters in this book focus on the processing and management of context data, whereas this chapter focuses on explicit user input along with explicit and peripheral outputs.

6.1.2 Inventing the Future

The time scale for the advancement and adoption of ubicomp systems and technologies may be unlike anything seen before in computing research and development. The adoption of this model of computation may reach a tipping point in key domains rapidly or it may be a slow process of features, functions, and indeed changes in expectations of what constitutes a computer, seeping into the general public consciousness. Weiser (1999) cites the symbolic representation of spoken language in written form as a ubiquitous technology. However, written language has developed over millennia and requires years of learning at a preschool, school, and even university level. Hence, the time scales involved in widespread adoption of a "ubicomp grammar" and hence interaction language may be longer than expected. Thankfully, researchers in academia, commercial laboratories, and industrial development have been producing ubicomp prototypes for more than 15 years. As a result, we have a rich assortment of research and experience to draw upon when considering UUIs.

Regardless of the eventual time scales of adoption, when planning a UUI for a particular ubicomp system, technology, or application, five fundamental questions typically arise:

1. Are the current generation of technologies applied to the problem sophisticated and aesthetic enough to afford a computationally enhanced experience that is realistic based on current practice?

2. Does the interface rely on well-understood metaphors and affordances? The WIMP metaphor has buttons, clicks, and drag-and-drop, but what can UUI designers and developers expect to use and build upon? In effect, what are the nouns, verbs, and grammar of ubicomp interfaces?

3. Does the proposed solution actually help people solve real problems or perform current tasks? Many of the technological scenarios used to motivate particular ubicomp methods, techniques, or technologies are aspirations of future needs and desires.

4. Will the costs or overheads involved prevent widespread adoption of these concepts into actual consumer use, that is, are the economics of this ubicomp solution sound?

5. When considering a UUI, if no one is expected to notice the presence of computation in the artifacts in their environment, then how are they expected to interact with them?

Even a cursory review of the ubicomp literature shows a range of technologies, methods, scenarios, and systems pushing the envelope of what is classically described as HCI. Some might describe many of these ubicomp systems as "solutions looking for a problem" but others classify them as attempts to invent the future. Alan Kay, the computer scientist, said, "Don't worry about what anybody else is going to do. The best way to predict the future is to invent it."

Current UUI development relies on a range of established and evolving user interface classes as described in Section 6.3. As new forms of sensing, actuation, technology, and interaction develops, we will see the creation of new interaction styles that may be hence classified as new classes of user interfaces themselves. Section 6.3 reviews the classical range of interaction styles considered in HCI and then describes three evolving classes of user interfaces that are central to the realization of UUIs. Table 6.1 describes

TABLE 6.1 Ten Rules for UUI Design

Rule	Meaning	Example
Bliss	Learning to interact with a new UUI should not require people to learn another skill or complex command language.	Good interaction design as discussed in Section 6.2
Distraction	Do not demand constant attention in a UUI. Inattention is the norm not the exception.	Ambient User Interfaces as discussed in Section 6.3.3
Cognitive Flow	Ubicomp systems that are everywhere must allow the user to retain total focus on the task at hand.	Multimodal interfaces as discussed in Section 6.4.3
Manuals	Do not require a user to read a manual to learn how to operate the current UUI. Do leverage prior experience.	Use of affordances (e.g., UI overlay) on real world in Figure 6.5
Transparency	Do not rely on users to hold application state in the mind to operate the UUI.	Tangible User Interfaces as discussed in Section 6.3.1
Modelessness	Avoid "modes" where the system responds differently to the same input stimulus dependent on some hidden state information.	State visible in SharePic as shown in Figure 6.8 (Section 6.3.2.1)
Fear of Interaction	Provide easy means to undo actions, otherwise users may become paralyzed with fear when using the UUI.	Use of well-understood design patterns as discussed in Section 6.2.4
Notifications	Feedback to the user can be piggybacked and layered into interactions with their physical environment.	Display of power usage as shown in Figure 6.12
Calming	Interfaces will support situated actions, interfaces will rely on a wide array of human inputs and human senses.	Surface interfaces as shown in Figure 6.6
Defaults	Good interfaces judiciously exploit what the system knows or can deduce.	Applications that reuse user input

the 10 key rules for UUI design. Each rule expresses either an aspect of the human state (e.g., maintaining bliss and not disturbing cognitive flow) or a base system requirement to ensure ease of use (e.g., no need of manuals and easy undo). Those developing a UUI should consider these rules from the start of the ideation process through to system design and development.

6.1.3 Chapter Overview

With all the artifacts in our environment imbibed with computation, input, and output, the future might become a very noisy, unpleasant, and controlling place. As such, it is important to study how to avoid the mistakes of the past in poor interface design while also relying on new technologies to help realize this vision of a useful future invisible "Everyware" (Greenfield, 2006) where we remain in control. Section 6.2 describes what is good interaction design for people in terms of affordances, metaphors, and the human action cycle. This is described in terms of user-centered design (UCD) in Section 6.2.1, systems design in Section 6.2.2, genius design in Section 6.2.3, and how patterns of interaction can be codified into design patterns in Section 6.2.4. The realization of a UUI can rely on a range of well-established user interface classes such as the GUI. Section 6.3 describes different classes of user interface. Three emerging classes of user interface that provide the natural interaction styles to allow interaction to remain in the periphery are described in Section 6.3.1 on tangible user interfaces (TUIs), Section 6.3.2 on surface user interfaces (SUIs) and particularly larger ones in Section 6.3.2.1, whereas ambient user interfaces (AUI) are discussed in Section 6.3.3. Three novel forms of input modality are described in Section 6.4, specifically sensor input in Section 6.4.1, gesture input in Section 6.4.2, and speech input in Section 6.4.3. We conclude this chapter with a brief discussion of some suggested UUI usability metrics in Section 6.5. These should be considered in the design, research, and development of any UUI.

6.2 INTERACTION DESIGN

Of particular note for UUI design is the field of *interaction design* (Shaffer, 2006). Interaction design is the discipline of defining the expected behavior of products and systems that a user can interact with. Donald Norman (2002) states, "far too many items in the world are designed, constructed, and foisted upon us with no understanding—or even care—for how we will use them." By contrast, interaction design aims to make products and systems usable and useful, and even engaging and fun. When considering the various aspects of interaction design for ubicomp, it is important to recall the breadth of the technological scenarios envisaged. Often, these scenarios revolve around people making connections to other people through ubicomp systems, not just connecting to the system itself. In addition, the breadth of the scenarios a UUI is required for suggests that interaction design should look to the fields of cinematography, kinesthesiology, and architecture.

Interaction design is a complex endeavor and draws on research, methods, techniques, and design guidelines from a range of overlapping and related fields and specialties including:

1. Cognitive psychology (metaphors, affordances, and mental models)
2. User experience design (storyboarding, personas, mockups)
3. Information architecture (shared data models, data stores)
4. Communication design (visual-auditory communication, graphic design)
5. User interface engineering (prototyping)
6. Human factors (human capability, ergonomics)
7. Industrial design (aesthetics)
8. HCI (new interface and interaction techniques)
9. Usability engineering (usability testing)

Good interaction design can be achieved in a number of ways depending on the complexity of the system proposed, its novelty, its degree of stability or ubiquity, and its cost. Design methodologies of interest to ubicomp include UCD, systems design, and genius design. The difference between these three approaches centers on the degree of user engagement in the process versus how much this can be abstracted away based on a whole system understanding or a confidence in the aesthetic of the designer. One approach to help bridge between these methods is to use documented design patterns. Such patterns can form a basis for exploiting lessons learned in each approach in the design of new systems based on a shared language for UUI design.

6.2.1 User-Centered Design

UCD focuses on the user's needs, problems, and goals to guide the design of a system. Users are involved at every stage of the process to help ensure that the system developed actually meets their needs and allows them to achieve their goals. As an approach to design, UCD dates back more than 30 years. It came from the realization that engineers frequently do not have the necessary skills required to develop user-friendly interfaces for computer systems. An important limitation of UCD is that most users typically cannot conceive of a radically different future and instead can only be relied on to inform design based on present needs. Henry Ford, the industrialist, is quoted as saying "if I had asked my customers what they wanted, they would have said a faster horse" (Jardim, 1970). UCD has been, and will continue to be, used in ubicomp research and development as a practical approach to entire systems design and user interface development (Fitton et al., 2005). UCD has its limitations in the face of evolving stated versus actual needs, user goals, technological shifts, or simply involving the wrong set or type of user in the process. Ubicomp research suggests that some of these limitations can be overcome by incorporating aspects of technology and cultural probes (Risborg and Quigley, 2003), contextual field research, intensive interviewing, and lag-sequential analysis (Consolvo et al., 2002) into the design process for ubicomp and its interfaces.

Many of the technological scenarios described in ubicomp literature are often beyond the current expectations, problems, and needs of users of today. From a design point of view, the solutions provided, given these scenarios, can be broadly classified as "systems design" or "genius design."

6.2.2 Systems Design

Systems design is a systematic and compositional approach to development, based on the combination of components to realize a solution in essence the development of a system of systems. A ubicomp system is typically composed of many systems including social systems (people), devices, appliances, computational artifacts, sensors, actuators, and services. In systems design, the documented or anticipated user needs and aims help set the *goal* of the system. Functional elements include computation, sensors, and actuators from the system itself. Inputs from the user are given from controls, which can be explicit (e.g., gesture) or implicitly defined (e.g., inference) on the system. Context data from the environment as a whole are sensed and matched with the goals, which drives the actuation of displays, services, or physical actuators. The actuation provides feedback (output) to the user, which should allow them to determine if the goal was met. If one removes the need for user control, then such a system can be described as self-governing or autonomic with the feedback loop helping to maintain control. Unlike desktop or Web application software development, systems design must consider and act upon a range of external factors from the real world as their inputs (disturbances, controls) and outputs (feedback, actuation). A systems design approach forces a designer to consider the entire environment in which the ubicomp system will be realized and not just one component of it. As the most analytical of the three design methodologies described here, it may be thought to appeal to ubicomp researchers who wish to ignore the user. This interpretation would be a mistake. As an analytical approach, it requires careful modeling and understanding of the implications of the user's goals, feedback, and controls provided.

6.2.3 Genius Design

Shaffer (2006) describes *genius design* as "the process of exclusively relying on the wisdom and experience of the designer to make all the design decisions." Based on their most informed opinion as to what users want and need, they design a product using their instinct and experience. This instinct and experience are developed over many successful projects and years of work. Many excellent devices, products, and systems in use today have come about from just this approach. Often, to maintain confidentiality, end users are not engaged in any manner, for example, in the design of the iPhone or indeed any Apple product. It is believed, but not confirmed, that designers in Apple must produce 10 versions of any new product feature for peer review and critique. We conjecture that this is how the majority of design decisions in ubicomp research and development are made today. A researcher invents a new physical component and interaction modality beyond the standard approach. Then, graduate students, academics, developers, or designers use their personal or collective skills to produce systems with a user interface. They do not have the time, resources, or inclination to engage a user cohort. Users may be involved at the end for testing or usability testing. Nielsen (2007) suggests that a quality user experience can be created with genius design *if* one starts by reducing risk and basing decisions on well-established usability data and studies. However, the contrived tasks often developed for usability testing are not suitable for the study of ubicomp systems and interfaces in authentic settings (Abowd and Mynatt, 2000). Instead, an iterative design process with evaluation and testing can remove the guesswork and inherent risk that the system and interfaces designed will fail. A description of the qualitative and quantitative user study techniques suitable for the evaluation and testing of a UUI can be found in Chapter 4.

6.2.4. Design Patterns

Design patterns are solutions to common design problems, tailored to the situation at hand (Tidwell, 2005). Introduced by Christopher Alexander and his colleagues in the field of architecture in *A Pattern Language: Towns, Buildings, Construction*, patterns represent a "shared language" for a discipline (Alexander et al., 1977). Patterns, now heavily used in software engineering (Gamma et al., 1994), are used to communicate common problems and appropriate solutions. Ubicomp researchers have started to describe common design patterns for this field (Landay and Borriello, 2003). Examples identified include Infrastructure (e.g., Proxies for devices), Privacy (e.g., Privacy Zones), Identification (e.g., Active Badge), Anticipation (e.g., "Follow-me Music"), Global Data (e.g., Interface teleporting), Discoverability, Capture and Access, Physical Space, and Location-based Services (e.g., Bluestar (Quigley et al., 2004)). A design pattern consists of 3–5 pages of text and provides general but descriptive methods for solving a particular problem. The text describes the background and how it might be fused with other patterns. A problem statement encapsulates a recurring domain specific problem. The answer identifies common solutions along with details, suggestions, and tradeoffs.

6.2.5 Discussion

With or without established patterns or an iterative process, UCD, systems design, and genius design approaches all have their strengths and weaknesses. Attempts to invent the future can be curtailed by the current generation of users who cannot imagine a future where this new device, service, or interface will be cheap, desired, or even required. However, this does not mean that the current generation of users should be ignored when researching and developing user interfaces for ubicomp systems. Instead, in each approach, user involvement should be layered in according to the assumptions around the eventual context of use for the system. This may be a UCD process where the ubicomp system solves a clear currently existing problem but relies on new techniques, methods, or infrastructure that are not yet commonplace or it may be an iterative systems design process where the ubicomp system affords users a new way to interact, relying on a novel combination of modalities with new appliances and affordances.

The process of design, or specifically interaction design, must draw on the available components, systems, interface elements, and modalities useful to realize a distraction-free, blissful, and calming experience. To understand the wide range of interface possibilities, we must first consider a range of "classes of user interface" currently available.

6.3 CLASSES OF USER INTERFACE

The user interface represents the point of contact between a computer system and a human, both in terms of input to the system and output from the system. The realization of this point of contact can exist in many forms and in many classes of user interface. In classical HCI texts, six classes are described that include command language, natural language, menu selection, form filling, direct manipulation, and anthropomorphic interfaces (Dix et al., 2003).

Direct manipulation embodied in the GUI has made computers accessible to many people. GUIs rely on visual and spatial cues, which are faster to learn, are easier to remember, and provide context for the user. Command languages require the user to remember concise yet cryptic commands. With rapid touch typing skills, these languages often appeal to experts but can strain one's power of recall. For example, consider the following UNIX commands

```
tail-n100 mynumbers|grep "^+353"
```

when issued at the command line will return any of the last 100 lines of the file mynumbers that start with string +353. Powerful indeed, but it requires detailed recall of command syntax to perfect.

Menu selection relies on recognition but not recall and can be easily supported on almost any visual device. However, this can create complex menu hierarchies and with audio systems users can easily get lost in (e.g., telephone menus). Form filling requires minimal training to use but is limited to data collection-style applications. Many of the most powerful web commerce applications are currently built around form fill-in interfaces. Research and usability has shown these reduce input errors and hence improve the user experience.

A natural language interface relies on a command language that is a significant subset of a language such as English. For example, IKEA offers an Online Assistant on their Web site that responds to typed English sentences. Related to natural language are humanlike or anthropomorphic interfaces that try to interact with people the same way people interact with each other. These are typically realized with interface characters acting as an embodied humanlike assistant or actors. Anna, the IKEA Online Assistant, has minimal anthropomorphic features through the use of 2-D animation.

From the WIMP paradigm developed in the Xerox Alto in the 1970s to the latest 3-D interface elements seen in Windows 7 and Mac OS X, the GUI has tended to dominate what is considered a user interface. However, clearly, a keyboard, screen, and mouse with GUI elements tied to every device affording computational interaction cannot be the future. The advent of the UUI will draw on elements from all these classes of interface and more.

For research and development, there are many issues in how elements of these well-established classes of user interface can be incorporated into future UUIs. There exist many further types of input technologies that we discuss in Section 6.4 which do not cleanly fit into any of these six classes because they rely on new devices. Examples include body movement in the form of gesture, speech, ambient feedback, surface interaction, and augmented reality (AR). Clearly, many new classes of user interface are being defined beyond these desktop-computing-bound six. Examples of this new class of interface include

- Tangible User Interface
- Surface User Interface
- Ambient User Interface

Alternate classes of user interface include gestural user interfaces, touch-based user interfaces, pen-based user interfaces, exertion interfaces (Mueller et al., 2007), and context-aware user interfaces. Two classes of interface, each deserving a chapter in their own right, which we do not explore in detail here, are AR user interfaces and multimodal user interfaces. An AR user interface (Figure 6.5) overlays computer graphics onto the image of a real-world scene typically with the aid of a supporting software interface framework (Sandor and Klinker, 2005). This scene can be viewed through a head-mounted display or a standard personal or desktop display. AR systems have been used to provide a graphical augmented-reality view of industrial equipment in the technician's immediate vicinity (Goose et al., 2003), Invisible Train Games (Wagner et al., 2005) as shown in Figure 6.5, or Human PacMan (Cheok et al., 2004). Multimodal interfaces attempt to take inputs from two or more of the inputs described here and "fuse" the disparate inputs into one. The motivation is to overcome the inherent limitations in any one modality with the expressiveness and error checking possible with multiple ones. For further details, we refer the reader to these guidelines for multimodal user interface design (Pavlovic et al., 1997; Reeves et al., 2004).

For the purposes of this UUI chapter, we limit our discussion to these three emerging areas (TUI, SUI, AUI) that rely on basic metaphors and

FIGURE 6.5 Augmented Reality invisible train game. (Accessed from Graz University of Technology, http://studierstube.org/handheld_ar/media_press.php)

well-understood affordances. These areas are both substantive and representative fields of user interface study for ubiquitous computing.

6.3.1 Tangible User Interface

Unlike a GUI, which presents manipulable elements virtually onscreen, a TUI integrates both representation and control of computation into physical artifacts. In essence, TUIs help provide physical form to computational artifacts and digital information. A "... TUI makes information directly graspable and manipulable with haptic feedback" (Ullmer and Ishii, 2000). The literal definition of the TUI as having both input and output factored into the same physical artifact has evolved over time (Fishkin, 2004). Accordingly, the accepted TUI paradigm is now: a user manipulates a physical artifact with physical gestures, this is sensed by the system, acted upon, and feedback is given. TUIs attempt to bridge the digital-physical divide by placing digital information in situ coupled to clearly related physical artifacts. The artifacts provide sites for both input, using well-understood affordances, and can act as output from the system. Unlike the GUI, a classical TUI device makes no distinction between the input devices and the output devices. Current TUI research and development attempt to form natural environments augmented with computation where the physical artifacts in that environments are digital embodiments of the state and information from a system.

FIGURE 6.6 Urp, architectural 3-D shapes with projected shadows beneath. (Copyright Tangible Media Group. Photograph taken by Dana Gordon.)

A stated aim for ubicomp is to make information, applications, and services available anywhere and at anytime in the human environment. Although TUIs are not yet ubiquitous in our daily lives, point examples drawn from both research and deployment can be discussed with reference to Weiser's original vision of "computation in basic objects, environments and the activities of our everyday lives in such a way that no one will notice its presence" (Weiser, 1999).

For example, "Urp" is a tangible workbench for urban planning and design (Underkoffler and Ishii, 1999) shown in Figure 6.6. In Urp, the basic objects are physical architectural models (houses, shops, schools) that can be placed by hand onto the workbench. The environment for Urp appears as a typical design desk with the computational infrastructure for the I/O Bulb (projector, vision system, computation) hidden away. Indeed, even in the case of no power, the desk and models still provide a view of the design space. Layered on top of the physical artifacts (desk and building models) are aspects of an underlying urban simulation including shadow, reflections, and wind flow. The seamlessness of this system relies on the natural interaction, rapid response, and in situ nature of the simulation data presented so that its presence is not noticed. The simulation data and interactions appear natural, obvious, and in keeping with the task at hand. Although not "visually invisible," Urp may be "effectively invisible" in action or behavior to the designer. This distinction between literal and effective invisibility is important to understand.

Posey is a computationally enhanced hub-and-strut kinematic construction kit (Weller et al., 2008). As with all types of construction kits such as Lego, Meccano, or Tinkertoy, Posey allows for a wide range of physical creations to be realized. In Posey, the basic objects are plastic childlike toy pieces, consisting of struts with a ball at each end and hubs with from one to four sockets. The environment for Posey can be a typical child's play area and the computational infrastructure for optocoupling with infrared (IR) light-emitting devices (LEDs) and photosensors and ZigBee wireless transmission remains hidden away. As with Urp, even without power, computation, or other digital-physical applications, Posey remains a usable toy. It is, after all, a TUI, which affords construction kit building activities. Aspects of the computationally enhanced system are currently rudimentary with Puppet Show outputs shown on a local display. The current version of Posey is not as seamless as proposed future versions, where the posable actions of Posey will be fluidly linked to the creation of animated character in a 3-D world for, example. Here, the creation of an animated bear puppet from physical interaction may allow future versions of this TUI to be invisible in action.

The field of TUI includes research and developments including areas such as audio systems (AudioCube, Audiopad (Patten et al., 2002), and Blockjam), construction toys (Topobo, Posey, FlowBlocks), physical tokens with interactive surfaces (Microsoft domino tags, metaDesk, Audiopad, Reactable, TANGerINE), toolKits (iStuff (Ballagas et al., 2003), Phidgets), "edutainment" (DisplayCube), and consumer products (I/O Brush, Nabaztag). Core methods, technologies, and techniques have been developed that are suitable for integration into physical artifacts, each of which contains many challenging research and engineering questions.

Many ubicomp scenarios are currently realized with GUIs on handheld computers, embedded displays, mobile phones, laptop, and desktop displays. In comparison, TUIs rely on devices concealed into everyday objects and everyday interaction modalities. These modalities rely on our basic motor skills and our physical senses. Rather than simply weaving GUI displays into our world with devices such as chumbys (http://www.chumby.com) or augmented refrigerators, TUIs are sites for computation, input, and output within objects in our physical environment. For example, Phicons are a TUI whose physical appearance suggests natural mappings to corresponding computational actions (Fishkin et al., 2002;

Ishii and Ullmer, 1997). The physical appearance serves as a metaphor for action. In high-end BMWs, the shape of a car seat control resembles the shape of the seat itself. Thus, pushing the physical button that is shaped like the bottom seat forward will cause the actual seat to move forward.

Fishkin [18] details two dimensions of *embodiment* and *metaphor* to describe a taxonomy with which to describe TUIs. Embodiment is the extent to which a user thinks of the state of computation as being embodied within physical housing of the artifact. This dimension includes four categories: full, nearby, environment, and distant (Fishkin, 2004).

1. With full, the output device is the input device.
2. With nearby, the output takes place near the input object, typically, directly proximate to it.
3. With environment, the output embodiment is around the user.
4. With a distant embodiment the output is "over there" on another device.

The dimension of metaphor relates physically afforded metaphors due to an artifact's physical tangibility. This dimension includes five categories: none, noun, verb, noun and verb, and full (Fishkin, 2004).

1. For none, the physical actions with the TUI are not connected to any real-world analogy (e.g., command line style UI).
2. For noun, the look of an input object is closely tied to the look of some real-world object but this is a superficial spatial analogy only.
3. For verb, the analogy is to the gesture used.
4. For noun and verb, the physical and virtual objects still differ but are related with appeal to analogy (e.g., Urp (Underkoffler and Ishii, 1999)).
5. Full gives a level of analogy where the virtual system is the physical system. Theses is no disconnect.

This work unifies several previous frameworks and incorporates a view from calm computing to classical GUI interfaces. So, expanding on

Fishkin's taxonomy, we define the five key characteristics of a TUI for a UUI to

1. Provide a clear coupling of physical artifact to relevant and related computation, state, and information

2. Ensure contextually appropriate physical affordances for computational interaction

3. Ensure contextually sensitive coupling of physical artifact to intangible representation (audio/graphics)

4. Support "invisibility in action" (not literal invisibility) and natural behavior

5. Ensure a grounding of the TUI interaction design in the fields of ethnography, industrial design, kinesthesiology, and architecture

Moving TUI from research to deployment requires careful interaction design to ensure that the affordances and computational support can be invisible in action and support natural behaviors. "… the appearance of the device must provide the critical clues required for its proper operation—knowledge has to be both in the head and in the world" (Norman, 2002). It is important to recall that even the original command line interface on terminal computers relied on tangible interfaces in the form of a physical keyboard and screen. Tangibility may be necessary to help realize many ubicomp scenarios, but by itself is not sufficient, as the outputs cannot change to fit the context as the GUI can.

6.3.2. Surface User Interface

An SUI is a class of user interface that relies on a self-illuminated [e.g., liquid crystal display (LCD)] or projected horizontal, vertical, or spherical interactive surface coupled with control of computation into the same physical surface (e.g., a touchscreen). As with a TUI, the outputs and inputs to an SUI are tightly coupled. They rely on computational techniques including computer vision, resistive membrane, capacitive and surface acoustic wave detection, to determine user input to the system. They are often used in public places (kiosks, ATMs) or small personal devices (PDA, iPhone) where a separate keyboard and mouse cannot or should not be used.

The scale of an SUI can range from small personal devices such as the iPhone or PDA, through a Tablet PC up to large public interactive

FIGURE 6.7 Microsoft surface detecting device and user touch.

surfaces such as the MERL DiamondTouch (Dietz and Leigh, 2001), as shown in Figures 6.7 and 6.9, or the Microsoft Surface as shown in Figures 6.2 and 6.8. An SUI can rely on a range of input types including passive stylus, active stylus, fingers, or tangible objects, or it may be tied to just one, as is the case with the Tablet PC with its active powered stylus. Whereas the input to the SUI is simply being used as a surrogate for a mouse input, many SUI applications function as classical GUIs do and as such they can be subject to many of the same design and usability studies and evaluations. In addition, large rendered or projected keyboards can further make such systems less ergonomic and functional than their GUI equivalents with attached keyboards and mice. Clearly, this is to be avoided.

As with all types of interface, SUIs must be fit for purpose and not be foisted on users as a gratuitous replacement for desktop computing. Fitts' law still applies! This law, although published more than 50 years ago, can be used to predict the time to move to a target area, as a function of the distance to the target and the size of that target. Equation (6.1) states that the movement time T (or MT as originally described), can be computed based on A, the distance to the center of the target from the starting position, and W, the target width. The empirically determined constants a and b are computed using a regression analysis on the movement time data.

FIGURE 6.8   User participants sharing photos with gestures in SharePic.

Refinements to this formulation to ensure a more stable model for the Fitts law have also been proposed.

$$T = a + b \log_2 \left(\frac{2A}{W} \right) \qquad (6.1)$$

As a result, small targets far away from where one's hands currently are will be difficult to acquire rapidly. Instead, the research, design, and development of an SUI can be seen as an opportunity to move beyond the classical desktop, keyboard, and mouse setting, and into the realization of more natural interaction styles *as appropriate*. The current generation of SUIs built into LCD displays or form factored into coffee tables cannot be considered a "basic object" in Weiser's vision. However, display technologies are now ubiquitous and if SUI interaction styles can be woven into the environments and activities of our everyday lives, then they will have achieved invisibility in action.

Basic SUIs have been commonplace for more than 20 years in the form of interactive kiosks, ATMs, and point-of-sale systems with a touch-sensitive surface overlaid with the display. These systems rely on touchscreen technology with simple button style interfaces. The basic technologies in many products such as the Nintendo DS with its resistive touchscreen technology were established in hardware research decades ago. The recent heightened interest in SUI products such as the iPhone generally stems from the low cost of production and the interaction styles they can afford beyond the desktop paradigm.

Due to their low cost and form factor, PDAs with rendered SUIs have formed components of many of the published ubicomp systems in the literature for the past 15 years, for example, the use of a PDA as memory aid for elders (Szymkowiak et al., 2005). Siren is a context-aware ubicomp system for firefighting (Jiang et al., 2004). Each firefighter carries a WiFi PDA on which the Siren messaging application is running. Each PDA has a Berkeley smart dust mote attached that is used as both a local sensing device and a communication facility to gather data from other motes embedded in the environment. In the Invisible Train system, an AR interface (SUI) is displayed on the screen of the PDA (Wagner et al., 2005). The attached camera allows the PDA to determine its relative 3-D location and rotation with respect to printed fiducial makers in the environment. The tracking and rendering overlay provides an SUI to a 3-D information space embedded within physical reality. In effect, the SUI acts as a "magic lens" onto the world. Interface elements that appear overlaid on the world such as the train speed button or moving track switches can all be operated with a tap on the rendered item on screen. Here, the small PDA acts as both input with stylus, touch, or gesture, and visual output to effect an interactive SUI.

Embedded, personal, or public displays provide opportunities for the development of an SUI. Smart phones can be used in ubicomp for taking and displaying context-aware notes tied to visual codes, photos, dictation, and text (Konomi and Roussos, 2007). It is important to note that coupling of input with output sometimes limits the tasks for which an SUI can be used. Consider the Audi Multi Media Interface (MMI), which provides controls for the company's high-end in-car navigation, radio, and car systems. The ergonomically designed remote control panel affords input while the two display outputs are in the driver's primary field of view. These displays use well-considered typography, shapes, shading, and opacity to help clearly present the information to the driver. However, this decoupling of input from output means the MMI is not a form of SUI unlike the more ubiquitous TomTom or Garmin navigation systems seen in-car.

The PDA has represented the first truly small, low-cost SUI platform that researchers experimented with. The interfaces developed have drawn heavily from GUI and mobile computing design and interaction styles. The availability of GUI toolkits has further rendered many of the interfaces on

PDAs displays for the "small screen" and the stylus or finger as a surrogate for the mouse. Character input, however, has moved beyond the GUI style with input languages such as Graffiti and Unistrokes (Castelluci and MacKenzie, 2008). PDAs typically support single-point interaction and the inputs they can accept in terms of tap, double tap, and gestures reflect this. Tablet PCs, although classified as SUIs, have featured less prominently in ubicomp research due to their size, form factor, and supported styles of interaction. Weiser (1999) developed inch-, foot-, and yard-scale displays called "tabs, pads, and boards." Tabs are akin to iPhones and PDAs, pads to the Tablet PC, and boards to larger SUIs.

6.3.2.1 Large SUIs
Larger interfaces that can act as an SUI have been researched and developed for more than 20 years. Although they can be classified in a number of ways, here we consider them from the standpoint of front-projected, rear-projected, and self-illuminated. For front-projected displays, the Digital Desk (Wellner, 1993) is a seminal ubicomp example incorporating an SUI. Both computer vision and audio sensing are used, and physical artifacts placed on the desk (under the gaze of a camera) act as interface elements. An exemplar application of a calculator has a paper sheet with printed buttons. Pressing these buttons is registered by the vision system and the current total is projected into the total square. Urp, described in Section 6.3.1, is a form of SUI relying on tangible objects to interact with it.

Various devices and applications rely on a single mouse input to operate. Single-touch interfaces can readily adapt to emulate single mouse input once issues of finger size and double tapping are considered. However, people typically have four fingers and one thumb on each of two hands. As a result, multitouch and multihanded interaction has emerged as an area of active research, and advances in both hardware and software solutions are supporting multitouch gestures, two-handed input, and multiperson input.

The DiamondTouch from MERL allows multiple, simultaneous users to interact with it as touch location information is determined independently for each user [16]. The DiamondTouch operates by transmitting a unique electrical signal to an array of rows and columns embedded in the table surface as shown in Figure 6.9. Each user is connected to a pad (standing, sitting, or touching) so that when they touch the surface with any part of their body, signals are capacitively coupled from the particular rows and columns touched. The DiamondTouch supports multipoint interaction and multiuser interaction, and is debris-tolerant and durable. SharePic is

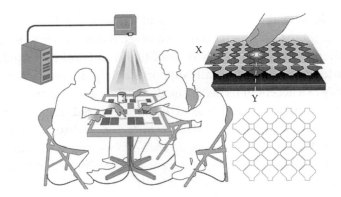

FIGURE 6.9 MERL Diamondtouch Multiuser input schematic. (Courtesy of Circle Twelve.)

a novel gesture-driven photo-sharing framework and application built on top of the DiamondTouch (Apted et al., 2006).

The Sony SmartSkin consists of a mesh of transmitter/receiver electrodes (e.g., copper wires) on the surface that forms a sensing layer (Rekimoto, 2007). The mesh can have a variable density, and it supports mouse emulation, shape-based manipulation, gestures, and capacitance tags on the bottom of tangible objects. When using capacitive sensing, both hand and finger interaction is supported along with physical dials and controls. AudioPad is a musical performance system using tangible music controls on a project surface (Patten et al., 2002). TANGerINE is a tangible tabletop environment, incorporating physical objects with embedded wireless sensors (Baraldi et al., 2007). For example, the Smart Micrel Cube (SMCube) is a wooden cube case with a matrix of IR emitter LEDs on each face. It embeds a WiMoCA node with a triaxial accelerometer and Bluetooth. The IR LEDs allow for computer vision methods to detect their position using simple noise removal, background subtraction, thresholding, and connected components analysis. Front-projected SUIs suffer from the problems of a person's body or hand occluding the display or blocking the line of sight for a camera system. In addition, the space required for the projector mounting can make this form of SUI difficult to manage. Short throw projectors like those used in commercial systems such as the latest SmartBoard go some way to overcoming some of these problems.

Rear-projected SUIs have made a large impression on the popular imagination in the past 5 years. They can be seen in nightly newscasts, during the Super Bowl, in the media, and new products from large multinational corporations such as Microsoft. The exact history of this part of the SUI field is both recent and deeply immersed in "patents pending." As such, we will review four representative systems with regard to the research and design methods behind them and interactions they support. The TouchLight relies on a rear-projected display onto a semitransparent acrylic plastic plane fitted with DNP HoloScreen material (Wilson, 2004). A touch image is computed on the plane situated between the user and a pair of cameras. An IR pass filter is applied to each camera and an IR cut filter to the projector if required. An IR illuminant is placed behind the display to illuminate the surface evenly in IR light. This configuration separates the projected light into the visible spectrum and the sensed light into the IR spectrum. Methods including depth mapping, sobel edge filters, mouse emulation algorithms, connected components algorithms, and template matching are used in TouchLight's computational support. The reacTable operates in a similar manner but it relies on a horizontal orientation and TUI physical objects with fiducial markers (Jordà et al., 2005). An example SUI is a collaborative electronic music instrument with the tabletop tangible multitouch interface. The Microsoft Surface operates in a similar manner to TouchLight, albeit with significant improvements in aspects such as industrial design, form factor, and software support. The IR lighting methods used here are typically referred to as *diffuse illumination*.

Fourier transform infrared (FTIR)-based displays such the multitouch work of Jeff Han (2005) relies on the total internal reaction (TIR) of light being frustrated. The frustration comes about when a object (e.g., finger) makes contact with the surface. The TIR is the same principal used in optical waveguides such as optical fibers. As with direct illumination, in FTIR IR illumination is used as the optical carrier. Here, a sheet of acrylic acts as a waveguide and is edge-lit by an array of high-power IR LEDs. IR light is emitted toward a pair of cameras situated behind the projected surface.

Self-projected systems are typically developed around LCD or plasma screen technologies. Given the space requirements for both front- and rear-projected SUIs, this class of device is drawing on both fundamental touchscreen technologies and the realization that alternate form factors are required. The digital vision touch relies on small cameras embedded in a device around the rim of the display. When an object enters the field of view, the angle within the field of view of the camera is calculated. Multiple cameras allow for triangulation and hence the location of the object (e.g., finger) onscreen to be determined. This is an area in its relative infancy as detecting multipoints as in FTIR or objects (e.g., the shape of a phone) as in TouchLight remains a significant research challenge for systems that do not have a clear view of the touch image from the back, or the front.

Regardless of the approach taken, the coupling of input with output into an SUI can cause a number of usability issues. These issues must be considered when developing an SUI component as part of a larger UUI:

1. The user's stylus, fingers, and hands may partially occlude the interface.
2. Interface elements may be difficult to select due to size of stylus, finger, or hand.
3. Users may suffer fatigue due to the range of human motion required.
4. The scale of the display may not be suitable for all aspects of the task at hand.
5. The screen surface can be damaged or dirty.
6. There is a lack of tactile feedback from passive screen surfaces.
7. Calibration between the display (projector or LCD) and the sensing elements can become misaligned.

Each of these problems can be addressed with careful design. For example, research is ongoing into the coupling of large displays with small personal displays (Terrenghi et al., 2009) to overcome the limitations of each type of SUI device. Other problems can be addressed with new technological advances such as haptic output to overcome the lack of tactile feedback on small SUIs. For example, Sony has developed tactile touch-screen feedback based on a very small actuator that bends the screen. The effect provides a haptic feedback along with a visual one to give the user the sense of buttons that actually click (Poupyrev and Maruyama, 2003).

Front, rear, or self-illuminated displays are becoming ubiquitous in our environment from ATMs to advertisements. Once coupled with support for HCI, they provide a key building block of the realization of always-on, always-available UUIs for ubicomp systems.

6.3.3 Ambient User Interfaces

SUIs require our engagement and involvement to operate correctly. Consider, however, a calm technology that can move easily from the periphery of our attention, to the center, and back again (Weiser, 1999). Ambient information displays or outputs are intended to be "ignorable" or "glanceable," allowing users to perceive the information presented in the periphery of their attention, but also to be bring this information (e.g., social reminders) into focus as required (Shannon et al., 2009). Although the periphery of our attention is a nebulous concept, it is grounded in the notion of our peripheral vision. Extensive research in vision science has demonstrated our abilities to recognize well-known structures and forms, to identify similar forms and particularly movements from outside our foveal line of sight. In practice, an ambient display, sound, movement, or even smell can convey background or context outputs from a ubicomp system (Heiner et al., 1999). Figure 6.10 demonstrates constant feedback through a tactile ambient display to express a mobile phone's state. Qualitative user studies demonstrated the ability to selectively "tune in and out" (Hemmert, 2009) status information was welcomed, even if the rudimentary form of actuation was not. In the future, a low-power heating element or other peripherally ignorable form of actuation could provide constant ambient feedback one could selectively attune to.

An ambient output (display or other) does not constitute a full user interface because it does not incorporate input. Many of the inputs we have described here would explicitly negate the glanceable nature of an AUI. As such, an AUI is a class of user interface where the output elements reside in the periphery of a user's awareness, moving to the center of attention only

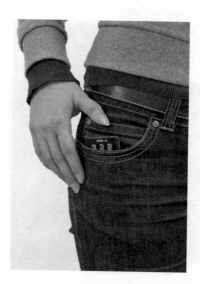

FIGURE 6.10 Ambient information provided by a tactile ambient display in a mobile phone. (Courtesy of Fabian Hemmert.)

when appropriate and desirable, and the inputs come from nonintrusive sensing or inference from other actions. Various dimensions of interest including a user's information capacity, suitable notification level, display representational fidelity, and aesthetic emphasis have been explored to understand AUIs and their limits (Pousman and Stasko, 2006). Additional studies have demonstrated a person's inability to recall the structure of an ambient information display, whereas the essential details and information conveyed can be recalled (Rashid and Quigley, 2009).

Fully realized AUIs as defined are not yet commonplace in our daily lives. However, ambient displays (e.g., digital advertisements and signage (Rashid and Quigley, 2009)) are commercially available and the subject of active research and development as part of larger information systems. Aspects of, or objects in, the physical environment become representations for digital information rendered as subtle changes in form, movement, sound, color, or light that are peripherally noticeable. Such displays can provide constant/situated awareness of information. Again, we can discuss these with reference to Weiser's original vision of "computation in basic objects, environments and the activities of our everyday lives in such a way that no one will notice its presence" (1999). Weiser himself discussed the Jeremjenko Dangling String called Live Wire showing network traffic in Xerox PARC as one of the earliest cited examples of an ambient display. Network activity caused the string to twitch, yielding a "peripherally noticeable indication of traffic."

The power-aware cord, as shown in Figure 6.11, is designed to visualize the energy flowing through it rather than hiding it (Gustaffson and

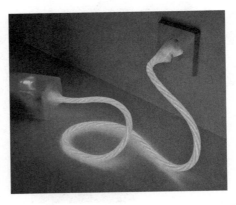

FIGURE 6.11 Power-aware cord, a power-strip designed to actively visualize the energy in use. (Photograph by Magnus Gyllenswärd. Accessed from http://tii.se/static/press.htm)

Gyllenswärd, 2005). This AUI aims to provide increased consumer awareness of their energy consumption. Switching off/on or unplugging devices acts as an input to the system and the display consists of glowing patterns produced by electroluminescent wires molded into the transparent electrical cord. This is a very basic object, present in our everyday environment, and with people ever more concerned with energy conservation, this has become a natural activity of our daily life, that is, consideration of energy usage. By relying on patterns of movement, this AUI can be attuned to in our periphery via peripheral vision. Longitudinal research studies of both the effectiveness and the disturbance of our visual field are required for this AUI. Would a moving pattern ever truly blend its way into our environments so as we might not notice it?

The information percolator (Heiner et al., 1999) is an "aesthetically pleasing decorative object." Here, the display is created by the control and release of air bubbles that rise up a series of tubes. The effect appears as a set of pixels, which can be used to convey a message that scrolls up the display. Sample applications include a personalized clock, activity awareness, a poetry display, and an interactive "bubble painting" application. The first three can be classed as AUIs as they require no explicit user input to operate and can be truly glanceable. The bubble painting treats the percolator as a large display with gestural input, thereby rendering it an SUI in this case. This blending of multiple interface classes will be typical and expected of future UUI developments. The percolator is not a basic object, but as a piece of installation art it can go unnoticed and thus remain glanceable.

By contrast, described as *information decoration*, the Datafountain displays relative currency rates using three water fountains side by side as shown in Figure 6.12. The relative heights of each fountain change with

FIGURE 6.12 Datafountain displaying relative currency rates.

respect to the currency changes (van Mensvoort, 2005). Here, the fountain is a basic object one would expect to find in day-to-day environments. However, unlike the information percolator, it remains only an output with inputs provided by a web service call. The InfoCanvas paints an appealing and meaningful representation of user-identified information on a virtual canvas. This ambient display allows people to monitor information in a peripheral manner (Miller and Stasko, 2001).

Finally, the suite of commercial systems for Ambient Devices including the Orb and Umbrella as shown in Figure 6.13 along with the Energy Joule represents the best examples of ambient displays, which have gained widespread deployment. The Ambient Orb is a frosted ball that glows in different colors to allow user to peripherally monitor changes of the status of everyday information. The color range and movement can be tailored via a website to the user's preference to show information including real-time stock market trends, pollen forecasts, weather, wind speed, and traffic congestion. The Ambient "Energy Joule" glows red, yellow, or green to indicate current cost of energy. Further details can be determined upon closer inspection with a left status bar detailing the cost, and right bar showing your current energy consumption. With the Ambient Umbrella, if rain is forecast, then the handle of the umbrella glows so you will not forget to take it. The Umbrella, as shown in Figure 6.13, is an everyday object, and deciding to take it or not each day is certainly an activity for those who live in rainy places. It remains an output only, but can move to the center of our focus if we peripherally notice the glowing handle. The Orb and Joule represent special purpose devices, not everyday objects that exhibit Ambient display features. The Orb may function as a component of a larger AUI or UUI as developed or designed in the future. With the

FIGURE 6.13   Ambient Orb and Ambient Umbrella. (Courtesy of Ambient Devices.)

Joule AUI turning on or off, an appliance can act as an input, whereas the color on the Joule provides ambient feedback that can remain glanceable.

Further AUI examples include the context-aware Visual Calendar (Neely et al., 2007), AuraOrbs and ambient trolley (Stasko et al., 2007), Information Art (Stasko et al., 2004), CareNet (Consolvo et al., 2004), BusMobile, and Daylight Display (Mankoff et al., 2003).

An AUI may exist in a device or environment for long periods without a conscious user action, yet the user is getting information from it, and by his/her actions or colocation, is providing it inputs. As a class of user interface, the ambient one conforms closest to Weiser's vision of calm technology. The design and evaluation of an AUI is particularly challenging from many standpoints including engaging users, conveying the utility, determining the effectiveness, efficiency, and usability (Mankoff et al., 2003), and collecting longitudinal efficacy data. In practice, an AUI may transition from implicit use to being an explicit SUI or from support for interaction with one to multiple users (Vogel and Balakrishnan, 2004). Clearly, there are limited types of information where the use of a dedicated single purpose AUI is useful. Hence, AUIs will form only part of a larger UUI for the system as a whole.

6.4 INPUT TECHNOLOGIES

A UUI relies on a broader range of inputs and outputs from the system than the classical GUI, TUI, or even an SUI. Examples of these inputs include physiological measurements, location, identity, video, audio, gesture, and touch. In addition, environmental sensors, personal/embedded sensors, data mining, historical data, inference, and preferences can all act as inputs to a system. Examples of the outputs that can be used include haptics, ambient displays, environmental updates, actuators, automated actions and personalized behaviors, and multiple audio/video channels. Such outputs are all reliant on our senses including sight, taste, smell, touch, hearing, and balance. As we have seen from our discussion on interface classes, some user actions will be interpreted via the UUI as an input to the system without the user being explicitly aware of it. Likewise, the UUI can provide outputs that are only intended for the periphery of the user's attention. Of course, many inputs and outputs will require explicit user action to form the input and to process the output provided. As we have already reviewed a wide range of output technologies, we focus on three categories of input namely sensor input, gesture input, and speech input for the purposes of concluding this chapter.

6.4.1 Sensor Input

A sensor is a device that can measure a physical property from the environment. Sensors can reside in the environment or on the body. Environmental sensors are becoming more widely deployed as components of systems to monitor traffic, air quality, water quality, and light pollution. More locally, sensors can be found embedded in doorways, security systems, weight monitors, and health systems. When sensors are collecting measurements about the person they are attached to, they are called physiological sensors (e.g., heart rate or galvanic skin response measurements); when not used as such, they are called mobile sensors (e.g., light levels, pollution monitoring, temperature). Computational measurements such as network traffic or memory usage can be determined by software sensors. The measurements from all these sensors are aspects of the context data that a ubicomp system can use to function correctly. By a user's regular actions, they can affect the environment in which the sensors operate (e.g., opening a window, turning on a stove, streaming a video). These actions can hence be viewed as implicit user inputs to a ubicomp system. For further details on sensors and context data, see Chapter 8 on context for ubicomp in this book.

The SensVest is a physiological monitoring system built into a wearable form factor. It measures aspects of human physical performance such as heart rate, temperature, and movement (Knight et al., 2005). The ring sensor, which measures blood oxygen saturation via pulse oximetry, is a component of a 24-hour patient monitoring system (Rhee et al., 1998). In both cases, these measurements form inputs to a larger system. Some commercial sensor systems rely on inertial technology such as the Moven motion capture suit, which can capture full body human motion.

As with many systems that rely on the collection of sensor inputs, the iHospital system is supported by a location and context-awareness infrastructure (Hansen et al., 2006). PCs can detect Bluetooth-enabled badges that the staff are wearing when they come within a 10 meter range. No explicit input is required; instead, the location data are sensed from the wireless environment, interpreted, and acted upon with various feedback in the form of reminders and surgery status details, on personal and public displays. Body-worn sensors or sensors embedded in devices can also collect various environmental measurements. Infrastructures can also be adapted to subtly alter the environmental conditions that allows for novel sensing systems to be developed. For example, an economical transmitter circuit can be realized by adding few components to a commercial electronic ballast circuit for fluorescent lamps

FIGURE 6.14 Magic Touch: use of RFID reader for tagged object identification and tracking. (Courtesy of Thomas Pederson.)

(Cheok and Li, 2008). A photo-receiver sensor can decode the pulse-frequency modulation that can allow this sensor to be localized. Once the location of the sensor, or more importantly the device or person it is attached to, can be determined, this can then provide an input to the system.

Active RFID systems can be sensed at a distance, whereas passive RFID requires closer contact and typically explicit user action for the sensor system to read the tag as shown in Figure 6.14. Examples of such large-scale sensor deployments include the Oyster card ticketing system used at the London Underground in the United Kingdom and retail applications deployed in Japan as examples, and other large-scale ubiquitous sensor deployments (Konomi and Roussos, 2007). For further details on such deployment, see Chapter 7.

6.4.2 Gesture Input

A gesture is the movement of a part of the body to express an idea of meaning. Typical gestures such as pointing, waving, or nodding are formed with the hand or the head as appropriate. For example, in a sign language a specific configuration and movement of the hands is considered a gesture. Figure 6.15 shows another common gesture, namely, a handshake, which can be detected and matched and used to infer social relationships (see Haddock et al., 2009). Both basic gestures such as pointing and complex gestures in sign language rely heavily on the appropriate cultural, geographical, or linguistic frame of reference for their interpretation. An innocuous gesture in one country can easily be interpreted as an insult in another. As such, gesture recognition is the process of interpreting human gestures using various inputs and computational processing. Almost any type of input device or system can be used to collect data

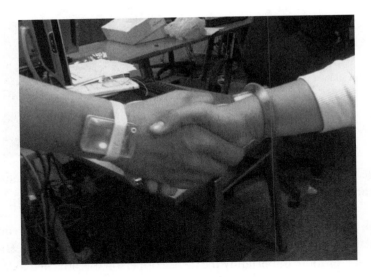

FIGURE 6.15 Matching accelerometer data for sensing handshakes for social network development.

on a user gesture. The difference between them is in the determination of how natural this movement actually is. Systems incorporating gesture recognition often suffer from the problems of feedback and visibility as described in Section 6.1.2—that is, how do I perform the gesture and how do I know what I just did was interpreted or not? Magic Touch, shown in Figure 6.16, is a novel gesture-driven object tracking and identification system (Pederson, 2001). Here, pages are tagged (either by the user or from the printer) and, as a user touches any page, the system develops a model that replicates the real-world organization of books, papers, and pages. Here, a gesture is simply the act of picking up and placing down a piece of paper, that is, very natural.

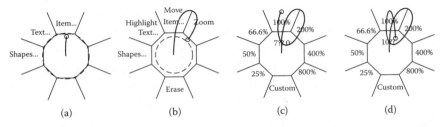

FIGURE 6.16 Flowmenu, single pen gesture to select option within a hierarchy.

Pen gestures using styli on PDAs and Tablets are the most ubiquitous example of gestural input to a computer system. Palm's Graffiti and Graffiti 2 are in widespread use and allow for rapid character and state changes (caps lock, shift) using simple gestures. However, they represent only a very small number of body movements. Pen input helped give rise to research on alternate forms of menu selection using gestural input, which do provide feedback and visibility. Circular pie, marking, and flow menus (Guimbretiére and Winograd, 2000) give each menu item a large target slice that is laid out radially and in different directions. Once you know the directions, you can quickly and reliably move ahead without looking. The user need not wait for the menus to render, so a single pen stroke can be a selection. Flow menus have a radial menu and overlaid submenus with eight octants. Experiments with it have shown that users can naturally learn gestures that require several strokes in one fluid gesture (Guimbretiére and Winograd, 2000). As single point inputs, all pen gestures can also be used with mouse input, although often with noticeable performance degradation.

Mouse gestures are commonly available for Web browsers such as Opera or FireFox via plugins. An example mouse gesture in Opera is for the user to hold down the right mouse button, move the mouse left, and release the right mouse button. This gesture encapsulates a series of movements (from current position to the back button) and a mouse click. Alternatively, a cheat sheet can be provided to a user on the make up of the pen and mouse gestures. In this case, the gesture can be considered a *command language* that must be memorized for efficient use. A specific problem with mouse gestures is the lack of standardization across applications, hence limiting the beneficial learning effect enjoyed by having visual widgets in a GUI.

The use of video input from a single or stereo camera array, relying on visible or near-visible light (e.g., IR) is a common method for collecting gestural input. This is a large field of research and development that is not limited to ubicomp scenarios. Classical UI interfaces augmented with video-based gestural input for gaming and sign language recognition are excellent examples of research in this field. The gestural input on SUIs such as the Microsoft Surface relies on the capture of IR light and subsequent processing with computer algorithms. Established algorithms and methods for processing images and video for gesture recognition include hidden Markov models (HMM) (Wilson and Bobick, 1999) and neural networks (Murakami and Taguchi, 1991). For further details, we refer the reader to the review by Pavlovic et al. (1997) on the visual interpretation of hand gestures. In ubicomp research, the FingerMouse (de la Hamette and Tröster 2008) relies on a button-sized camera that can be worn on the body, capturing the user's hand and supporting real-time segmentation and processing. Although not an everyday object, it is not a stretch to imagine such inputs factored into jewelry and other items commonly worn on the body.

Physical artifacts can be instrumented with sensors to provide accelerometer data. Body movement with the artifact can form an input that can be interpreted as a gesture such as shake, bump, tap, bounce, and rotate. TiltType relies on the combination of tilting actions with button presses to input characters to the device (Fishkin et al., 2002).

The Display Cube provides gesture recognition and output through six displays mounted at the sides of the cube (Kranz et al., 2005). A form of TUI, this is a playful learning interface for children. Simple simultaneous button presses affords SyncTap the ability to seamlessly authenticate two devices (Rekimoto, 2004). In contrast, simply holding two devices together and shaking them provides a method for automatic authentication in Shake Well Before Use (Mayrhofer and Gellersen, 2007). Smart-Its Friends are small embedded devices that become connected when a user holds them together and shakes them (Holmquist et al., 2001). The Wiimote is a handheld wireless game controller that can sense motion. Although this is a special purpose device and not an everyday object, its widespread adoption and ease of use can make the combination of game input and output invisible in action. The number of Wiimote devices that have ended up embedded in display screens is a testament to the fact that people forget it is a physical object that can damage other things and people!

6.4.3 Speech Input

Speech is the ability to produce articulate sounds to express an idea or meaning, and is typically realized through a spoken language. Speech recognition is the process of interpreting human speech using a variety of audio inputs and computational processing. As with video-based input processing, speech recognition is represented by a large body of research, development, and commercial deployment. The most widespread deployment of speech recognition can be found in phone call routing systems relying on a limited context sensitive vocabulary of words such as yes/no, numbers, and city names. The next most common example is in commercial dictation systems such as Dragon NaturallySpeaking. These rely on training periods with the user and a relatively noise-free environment in which to achieve high levels of speech recognition. Many high-end cars

FIGURE 6.17 The MEMENTO speech agent in a user's personal area network.

use commercial systems such as IBM's ViaVoice for embedded speech recognition in automotive telematics and other basic controls.

Speech represents a popular view of how humans will interact with computers as evidenced in literature and film. From natural interaction with next-generation robotic systems to HAL seen 40 years ago in Arthur C. Clarke's *2001: A Space Odyssey*, speech represents one view of how people can interact naturally with future ubicomp systems. Speech recognition and natural speech output represent the backbone of natural language and anthropomorphic interfaces. In addition, speech is often used as a secondary input modality to multimodal systems. For example, Memento (West et al., 2007) uses a combination of speech and pen input for the creation of a digital-physical scrapbook as shown in Figure 6.17. Speech is recognized by Sphinx components (Huang et al., 1993), and gestures by the Microsoft handwriting and gesture recognition libraries. The fused input forms the basis for action by Memento, which in this case is a digital-physical scrapbook for memory sharing.

Speech forms a central mode of input in infrastructure-rich visions of ubicomp as in Project Aura (Garlan et al., 2002). Here, the environment consists of many agents with which one can communicate using natural language. Aura represents a class of Intelligent Environment research popular in ubicomp relying on a rich range of seamlessly coupled inputs and outputs. Beyond such next-generation environments, today there are many work environments where speech can provide a more natural interface. These are environments where the classic GUI with keyboard and mouse or even next-generation SUI or TUI cannot provide the required computational support for the task at hand in a natural manner. Environments such as surgeries, dangerous work environments, and driving all represent environments where UUIs with speech as input have been researched and developed. Ubicomp scenarios incorporating in-car elements are a natural environment for the exploration of speech input in a UUI (Graham and Carter, 2000). SEAR (speech-enabled augmented reality) offers a technician a context-sensitive speech dialogue concerning industrial equipment in the technician's immediate vicinity (Goose et al., 2003).

Speech recognition systems using statistically based algorithms such as HMMs rely on both an acoustic and language model. For a UUI researcher and developer, there exist a number of toolkits that can speed up the rapid prototyping and iterative design process with speech. VoxForge can be used to create acoustic models for speech recognition engines. Designed to collect transcribed speech, VoxForge aims to create multiuser acoustic models that can be used without training. The Julius large vocabulary continuous speech recognition engine is by the Continuous Speech Recognition Consortium in Japan (Kawahara et al., 2004). Julius can be incorporated into a UUI, although it requires the computation power found in modern desktop PCs to operate. The CMU Sphinx Group produces a series of Open Source Speech Recognition components (Huang et al., 1993). These include decoders, acoustic model training, and language model training. Other toolkits include the Internet-Accessible Speech Recognition Technology C++ libraries and the HMM Toolkit (Young et al., 2002). Other groups are attempting to realize the standardization seen in GUIs across "Universal Speech Interfaces." The learning from interacting with one such USI can be transferred to future experiences with others (Rosenfeld et al., 2001).

Although speech input is key, audio can be used to provide speech output of verbal instructions in a UUI (Bradley and Dunlop, 2005). ONTRACK provides navigation cues to a user navigating an environment by adapting the audio he or she is listening to. This is achieved by modifying the spatial balance and volume to help guide the user to his or her destination (Jones et al., 2007). NAVITIME, used by more than 1.82 million people in Japan, provides map-based routes, text routes, and provides turn-by-turn voice guidance for pedestrians on handheld mobile phones and PDAs (Arikawa et al., 2007). Gesture and speech interaction can be provided with patient information systems (O'Neill et al., 2006). Bluetooth headsets can be used to provide direct audio and speech outputs to the user, as in the feedback provided in Memento, where the headset operates for both speech input and output in a personalized manner to the user (West et al., 2004).

6.5 INTERFACE USABILITY METRICS

Usability is a quality attribute that assesses how easy user interfaces are to use. The word "usability" also refers to methods for improving ease-of-use during the design process. Usability is defined by five quality components: learnability, efficiency, memorability, errors, and satisfaction.

For a UUI, we can describe a set of interface usability metrics, as shown in Table 6.2.

TABLE 6.2 Seven Key UUI Usability Metrics

Metric	Meaning
Conciseness	A few simple actions in a brief time can achieve a task. This can be measured by time (attention or gaze), keystrokes, gestures, and taps.
Expressiveness	Does a combination of actions not anticipated give consistent results?
Ease	How much does a user need to learn or recall just to start using the interface?
Transparency	How much does a user need to remember about the state of his or her problem while using the interface telephone speech interface versus a GUI?
Discoverability	Can the user easily understand and form a mental model of the interface functionality?
Invisibility	How much does the interface make itself know when it could have inferred, deduced, or waited for the data required?
Programmability	Can the application, device, or service be used in repetitive tasks or can it become a component in a larger system?

6.6 CONCLUSIONS

This chapter has provided an overview of the classes of interface both new and old that may be suitable to realize interfaces of applications in environments imbibed with computation, input, and output. This chapter started by setting the context for what a UUI is and hence detailing 10 rules for UUI design (Table 6.1). We have shown contrasting approaches to interaction design and how these have impacted on each other and the relative merits of each approach in Section 6.2. Section 6.3 provided a rich description of different classes of user interface. Three novel forms of input modality are described in Section 6.4. This chapter concludes with interface usability metrics described in Table 6.2 for UUIs in Section 6.5 and how these may be applied to future interface design.

Although this chapter offers a snapshot of the state of the art in UUI development, it is best viewed as an entry point to the research papers cited here. Considerable research and development has been undertaken in both ubicomp and the research areas that underpin the subject. Developing a UUI as a prototype or proof of concept to demonstrate an idea or run some small-scale user trials is fundamentally different to the realization of a UUI and system for use in the wild. In the wild, an interface is expected to have all the properties shown in Table 6.1 and more. Anything less constitutes risk that people will not purchase, adopt, or use your system. General usability metrics and our ubicomp interface usability metrics are described in Table 6.2. When and where they are to be applied should be considered and reflected upon time and again in the research and development process.

The ultimate goal for ubicomp is to have interfaces and hence systems that seamlessly support the actions of their users. In the future, enhanced computational artifacts, environments, and full ubicomp systems will become so commonplace in restaurants, colleges, workplaces, and homes that no one will notice their presence. Our job as researchers, designers, developers, and ultimately users is to constantly question this vision to ensure we end up with calm and blissful digital-physical experiences.

ACKNOWLEDGMENTS

The authors thank Ross Shannon, John Krumm, and Marc Langheinrich for helpful comments and discussions on this text.

REFERENCES

Gregory D. Abowd and Elizabeth D. Mynatt. Charting past, present, and future research in ubiquitous computing. *ACM Transactions on Computer-Human Interaction* 7(1):29–58, 2000.

Christopher Alexander, Sara Ishikawa, and Murray Silverstein. *A Pattern Language: Towns, Buildings, Construction (Center for Environmental Structure Series)*. Oxford University Press, Oxford, 1977.

Trent Apted, Judy Kay, and Aaron Quigley. Tabletop sharing of digital photographs for the elderly. In *CHI '06: Proceedings of the SIGCHI Conference on Human Factors in Computing Systems*, pp. 781–790, New York, NY, USA, April 2006. ACM Press.

Masatoshi Arikawa, Shin'ichi Konomi, and Keisuke Ohnishi. Navitime: Supporting pedestrian navigation in the real world. *IEEE Pervasive Computing* 6(3):21–29, 2007.

Rafael Ballagas, Meredith Ringel, Maureen Stone, and Jan Borchers. iStuff: A physical user interface toolkit for ubiquitous computing environments. In *CHI '03: Proceedings of the SIGCHI Conference on Human Factors in Computing Systems*, pp. 537–544, New York, NY, USA, 2003. ACM.

Stefano Baraldi, Alberto Del Bimbo, Lea Landucci, Nicola Torpei, Omar Cafini, Elisabetta Farella, Augusto Pieracci, and Luca Benini. Introducing tangerine: A tangible interactive natural environment. In *MULTIMEDIA '07: Proceedings of the 15th International Conference on Multimedia*, pp. 831–834, New York, NY, USA, 2007. ACM Press.

Nicholas Bradley and Mark Dunlop. An experimental investigation into wayfinding directions for visually impaired people. *Personal and Ubiquitous Computing* 9(6):395–403, 2005.

Steven J. Castellucci and I. Scott MacKenzie. Graffiti vs. unistrokes: An empirical comparison. In *CHI '08: Proceeding of the Twenty-Sixth Annual SIGCHI Conference on Human Factors in Computing Systems*, pp. 305–308, New York, NY, USA, 2008. ACM Press.

Adrian Cheok and Yue Li. Ubiquitous interaction with positioning and navigation using a novel light sensor-based information transmission system. *Personal and Ubiquitous Computing* 12(6):445–458, 2008.

Adrian David Cheok, KokHwee Goh, Wei Liu, Farzam Farbiz, SiewWan Fong, SzeLee Teo, Yu Li, and Xubo Yang. Human pacman: A mobile, wide-area entertainment system based on physical, social, and ubiquitous computing. *Personal and Ubiquitous Computing* 8(2):71–81, 2004.

Sunny Consolvo, Larry Arnstein, and Robert B. Franza. User study techniques in the design and evaluation of a ubicomp environment. *UbiComp 2002: Ubiquitous Computing*, pp. 281–290, 2002.

Sunny Consolvo, Peter Roessler, and Brett E. Shelton. The CareNet display: Lessons learned from an in home evaluation of an ambient display. *UbiComp 2004: Ubiquitous Computing*, pp. 1–17, 2004.

Joëlle Coutaz, James L. Crowley, Simon Dobson, and David Garlan. Context is key. *Communications of the ACM* 48(3):49–53, 2005.

Patrick de la Hamette and Gerhard Tröster. Architecture and applications of the fingermouse: A smart stereo camera for wearable computing HCI. *Personal and Ubiquitous Computing* 12(2):97–110, 2008.

Paul Dietz and Darren Leigh. Diamondtouch: A multi-user touch technology. In *UIST '01: Proceedings of the 14th Annual ACM Symposium on User Interface Software and Technology*, pp. 219–226, New York, NY, USA, 2001. ACM Press.

Alan Dix, Janet E. Finlay, Gregory D. Abowd, and Russell Beale. *Human-Computer Interaction, 3rd Edition*. Prentice Hall, Inc., Upper Saddle River, NJ, USA, 2003.

Kenneth P. Fishkin. A taxonomy for and analysis of tangible interfaces. *Personal and Ubiquitous Computing* 8(5):347–358, 2004.

Kenneth P. Fishkin, Kurt Partridge, and Saurav Chatterjee. Wireless user interface components for personal area networks. *IEEE Pervasive Computing* 1(4):49–55, 2002.

D. Fitton, K. Cheverst, C. Kray, A. Dix, M. Rouncefield, and G. Saslis-Lagoudakis. Rapid prototyping and user-centered design of interactive display-based systems. *Pervasive Computing, IEEE* 4(4):58–66, 2005.

Erich Gamma, Richard Helm, Ralph Johnson, and John Vlissides. *Design Patterns*. Addison-Wesley Publishing, Reading, MA, 1994.

David Garlan, Dan Siewiorek, Asim Smailagic, and Peter Steenkiste. Project aura: Toward distraction-free pervasive computing. *IEEE Pervasive Computing* 1(2):22–31, 2002.

Stuart Goose, Sandra Sudarsky, Xiang Zhang, and Nassir Navab. Speech-enabled augmented reality supporting mobile industrial maintenance. *IEEE Pervasive Computing* 2(1):65–70, 2003.

Robert Graham and Chris Carter. Comparison of speech input and manual control of in-car devices while on the move. *Personal and Ubiquitous Computing* 4(2):155–164, 2000.

Adam Greenfield. *Everyware: The Dawning Age of Ubiquitous Computing*. Peachpit Press, Berkeley, CA, USA, 2006.

François Guimbretiére and Terry Winograd. Flowmenu: Combining command, text, and data entry. In *UIST '00: Proceedings of the 13th Annual ACM Symposium on User Interface Software and Technology*, pp. 213–216, New York, NY, USA, 2000. ACM Press.

Anton Gustafsson and Magnus Gyllenswärd. The power-aware cord: Energy awareness through ambient information display. In *CHI '05: CHI '05 Extended Abstracts on Human Factors in Computing Systems*, pp. 1423–1426, New York, NY, USA, 2005. ACM Press.

David Haddock, Aaron Quigley, and Benoit Gaudin. Sensing handshakes of social network development. In *Proceedings of the Artificial Intelligence and Cognitive Science 20th Irish International Conference*. AICS, 2009.

Jefferson Y. Han. Low-cost multi-touch sensing through frustrated total internal reflection. In *UIST '05: Proceedings of the 18th Annual ACM symposium on User Interface Software and Technology*, pp. 115–118, New York, NY, USA, 2005. ACM Press.

Thomas Riisgaard Hansen, Jakob E. Bardram, and Mads Soegaard. Moving out of the lab: Deploying pervasive technologies in a hospital. *IEEE Pervasive Computing* 5(3):24–31, 2006.

Jeremy M. Heiner, Scott E. Hudson, and Kenichiro Tanaka. The information percolator: Ambient information display in a decorative object. In *UIST '99: Proceedings of the 12th Annual ACM Symposium on User Interface Software and Technology*, pp. 141–148, New York, NY, USA, 1999. ACM Press.

Fabian Hemmert. Life in the pocket: The ambient life project life-like movements in tactile ambient displays in mobile phones. *International Journal of Ambient Computing and Intelligence* 1(2):13–19, 2009.

Lars Holmquist, Friedemann Mattern, Bernt Schiele, Petteri Alahuhta, Michael Beigl, and Hans W. Gellersen. Smart-Its friends: A technique for users to easily establish connections between smart artefacts. *Ubicomp 2001: Ubiquitous Computing*, pp. 116–122, 2001.

Xuedong Huang, Fileno Alleva, Hsiao-Wuen Hon, Mei-Yuh Hwang, and Ronald Rosenfeld. The SPHINX-II speech recognition system: An overview. *Computer Speech and Language* 7(2):137–148, 1993.

Hiroshi Ishii and Brygg Ullmer. Tangible bits: Towards seamless interfaces between people, bits and atoms. In *CHI '97: Proceedings of the SIGCHI Conference on Human Factors in Computing Systems*, pp. 234–241, New York, NY, USA, 1997. ACM Press.

Anne Jardim. *The First Henry Ford: A Study in Personality and Business. Leadership*. MIT University Press, Cambridge, MA, 1970

Xiaodong Jiang, Nicholas Y. Chen, Jason I. Hong, Kevin Wang, Leila Takayama, and James A. Landay. Siren: Context-aware computing for firefighting. *Pervasive Computing*, pp. 87–105, 2004.

Matt Jones, Steve Jones, Gareth Bradley, Nigel Warren, David Bainbridge, and Geoff Holmes. Ontrack: Dynamically adapting music playback to support navigation. *Personal and Ubiquitous Computing*, 2007.

Sergi Jordà, Martin Kaltenbrunner, Günter Geiger, and Ross Bencina. The reactable*. In *Proceedings of the International Computer Music Conference (ICMC 2005)*, Barcelona, Spain, 2005.

Tatsuya Kawahara, Akinobu Lee, Kazuya Takeda, Katsunobu Itou, and Kiyohiro Shikano. Recent progress of open-source LVCSR Engine Julius and Japanese model repository and Japanese model repository—software of continuous speech recognition consortium. In *INTERSPEECH-2004*, pp. 3069–3072. http://julius.sourceforge.jp/, 2004.

James F. Knight, Anthony Schwirtz, Fotis Psomadelis, Chris Baber, Huw W. Bristow, and Theodoros N. Arvanitis. The design of the SensVest. *Personal and Ubiquitous Computing* 9(1):6–19, 2005.

Shin'ichi Konomi and George Roussos. Ubiquitous computing in the real world: Lessons learnt from large scale RFID deployments. *Personal and Ubiquitous Computing* 11(7):507–521, 2007.

Matthias Kranz, Dominik Schmidt, Paul Holleis, and Albrecht Schmidt. A display cube as tangible user interface. In *Adjunct Proceedings of the Seventh International Conference on Ubiquitous Computing (Demo 22)*, September 2005.

James A. Landay and Gaetano Borriello. Design patterns for ubiquitous computing. *Computer* 36(8):93–95, 2003.

Jennifer Mankoff, Anind K. Dey, Gary Hsieh, Julie Kientz, Scott Lederer, and Morgan Ames. Heuristic evaluation of ambient displays. In *CHI '03: Proceedings of the SIGCHI Conference on Human Factors in Computing Systems*, pp. 169–176, New York, NY, USA, 2003. ACM Press.

Rene Mayrhofer and Hans Gellersen. Shake Well Before Use: Authentication based on accelerometer data. In *Proceedings of 5th International Conference on Pervasive Computing*, pp. 144–161, 2007.

Todd Miller and John Stasko. The InfoCanvas: Information conveyance through personalized, expressive art. In *CHI '01: CHI '01 Extended Abstracts on Human Factors in Computing Systems*, pp. 305–306, New York, NY, USA, 2001. ACM Press.

Florian Mueller, Gunnar Stevens, Alex Thorogood, Shannon O'Brien, and Volker Wulf. Sports over a distance. *Personal and Ubiquitous Computing* 11(8):633–645, 2007.

Kouichi Murakami and Hitomi Taguchi. Gesture recognition using recurrent neural networks. In *CHI '91: Proceedings of the SIGCHI Conference on Human Factors in Computing Systems*, pp. 237–242, New York, NY, USA, 1991. ACM Press.

Steve Neely, Graeme Stevenson, and Paddy Nixon. Assessing the suitability of context information for ambient display. In William R. Hazlewood, Lorcan Coyle, and Sunny Consolvo, Eds., *Proceedings of the 1st Workshop on Ambient Information Systems. Colocated with Pervasive 2007*, Toronto, Canada, volume 254 of CEUR Workshop Proceedings, ISSN 1613–0073, May 2007.

Jakob Nielsen. The Myth of the Genius Designer, http://www.useit.com/alertbox/genius-designers.html, May 2007.

Donald A. Norman. *The Design of Everyday Things*. Basic Books, New York, NY, 2002.

Eamonn O'Neill, Manasawee Kaenampornpan, Vassilis Kostakos, Andrew Warr, and Dawn Woodgate. Can we do without GUIs? Gesture and speech interaction with a patient information system. *Personal and Ubiquitous Computing* 10(5):269–283, 2006.

James Patten, Ben Recht, and Hiroshi Ishii. Audiopad: A tag-based interface for musical performance. In *NIME '02: Proceedings of the 2002 Conference on New Interfaces for Musical Expression*, pp. 1–6, Singapore, Singapore, 2002. National University of Singapore.

Vladimir I. Pavlovic, Rajeev Sharma, and Thomas S. Huang. Visual interpretation of hand gestures for human-computer interaction: A review. *IEEE Transactions on Pattern Analysis and Machine Intelligence* 19(7):677–695, 1997.

Thomas Pederson. Magic touch: A simple object location tracking system enabling the development of physical-virtual artefacts in office environments. *Personal and Ubiquitous Computing* 5(1):54–57, 2001.

Ivan Poupyrev and Shigeaki Maruyama. Tactile interfaces for small touch screens. In *UIST '03: Proceedings of the 16th Annual ACM Symposium on User Interface Software and Technology*, pp. 217–220, New York, NY, USA, 2003. ACM Press.

Zachary Pousman and John Stasko. A taxonomy of ambient information systems: Four patterns of design. In *AVI '06: Proceedings of the Working Conference on Advanced Visual Interfaces*, pp. 67–74, New York, NY, USA, 2006. ACM Press.

Aaron Quigley, Belinda Ward, Chris Ottrey, Dan Cutting, and Robert Kum-merfeld. Bluestar, a privacy centric location aware system. In *Position Location and Navigation Symposium (PLAN 2004)*, pp. 684–689, April 2004.

Umer Rashid and Aaron Quigley. Ambient displays in academic settings: Avoiding their underutilization. *International Journal of Ambient Computing and Intelligence* 1(2):31–38, 2009.

Leah M. Reeves, Jennifer Lai, James A. Larson, Sharon Oviatt, T. S. Balaji, Stéphanie Buisine, Penny Collings, Phil Cohen, Ben Kraal, Jean-Claude Martin, Michael McTear, T. V. Raman, Kay M. Stanney, Hui Su, and Qian Ying Wang. Guidelines for multimodal user interface design. *Communications of the ACM* 47(1):57–59, 2004.

Jun Rekimoto. Smartskin: An infrastructure for freehand manipulation on interactive surfaces. In *CHI '02: Proceedings of the SIGCHI Conference on Human Factors in Computing Systems*, pp. 113–120, New York, NY, USA, 2002. ACM Press.

Jun Rekimoto. Synctap: Synchronous user operation for spontaneous network connection. *Personal and Ubiquitous Computing* 8(2):126–134, 2004.

Sokwoo Rhee, Boo-Ho Yang, Kuowei Chang, and H. H. Asada. The ring sensor: A new ambulatory wearable sensor for twenty-four hour patient monitoring. *Engineering in Medicine and Biology Society, 1998. Proceedings of the 20th Annual International Conference of the IEEE*, 4:1906–1909, 1998.

Peter Risborg and Aaron Quigley. Nightingale: Reminiscence and technology—from a user perspective. In *OZeWAI 2003: Australian Web Accessibility Initiative*. La Trobe University, Victoria, Australia, December 2003.

Ronald Rosenfeld, Dan Olsen, and Alex Rudnicky. Universal speech interfaces. *Interactions* 8(6):34–44, 2001.

Christian Sandor and Gudrun Klinker. A rapid prototyping software infrastructure for user interfaces in ubiquitous augmented reality. *Personal and Ubiquitous Computing* 9(3):169–185, 2005.

Dan Shaffer. *Designing for Interaction: Creating Smart Applications and Clever Devices*. Peachpit Press, Berkeley, CA, 2006.

Ross Shannon, Eugene Kenny, and Aaron Quigley. Using ambient social reminders to stay in touch with friends. *International Journal of Ambient Computing and Intelligence* 1(2):70–78, 2009.

John Stasko, Myungcheol Doo, Brian Dorn, and Christopher Plaue. Explorations and experiences with ambient information systems. In William R. Hazlewood, Lorcan Coyle, and Sunny Consolvo, Eds., *Proceedings of the 1st Workshop on Ambient Information Systems. Colocated with Pervasive 2007*, Toronto, Canada, volume 254 of CEUR Workshop Proceedings ISSN 1613-0073, May 2007.

John Stasko, Todd Miller, Zachary Pousman, Christopher Plaue, and Osman Ullah. Personalized peripheral information awareness through information art. *UbiComp 2004: Ubiquitous Computing*, pp. 18–35, 2004.

Andrea Szymkowiak, Kenny Morrison, Peter Gregor, Prveen Shah, Jonathan J. Evans, and Barbara A. Wilson. A memory aid with remote communication using distributed technology. *Personal and Ubiquitous Computing* 9(1):1–5, 2005.

Lucia Terrenghi, Aaron Quigley, and Alan Dix. A taxonomy for and analysis of coupled displays. *Personal and Ubiquitous Computing*, 2009.

Jenifer Tidwell. *Designing Interfaces: Patterns for Effective Interaction Design.* O'Reilly Media, Inc., November 2005.

Brygg Ullmer and Hiroshi Ishii. Emerging frameworks for tangible user interfaces. *IBM Systems Journal* 39(3–4):915–931, 2000.

John Underkoffler and Hiroshi Ishii. Urp: A luminous-tangible workbench for urban planning and design. In *Proceedings of the SIGCHI Conference on Human Factors in Computer Systems: The CHI Is the Limit*, pp. 386–393, Pittsburgh, VA, May 15–20, 1999.

Koert van Mensvoort. Datafountain: Money translated to water, 2005.

Daniel Vogel and Ravin Balakrishnan. Interactive public ambient displays: Transitioning from implicit to explicit, public to personal, interaction with multiple users. In *UIST '04: Proceedings of the 17th Annual ACM Symposium on User Interface Software and Technology*, pp. 137–146, New York, NY, USA, 2004. ACM Press.

Daniel Wagner, Thomas Pintaric, Florian Ledermann, and Dieter Schmalstieg. Towards massively multi-user augmented reality on handheld devices. In *Proceedings of the 3rd International Conference on Pervasive Computing*, pp. 208–219, 2005.

Mark Weiser. The computer for the 21st century. *SIGMOBILE Mobile Computing and Communications Review* 3(3):3–11, 1999.

Michael Philetus Weller, Ellen Yi-Luen Do, and Mark D. Gross. Posey: Instrumenting a poseable hub and strut construction toy. In *TEI '08: Proceedings of the 2nd International Conference on Tangible and Embedded Interaction*, pp. 39–46, New York, NY, USA, 2008. ACM Press.

Pierre Wellner. Interacting with paper on the digitaldesk. *Communications of the ACM* 36(7):87–96, 1993.

David West, Trent Apted, and Aaron Quigley. A context inference and multimodal approach to mobile information access. In *Artificial Intelligence in Mobile Systems 2004 (workshop of UbiComp 2004)*, September 2004.

David West, Aaron Quigley, and Judy Kay. Memento: A digital physical scrapbook for memory sharing. *Journal of Personal and Ubiquitous Computing* 11(4):313–328, 2007. Special Issue on Memory and sharing of experiences.

Andrew D. Wilson. TouchLight: An imaging touch screen and display for gesture-based interaction. In *ICMI '04: Proceedings of the 6th International Conference on Multimodal Interfaces*, pp. 69–76, New York, NY, USA, 2004. ACM Press.

Andrew D. Wilson and Aaron F. Bobick. Parametric hidden Markov models for gesture recognition. *IEEE Transactions on Pattern Analysis and Machine Intelligence* 21(9):884–900, 1999.

Steve J. Young, Gunnar Evermann, Thomas Hain, Dan Kershaw, Gareth L. Moore, Julian J. Odell, Dave Ollason, Dan Povey, Valtcho Valtchev, and Phil C. Woodland. *The HTK Book (Version 3.2.1)*. HTK, 2002.

CHAPTER 7

Location in Ubiquitous Computing

Alexander Varshavsky and Shwetak Patel

CONTENTS

7.1	Introduction	286
7.2	Characterizing Location Technologies	288
	7.2.1 Location Representation	288
	7.2.2 Infrastructure and Client-Based Location Systems	289
	7.2.3 Approaches to Determining Location	290
	7.2.3.1 Proximity	291
	7.2.3.2 Trilateration	291
	7.2.3.3 Hyperbolic Lateration	294
	7.2.3.4 Triangulation	295
	7.2.3.5 Fingerprinting	295
	7.2.3.6 Dead Reckoning	297
	7.2.4 Error Reporting	298
	7.2.4.1 Sources of Errors	298
	7.2.4.2 Reporting Error	299
7.3	Location Systems	300
	7.3.1 Global Positioning System	302
	7.3.2 Active Badge	304
	7.3.3 Active Bat	305
	7.3.4 Cricket	306
	7.3.5 UbiSense	307
	7.3.6 RADAR	308

7.3.7 Place Lab	310
7.3.8 PowerLine Positioning	311
7.3.9 ActiveFloor	312
7.3.10 Airbus	314
7.3.11 Tracking with Cameras	315
7.4 Conclusions and Challenges	316
References	317

7.1 INTRODUCTION

This chapter discusses the fundamentals of location technologies and gives an overview of both historical and current location systems. Location technologies have been an important part of ubiquitous computing (ubicomp) and have been an active topic of research for the past decade (LaMarca and de Lara, 2008). The ability to determine a user's location enables a variety of ubicomp applications that provide services and functionality appropriate to the specific location and context. In other words, location-aware applications use the location of the target to add value to the services they provide. For example, some of the earlier location-aware applications were routing phone calls to the phone closest to the user's current location (Want et al., 1992), sending printouts to the nearest printer, and displaying files and programs specific to the user's location (Schilit et al., 1999). Since the early days, location-aware applications have grown in sophistication and utility, and people rely more and more on these applications. Today, people use location-aware applications in almost any life domain, including entertainment, navigation, asset tracking, health care monitoring, and emergency response. The number of location-aware applications is still growing fast, with the annual market for global positioning system (GPS) and navigation services and products alone projected to grow to U.S.$200 billion by 2015 (Rizos et al., 2005).

Location is one of the most important components of user context (Schilit et al., 1999). In addition to being useful in its own right, location information can also be used to infer additional pieces of context, such as user activity, mode of transportation, and social relationship. For example, spending time at a gym is indicative of exercising, changing location at a speed of 65 miles per hour is indicative of driving, and driving someone every morning to and from work is indicative of a close relationship. We refer the reader to Chapter 8 for an in-depth discussion on the use of contextual information in ubicomp.

Location information can be conveyed in absolute, relative, or symbolic form. An absolute location describes an exact position, such as an address or geographic coordinates. For example, AT&T Labs is located at 180 Park Avenue, Florham Park, NJ. A relative location describes a position of an object relative to another absolute location. For instance, Dunn Gardens are located approximately 10 miles north of downtown Seattle, WA. A symbolic location is a descriptive name of a place, such as "home," "work," or "bedroom."

Despite the importance of location information, there is no single location technology that is accurate, low-cost, easy to deploy, and ubiquitous. Instead, there is a collection of location technologies each of which is best suited for a particular situation and need, ranging from accuracy of 1 millimeter using magnetic fields (Ascension Technology) to tens of kilometers using FM radio signals (Krumm et al., 2003). Since location technologies generally trade off accuracy for coverage and cost, one should choose the location system that satisfies the accuracy requirement of the particular location-aware application of interest. For example, printing a document on the nearest printer will, in most cases, work as well with 1 meter accuracy as it will with 1 millimeter accuracy. This chapter covers a variety of location systems with different accuracy, coverage, and cost trade-offs. In addition, most location-aware applications make an implicit assumption that location of a device, such as a mobile phone, is a good proxy for the location of the human using the device. Although, in practice, it is not always the case (Patel et al., 2006a), for the ease of presentation, this chapter refers to the location of the device and the location of the user using the device interchangeably.

The outline of the rest of the chapter is as follows. Section 7.2 discusses various aspects of location technologies, including ways to specify location, differences between client-based and network-based location systems, common approaches for determining location, and ways to describe location error. Section 7.3 gives an overview of both historical and current location systems, while helping distill why certain design decisions where made for each system. Section 7.4 concludes the chapter and lists some of the remaining challenges in the field of location tracking.

Note that this chapter does not cover the important topics of location modeling, issues related to location privacy, and more advanced stochastic methods for inferring location. See Chapter 3 for a discussion on privacy and Chapter 9 for a description of how to process sequential sensor data.

7.2 CHARACTERIZING LOCATION TECHNOLOGIES

Location technology is a combination of methods and techniques for determining a physical location of an object or a person in the real world. This section describes various aspects of location technologies. The section starts with discussing ways to represent absolute, relative, and symbolic locations. Next, Section 7.2.2 describes the differences between client-based, network-based, and network-assisted location systems. Section 7.2.3 surveys popular technologies for location determination. Finally, Section 7.2.4 discusses common sources of location error and ways to report the accuracy of a location system.

7.2.1 Location Representation

Location is a position in a physical space and it can be represented in absolute, relative, or symbolic form. The most common means of specifying a precise absolute location is using the points' degrees of latitude and longitude on the surface of the Earth, as defined by the geographic coordinate system. If Earth were a perfect ellipsoid, the latitude would measure the angle between the point and the equatorial plane from the center of Earth. In reality, however, the latitude, or the geodetic latitude, measures the angle between the equator and a line that is normal to the reference ellipsoid, which approximates the shape of Earth. The longitude measures the angle along the Equator to the point. A line that passes near the Royal Observatory, Greenwich, England, is accepted as the zero-longitude point and it is called the Prime Meridian. Lines of constant latitude are called parallels and lines of constant longitude are called meridians. Meridians, unlike parallels, are not parallel and all intersect at the North and South Poles. This form of representation is often used in outdoor location systems such as GPS. See Figure 7.1 for an example.

One can specify any location on the surface of the Earth using latitude and longitude. For example, Seattle, WA, has a latitude of 47.60°N and a longitude of 122.33°W. Thus, a vector from the center of the Earth to a point 47.60° north of equator and 122.33° west of Greenwich, England, will pass through Seattle. By adding a vertical distance from the center of the Earth or, more commonly, from the mean sea level at a given point, it is possible to specify any location below or above the surface of the Earth.

Although geographic coordinates are useful for specifying a precise absolute location, they are not convenient to use in the types of applications that involve reasoning with location information by humans. One

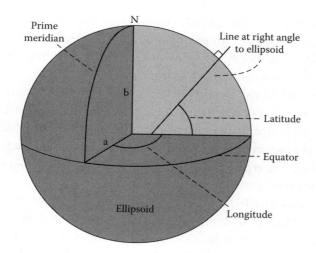

FIGURE 7.1 An example of a latitude and longitude angles to a point on Earth. (Accessed from http://home.online.no/~sigurdhu/artimages/Lat-Long.gif)

can rarely overhear a person saying: "I am at 40.7755 North, 74.4135 West. Do you want to grab a coffee at 40.789 North, 74.393 West in half an hour?" As an alternative, it is typical to use an address to specify a location (e.g., 40 St. George Street), a symbolic name of a place that is familiar to both parties (e.g., the mall) or a relative coordinate (e.g., 100 meters north of Main and Ridgedale). A geocoder, such as Microsoft's Virtual Earth or AT&T's Yellow Pages, can be used to translate an address or a zip code into a geographic coordinate. Translating a geographic coordinate into an address or a zip code can be performed using a reverse geocoder.

Describing location within an indoor space is similarly challenging. A system may represent a location using a local coordinate system within the building by specifying an X and Y distance from a fixed corner of the building. For a multistory building, an anchor point could be specified for each floor. Although these representations are useful at a system level, they are usually also mapped to higher-level relative or symbolic representations, such as "living room," "bedroom," "Joe's office," or "next to the coffee pot."

7.2.2 Infrastructure and Client-Based Location Systems

This section describes the differences between three classes of location systems: client-based, network-based, and network-assisted. In a client-based location system, a device computes its own location without relying on the network infrastructure. An example of a client-based location system is GPS, in which a device equipped with a GPS chip calculates its own location using signals received from at least four GPS satellites.

In a network-based location system, the network infrastructure calculates the position of a device. An example of a network-based location system is the Active Badge system (Want et al., 1992), in which a badge carried by the user emits infrared (IR) signals captured by the IR receivers in the ceiling. The receivers, in turn, transmit signal data to a networked processor that computes the badge's location.

In a network-assisted location system, both the device and the infrastructure participate in computing the location of the device. An example of a network-assisted location system is the Assisted GPS, in which a device calculates its own location based on its GPS measurements and additional information about the GPS constellation received over the cellular link from the cellular network infrastructure. The additional information allows the device to calculate its location even if fewer than four satellites are in view and it reduces the time from turning on of the device to the initial location acquisition.

The main advantage of a client-based location system is that it preserves the location privacy of the device. Since the mobile device simply listens to the beacons from the infrastructure without transmitting data, the infrastructure has no way of determining the location of the device, unless the device is willing to share the data. On the other hand, calculating the location on a device may reduce its battery life and it adds requirements on the device's processing and storage capabilities. Note that the network infrastructure may learn the device's location when the device requires additional information or services based on its location. For example, requesting maps or nearby restaurants requires the device to disclose its location with a certain degree of accuracy. For more information about preserving a user's privacy, see Chapter 3.

7.2.3 Approaches to Determining Location

This section describes six fundamental techniques for determining the location of a device: proximity, trilateration, hyperbolic lateration, triangulation, fingerprinting, and dead reckoning. Some of these techniques assume a presence of one or more reference points, whose precise location is known in advance. Examples of a reference point include a GPS satellite, a WiFi access point (AP), or a cellular tower.

7.2.3.1 Proximity

Proximity sensing is the simplest location technique. It uses the closeness of a device to a reference point to estimate the location of the device. The device's location is typically estimated to be the location of the reference point. Either the device or the reference point can sense the proximity. Note that detecting a device in close proximity does not necessarily reveal the identity of the device. Therefore, a separate identity detection mechanism may be necessary for recognizing the identity of the device.

Proximity can be detected through either direct physical contact or detection of the device being in range of one or more reference points. For example, stepping on a pressure sensor reveals the presence of an individual on the sensor and communicating with a WiFi AP indicates that the device is near the AP. In the latter case, proximity sensing relies on a limited range of coverage of the underlying wireless communication technology. For instance, the range of a near-field communication device is a few centimeters, the range of a Bluetooth device is tens of meters, the range of a WiFi device is hundreds of meters and a cellular phone may receive signals kilometers away. If the device happens to be in a range of several reference points, it is possible to compute a more accurate location estimate. For instance, it is possible to estimate the location of the device as an average of reference point positions. If the strength of the signal with which the reference points can overhear the device is available, a more precise location can be obtained using the weighted average of the reference point positions. A more advanced technique that uses the actual distances between a device and reference points is called trilateration, which is discussed in the next section.

7.2.3.2 Trilateration

Trilateration is a location technique that computes the position of a device by measuring the distance between the device and a number of reference points at known locations. The number of reference points required for computing the location is one greater than the number of the physical space dimensions. For example, calculating the device's location in two dimensions (2-D) requires three noncoplanar reference points, whereas calculating the device's location in 3-D requires four reference points.

Figure 7.2 shows an example of trilateration in 2-D. Each black dot represents a reference point and it defines a center of a circle with the radius equal to the estimated distance to the device. Thus, estimating the distance to a single reference point yields an infinite number of possible locations

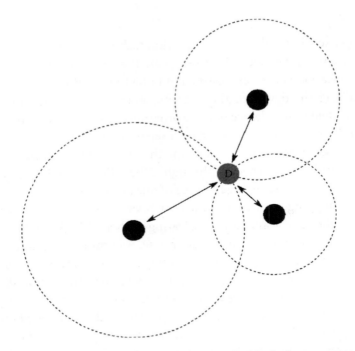

FIGURE 7.2   Example of trilateration in 2-D. The black dots represent reference points. The gray dot represents the location of the device.

of the device on the perimeter of the circle; estimating the distances to two reference points yields two possible locations of the device at the intersections of the two circles; and estimating the distances to three reference points uniquely defines the device's location.

To estimate the distance between a device and a reference point, it is common to either measure the time-of-flight of the signal or to measure the attenuation of the strength of the signal at the receiver. The next two sections cover these distance estimation techniques.

7.2.3.2.1 Time of Flight   Estimating the time-of-flight of a signal between a device and a reference point is possible because the speed of sound and the speed of light are known quantities (344 meters per second in 21°C air for sound and 299,792,458 meters per second for light). Therefore, by measuring the time it takes for the signal to travel the distance between the device and the reference point and multiplying it by the speed of travel, it is possible to estimate the traveled distance. For instance, if an ultrasonic pulse sent from a device to a reference point reached its destination in

29 milliseconds, the distance between the two is 10 meters. A radio or light signal would cover the same distance in 33.4 nanoseconds.

Measuring the time-of-flight requires precise clock synchronization between the device and the reference point. This is especially true when estimating the time-of-flight of a radio or light signal because small skews in the clocks will result in large measurement errors. To avoid the clock synchronization problem, instead of measuring the time-of-flight between two devices, some systems measure the round-trip delay and divide it by 2. This eliminates the need to synchronize clocks on two devices because the same device both transmits and receives the signal.

The time-of-flight can be measured either by the device or the network infrastructure. In the former, each reference point transmits a signal that is being received and decoded by the device. In the latter, the device sends a signal that is being received by all reference points. In both cases, the reference points need to have synchronized clocks for the precise time-of-flight calculation.

7.2.3.2.2 Signal Strength Attenuation   Another approach to estimate the distance between a device and a reference point is based on the ability to estimate the decrease in the strength of a signal as it travels away from its source. The formula that estimates the strength of the signal at a certain distance from the source is called the signal attenuation model.

A signal attenuation model depends on a multitude of factors, including the distance from the source, terrain contours, physical environment, propagation medium, and the height of the antennas at the source and the destination. For instance, the signal attenuation model for free space far-field radio frequency (RF) states that the strength of a radio signal decreases by a factor of $1/r^2$, where r is the distance from the radio source. For near-field inductively coupled communication, the attenuation of the signal could decrease by a factor of as high as $1/r^6$. Thus, given the correct signal attenuation model and the strengths of the signal at the source and at the destination, it is possible to estimate the distance between the source and the destination.

Unfortunately, in more complex physical environments, such as an indoor office space, radio attenuation models have trouble estimating the distance accurately due to complex interactions of the signal with objects in the physical space. As the signal travels, it may be reflected, refracted, or diffracted, which may cause the signal to change direction, reach areas that would not be possible to reach if the radio signals traveled in a direct line, or arrive at the destination by two or more paths. These complex interactions present a formidable challenge for creating accurate signal attenuation models.

7.2.3.3 Hyperbolic Lateration

Hyperbolic lateration uses the difference between the signal arrival times from a device to three or more reference points, instead of using the signal travel time itself. The hyperbolic lateration technique is also applicable when a device receives signals that were simultaneously transmitted by three or more reference points. For the ease of presentation, this section explains hyperbolic lateration in 2-D.

The signal transmitted from a device will be received at different times by reference points located at different distances from the device. The difference between the signal arrival times at two reference points restricts the possible location of the device to be along a hyperbolic line, with the two reference points serving as the foci of the hyperbola. In other words, transmitting a signal while being located at any point on the hyperbola will result in the same time difference of arrival of the signal to the two reference points. Adding a third reference point gives two more pairs of TDOA and therefore two more hyperbolas. Intersection of any two of the hyperbolas defines the unique possible location of the device. Figure 7.3 shows an example of hyperbolic lateration in two dimensions. The intersection of the two hyperbolas uniquely defines the location of a device.

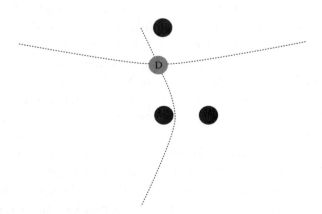

FIGURE 7.3 An example of hyperbolic lateration in 2-D. The black dots represent reference points. The gray dot represents the location of the device.

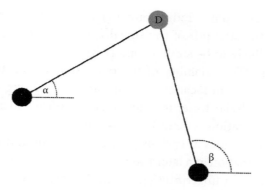

FIGURE 7.4 An example of triangulation in 2-D. The black dots represent reference points. The gray dot represents the location of the device.

7.2.3.4 Triangulation

Triangulation uses the angle of arrival (AOA) of signals traveling from a device to reference points to estimate the device's location. See Figure 7.4 for an example of triangulation in 2-D. Measuring the angle at which the signal arrives from the device (represented as the gray dot) to a reference point (represented as a black dot) restricts the position of the device along the line that passes through the reference point along the AOA. Measuring angles from two reference points results in two lines that uniquely define the device's location at the point of intersection. Thus, it is enough to have angle measurements from only two reference points to determine the location of the device in two dimensions; in practice, however, more than two reference points are used to reduce angle measurement errors.

To estimate the AOA of a signal, either a directional antenna or an antenna array is needed. Since neither is typically available on a mobile device, most existing location systems based on triangulation choose to measure the AOA at the reference points.

7.2.3.5 Fingerprinting

Fingerprinting is a location technology that uses pattern matching techniques to estimate the location of a device. For the ease of presentation, this section describes RF fingerprinting; however, the techniques presented in this section are applicable to other signal sources as well, such as sound and colored light.

RF fingerprinting relies on two properties of radio signals: temporal stability and spatial variability. Temporal stability refers to the stability of a radio signal from a radio source at any given location over time. For example, the strength of the signal from a nearby cellular tower at one's office is likely to be similar tomorrow and next week. Spatial variability refers to the variability of the radio signal from the same radio source at two different locations. For instance, the strength of the signal from a nearby cellular tower is different at the office and at the cafeteria. Fingerprinting location systems take advantage of these two properties by capturing radio profiles at various physical locations and using them for location determination at a later time.

The accuracy of a fingerprinting system is closely tied to the degree of spatial variability of the signal. For example, signal strength from WiFi APs exhibits spatial variability at the 1 to 10 meter level, or, in other words, a given WiFi AP may be heard stronger or not at all a few meters away. This allows for WiFi-based fingerprinting systems with about 1 meter of spatial error.

Fingerprinting relies on a training phase to build a radio map of the target environment before it can be used for location determination. During the training phase, a device moves through the environment, taking measurements of the strength of signals emanating from a group of radio sources (e.g., WiFi APs). An example of a radio measurement is shown in Table 7.1. The table shows a list of WiFi APs and the signal strengths as received from each of the APs.

At the end of the training process, the fingerprinting system has a radio profile for a multitude of locations in the target physical space. Since fingerprinting does not model radio propagation, a fairly dense grid of radio measurements needs to be collected to achieve good accuracy. The original fingerprinting system (Bahl and Padmanabhan, 2000), for example, collected measurements of WiFi signal strengths about 1 m apart.

Once the training phase is complete, a device can estimate its location by collecting a measurement and feeding it to the fingerprinting system, which may reside either on the device itself or in the network infrastructure. The fingerprinting system estimates the location of the device based

TABLE 7.1 An Example of a Measurement That Includes Names, MAC Addresses, and Received Signal Strength Indicator Values of Three WiFi Access Points

SSID (Name)	BSSID (MAC Address)	Signal Strength (RSSI)
linksys	00:0F:66:2A:61:00	18
starbucks	00:0F:C8:00:15:13	15
newark wifi	00:06:25:98:7A:0C	23

on the similarity between the current measurement and the measurements recorded during the training phase.

The similarity between measurements can be computed in a variety of ways, but it is common to use the Euclidean distance in signal space. For example, if one WiFi measurement contains signal strengths for N APs $(S_1,\ldots,S_n)$ and another measurement contains signal strengths for the same N sources $(R_1,\ldots,R_n)$, then the Euclidean distance between the two fingerprints will be calculated as:

$$E = \sqrt{(S_1 - R_1)^2 + (S_2 - R_2)^2 + \cdots + (S_N - R_N)^2}$$

If an AP is not present in one of the measurements, it is common to substitute its signal strength with either the minimal signal strength found in this measurement or with a fixed predetermined value.

There are several variations of the fingerprinting algorithm. A fingerprinting system based, for example, on the nearest neighbor (NN) algorithm estimates the location of a device to be the location of the measurement in the training radio profile with the smallest Euclidean distance to the current measurement. A K-NN algorithm produces an estimate of the device's location by averaging the locations of the K measurements in the training profile with the smallest Euclidean distance to the current measurement. The ideal K value can be determined through experimentation in a representative environment, but a small value of 3 or 4 was shown to work well in practice (Cheng et al., 2005).

7.2.3.6 Dead Reckoning

Dead reckoning is a location technique that computes the location of a device based on its previously known location, or fix, elapsed time, direction, and average speed of movement. The assumption behind dead reckoning is that the direction and the average speed of movement since the last fix is either known or can be estimated. Figure 7.5 illustrates the principle behind dead reckoning. The black dot represents the last known location of the device. Knowing the last fix, the direction, and the average speed, it is possible to estimate the new location of the device at the location of the gray dot.

Since dead reckoning calculates the *relative* position since the last fix, it is being used in combination with another location technology capable of calculating the absolute location of the device. Dead reckoning is thus often used to refine the estimates of another location system

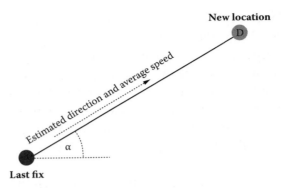

FIGURE 7.5 An example of dead reckoning in 2-D. The black dot represents the position of the last fix. The gray dot represents the estimated location of the device.

or to calculate the estimates when the other location system becomes temporary unavailable (e.g., when a car enters a tunnel and loses signals from GPS satellites).

The accuracy of dead reckoning depends on the quality of estimation of the speed and direction of movement. These can be estimated either through extrapolation from two or more previous fixes or measured by the sensors on the device itself. Some of the sensors that are often used for dead reckoning are accelerometers, which can be used to measure the acceleration of the device; odometers, which can be used to measure the distance traveled by a car; and gyroscopes, which can be used to measure the direction of movement.

7.2.4 Error Reporting

The goal of a location system is to produce accurate location estimates. Unfortunately, in practice, location systems often produce inaccurate estimates due to a variety of reasons. This section surveys some of the common reasons for errors in location systems and describes the common procedure for reporting location errors.

7.2.4.1 Sources of Errors

Location systems are designed to produce accurate location estimates given that the measurements that the location system uses are accurate as well. Unfortunately, there are several factors that introduce errors into location systems.

Incorrect reference point coordinates. Location systems that require the precise location of reference points produce location errors when the given locations are incorrect. This problem can be mitigated or eliminated for stationary reference points by carefully mapping the location of the reference points. However, for reference points that are mobile (e.g., GPS satellites) this may be a difficult problem due to unexpected factors (e.g., solar winds) that may alter a reference point's location.

Ionospheric and tropospheric delay. Signals traveling through the ionosphere and troposphere experience delays due to interactions with the Earth's atmosphere. Although there are mathematical models that try to estimate the delay, it still accounts for the major part of error in GPS-based positioning.

Clock synchronization. Precise time measurement requires tight clock synchronization between the sender and the receiver of the signal, or between devices that transmit signals simultaneously. Existing synchronization algorithms reduce the effect of clock skews, but do not eliminate them completely. Clock skews are a common error source for all location systems that use time measurements for location determination.

Multipath. A signal traveling through space may arrive at the destination along several paths due to interactions with obstacles along the way. Copies of the signal may overlay at the receiver, causing distortions of the amplitude and phase of the signal. Having no line of sight between the sender and the receiver exacerbates the multipath problem, making measurement errors more severe.

Geometry. The configuration geometry of the reference points has an effect on accuracy. Positioning the reference points too close to each other or on a line typically results in large location errors.

7.2.4.2 Reporting Error

The quality of location estimates produced by a location system varies depending on many factors, including the physical location, the time of day, the current weather, and the environment. Therefore, to fully understand the quality of a location system, it is necessary to collect a large number of location estimates under various conditions. The procedure for reporting location errors depends on whether the location system produces symbolic or absolute locations.

Location systems estimating symbolic locations, such as home or work, produce estimates that are either correct or not. In this case, the most common means of expressing the accuracy of a location system is to present it as a percentage. For instance, a location system may determine the room in a building correctly 85% of the time. Repeating the experiment several times allows for the calculation of confidence intervals of the accuracy of the location system.

For location systems estimating absolute locations, such as geographic coordinates, it is typical to show a cumulative distribution function of the location error. In this case, a location error is defined as the distance between the true and the estimated locations. It is also common to specify the 50th and the 95th percentile of the location error, both of which can be derived from the cumulative distribution function. If the location system performs differently along horizontal and vertical dimensions, then it is common to specify the location error separately for each dimension.

7.3 LOCATION SYSTEMS

The first part of this chapter discussed techniques and general concepts for building location systems. The remainder of this chapter discusses specific commercial and research location systems that use these principles and techniques. In addition, this chapter also attempts to highlight important characteristic that should be considered when applying a particular technology to an application.

Localization has been a very active research problem in the ubiquitous computing community in the preceding decade. Several characteristics distinguish the different solutions, such as the underlying signaling technology (e.g., IR, RF, load sensing, computer vision, or audition), line-of-sight requirements, accuracy, and cost of scaling the solution over space and over the number of objects. This section provides an overview of some of the historically important and current location systems and highlights the different characteristics of each system. The intent of this section is not to present an entire survey of location systems, but to highlight important historic and current systems that have addressed the general problem of location tracking in variety of ways.

It is important to note that there is no one perfect location system. Each system must be evaluated based on the intended application across a variety of dimensions such as its accuracy, the infrastructure requirements, the ability to scale, etc. Table 7.2 summarizes location systems covered in this section and dimensions that one should consider when evaluating, building, and using a location system. An important consideration is the performance or accuracy of the system and its resolution (e.g., low resolution for weather forecasts and high resolution for indoor navigation). At the same time, one must consider the infrastructure requirements to

TABLE 7.2 Location Tracking Technologies across a Collection of Factors Used to Evaluate a Particular Location System

	Location Type	Resolution, Accuracy	Infrastructure Requirements	Location Data Storage	Spectral Requirements	Location System Type
Active Badge	Symbolic Indoor	Room level	IR Sensors and customs tag	Central	IR	Custom active tagging
ActiveBat	Absolute Indoor	3 cm, 90%	Ultrasonic (US) receivers and transmitters	Central	30 kHz ultrasound and 900 MHz RF	Custom active tagging
ActiveFloor	Symbolic Indoor	1 m, 91%	Custom floor tiles	Central	Load sensor	Passive
Airbus	Symbolic Indoor	Room level, 88%	Single sensor in HVAC	Central	Pressure sensor	Passive
Cricket	Absolute Indoor	3 cm, 90%	US receivers and transmitters	Local	30 kHz ultrasound and 900 MHz RF	Custom active tagging
GPS	Absolute Outdoor	10 m, 50%	GPS receiver	Local	1500 MHz RF	Custom active tagging
PlaceLab (GSM)	Symbolic Indoor/Outdoor	20 m, 90% 5 m, 50%	Existing GSM towers	Local	900–2000 MHz RF	Active tagging
LaceLab (WiFi)	Symbolic Indoor/Outdoor	20 m, 50%	Existing WiFi APs	Local	2.4 GHz RF	Active tagging
PowerLine Positioning	Symbolic Indoor	2 m, 93% 0.75 m, 50%	2 plug-in module and custom tag	Local or central	300–1600 kHz RF	Custom active tagging
RADAR	Symbolic Indoor	6 m, 90% 2–3 m, 50%	3–5 WiFi APs	Local	2.4 GHz RF	Active tagging
Ubisense	Absolute Indoor/Outdoor	15 cm, 90%	Custom sensors and tags	Central	2.5 GHz and 6–8 Ghz wideband RF	Custom active tagging
Vision	Absolute Indoor/Outdoor	1 m, 50–80% (varies by camera density)	Multiple cameras	Central	RF for wireless cameras	Passive

Note that the accuracy is reported as a percentile.

evaluate the ease of deployment, cost and installation, and maintenance burden. For example, targeting location systems for the home presents several challenges. One major challenge is cost. In a commercial setting, more resources are typically available for disposal, and thus a company can justify the investment based on added productivity and reduction of other costs. On the other hand, the average homeowner would have difficulty justifying a high cost. Also, consider a researcher wanting to install location systems in various homes for a study. The cost of simultaneously deploying a system in multiple homes is much greater than a single, larger commercial building, such as an office building or a hospital, because parts of the infrastructure have to be replicated for each home being studied. Other important considerations are the spectral requirements of the location system. For example, certain parts of a hospital have very strict regulations on RF emission. Thus, in these environments, one may choose an IR-based solution. Another important consideration may be whether it is practical to have an individual carry a location tag. Finally, certain applications may require the protection of one's privacy, thus requiring a location system that computes its location locally as opposed to at a central server.

7.3.1 Global Positioning System

Currently, GPS is the most popular outdoor location tracking system worldwide. GPS first originated for military applications, but today, GPS-based solutions permeate throughout many civilian and consumer applications, such as in-car navigation systems, marine navigation, and fleet management services. Civilian GPS has a median accuracy of 10 meters outdoor, but areas with substantial occlusions, such as tall buildings and large mountains can reduce the accuracy of the system. GPS typically does not work well in most indoor settings, because of constant occlusions from the GPS satellites.

GPS consists of receivers that passively receive signals being transmitted from a subset of at least 24 geosynchronous satellites orbiting the Earth. Each GPS satellite transmits data that contains its location and the current time. Although the signals transmitted by the satellites are synchronized, they arrive at the receiver at different times due to the difference in distance between the satellites and the receiver. Thus, the distance to the GPS satellites can be determined by estimating the amount of time it takes for their signals to reach the receiver. At least four GPS satellites are needed to calculate the position of the receiver.

These GPS satellites transmit data over various radio frequencies, designated as L1, L2, etc. Civilian GPS uses the L1 frequency of 1575.42 MHz in the ultrahigh frequency band. This signal consists of three different pieces of information—a pseudorandom ID code, ephemeris data, and almanac data. The pseudorandom code is a simple ID code that identifies which satellite is transmitting information. The ephemeris data indicates to the GPS receiver where each GPS satellite should be located (orbital data) at a given time in the day. Finally, the almanac data contains information about the status of the satellite (healthy or unhealthy), current date, and time.

Unlike the GPS satellites, GPS receivers do not have atomic clocks and are not synchronized with the GPS satellites. Therefore, a GPS receiver calculates the time difference of arrival (TDOA) using the timing slack required to synchronize the GPS receiver's generation of a pseudorandom ID code with those being transmitted by the satellite to determine the signals' travel time. To determine its location, the receiver applies hyperbolic lateration in 3-D using the estimated TDOA values. In addition, a fourth satellite is required to correct any synchronization errors.

A GPS receiver also takes into account a variety of correction factors that may impact the signal delay. Here are some factors the can degrade the quality of the GPS signal originating from the satellites:

- Multipath—occurs when the GPS signal is reflected off tall buildings, thus increasing the time-of-flight of the signal.

- Too few satellites visible—occurs when there are major obstructions (e.g., GPS does not work well indoors or underground).

- Atmospheric delays—signals can slow as they pass through the atmosphere.

There are several ways to minimize some of these errors. One way is to predict and model the atmospheric delays and apply a constant correction factor to the received signal. The other strategy is to increase the number of channels in the receiver to allow for more satellite signals to be seen. A recent system, called differential GPS, uses a collection of terrestrial beacons to emit correction codes (using long wave radio between 285 and 325 kHz) in multipath-prone areas. The accuracy of differential GPS has been shown to be 1.8 meters at least 95% of the time (LaMarca and de Lara, 2008). Another approach called Real-Time Kinematic GPS uses phase measurements from existing GPS signals to provide receivers with real-time corrections.

FIGURE 7.6 Original Active Badges used at Olivetti (http://koo.corpus.cam.ac.uk/projects/badges/index.html).

7.3.2 Active Badge

The Active Badge (Want et al., 1992) was first introduced by the Olivetti Research Laboratory in Cambridge and was one of the first indoor location tracking systems developed. The Active Badge is designed to be worn by visitors and employees of an organization to allow a central database to keep track of their location within the building (see Figure 7.6). The badge transmits a unique code via a pulse-width modulated IR signal to networked sensors/receivers deployed throughout a building. The Active Badge uses 48-bit ID codes and is capable of two-way communication.

The badge periodically beacons the unique code (approximately every 10–15 seconds), and the information regarding which sensors detected this signal is stored in a central database. The IR-based solution is designed to operate up to 6 meters away from a sensor. The IR signal is strong enough to be reflected off walls and ceiling, so that sensors can detect these signals in a small room without line-of-sight operation. Since the IR signal does not travel through walls, the sensors are deployed throughout the space. The walls of the room can also be used as a natural boundary to contain IR signals, thus enabling a receiver to identify the badge within a room. The density and the strategic placement of the sensors also dictate the resolution of the location tracking. For example, multiple sensors may be deployed in a large conference room to detect if the individual is near the podium or sitting at the table. The Active Badge system uses a lookup table in the central server for determining the location of the badge, based on which sensors are detecting the badge. This physical location is associated to each sensor with an initial setup and installation phase.

Based on the Active Badge system, other IR-based location tracking solutions have also been developed, such as the SPECs project from HP Labs and the Versus system (http://www.versustech.com/). Although the disadvantage of these solutions is that they require line-of-sight operation, the use of IR allows for a low-cost and simple tag and receiver design.

7.3.3 Active Bat

Active Bat (Ward et al., 1997) is an ultrasonic-based location tracking systems consisting of ultrasound receivers dispersed in a space and location tags that emit ultrasonic pulses. Active Bat tags emit short pulses of ultrasound and are detected by receivers mounted at known points on the ceiling, which measure the time-of-flight of each pulse. Using the speed of sound, the distance from the tag to each receiver is calculated. Given three or more measurements to the receivers, the 3-D position of the tag can be determined using trilateration. One key concept of Active Bat is the use of an RF signal to cue the tag to transmit its ultrasonic pulse. The RF cue gives the receivers in the environment a starting point for timing the received ultrasonic pulse. Since the speed of light is significantly faster than the speed of sound, the RF signal delay is negligible and does not need to be considered for calculating the time-of-flight of the acoustical signal.

The information about the location of Active Bat tags is managed by a central server. This coordination is essential in garnering efficient use of the available ultrasound bandwidth among all the tags. Multiple tags must coordinate their pulses so as not to interfere with each other's time-of-flight calculations. In addition, multiple receivers are needed to ensure line-of-sight operation and to reduce multipath problems. Active Bat has a location accuracy of 90% at 3 centimeters. The system supports 75 tags being tracked in a 1000 square meters space consisting of 720 receivers. Although Active Bat offers precise indoor location tracking, it does require significant instrumentation to the space.

One of the drawbacks of the Active Bat architecture is its active approach of the tag beaconing, as opposed to using a passive approach, where the tag listens to pulses emanating from the environment (see Figure 7.7). The

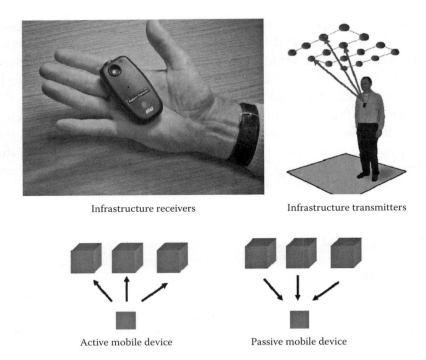

FIGURE 7.7 Top: Active Bat tags and example placement of receivers (http://research.microsoft.com/en-us/people/shodges/past.aspx). Bottom: Active (e.g., Active BAT) versus passive (e.g., Cricket) approach to location tracking systems.

passive architecture scales better than the active architecture as the density of location tags increase, because the RF and acoustical channels use is independent of the number of tags in the environment. In addition, the active mobile architecture requires a significant network infrastructure to connect the deployed receivers to a central server. This also leads to privacy concerns, because the central server knows the position of all tags in the systems. In contrast, the passive architecture allows a mobile device to estimate its location locally on each tag (see Cricket).

7.3.4 Cricket

The Cricket location system (Priyantha et al., 2000), unlike ActiveBat or ActiveBadge, does not rely on a centralized architecture to compute location information. Each Cricket tag is a small platform incorporating an RF transceiver, a microcontroller, and hardware receiving ultrasonic signals.

FIGURE 7.8 Cricket tags and example placement of the transmitters (http://nms.lcs.mit.edu/projects/cricket/).

Cricket beacons are affixed to known locations and are typically attached to the ceilings or walls of a building (see Figure 7.8). Each beacon periodically transmits a wireless (RF) message while at the same time sending a short ultrasonic pulse that allows the receiver tag to measure the distance from the beacon using the time-of-flight of the ultrasonic signal (similar to the technique used in Active Bat). The tag determines the time-of-flight of the acoustical signal and computes the position of the tags relative to the nearby beacons using trilateration. Each tag then uses these distance measures and the beacon position information contained in the RF messages to compute their location relative to the space. Since the tags do not actively transmit data, the scaling of the system is independent of the number of tags in the environment.

Unlike the ActiveBat system, Cricket is decentralized, so it preserves privacy by performing location calculations directly on the tag itself. In addition, the beacons deployed in the space do not have to be networked together, reducing some of the installation burden. Similar to the ActiveBat, line-of-sight operation is needed between each tag and at least three beacons, which requires sufficient installation of Cricket beacons to ensure full coverage in a space.

7.3.5 UbiSense

Ubisense (http://www.ubisense.net) is a commercial location tracking system using an ultrawideband (UWB) signal for localization (see Figure 7.9). Ubisense offers high precision at about 15 cm (at 90th percentile) by triangulating the location of active tags (called Ubitags) from a collection of networked sensors (called Ubisensors) distributed in a space. Each Ubitag incorporates a conventional RF radio (2.4 GHz) and a UWB radio (6–8 GHz). The conventional radio is used to coordinate and schedule when a

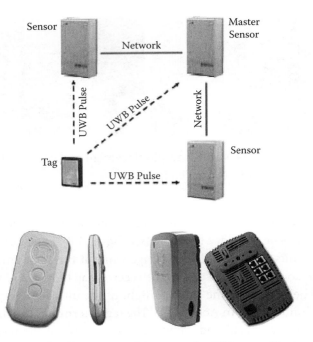

FIGURE 7.9 Top: Infrastructure architecture for Ubisense. Bottom: Ubitags are shown on the left and Ubisensor on the right (http://www.ubisense.net).

particular Ubitag should transmit its UWB pulse. The Ubisensors consist of a collection of phased array antennas.

After a tag is queried to transmit its UWB pulse, the Ubisense system uses TDOA and AOA to triangulate the location of the tag. Thus, at least two Ubisensors are needed to calculate the 3-D position of a Ubitag. The TDOA information is computed from sensors connected together with a physical timing cable. The advantage of using UWB pulses is that it is easier to filter multipath signals and can endure some occlusion. However, Ubisense does require line-of-sight operation for optimal performance. Since a timing cable is required to each Ubisensor, the installation process can be challenging in certain environments.

7.3.6 RADAR

The cost and effort of installation of the necessary infrastructure is a major drawback to wide-scale deployment of a location system. Thus, there have been efforts in developing location-based systems that reuse existing infrastructure to ease the burden of deployment and lower the cost. The RADAR system implements a location service using the information obtained from an already existing 802.11 WiFi network (Bahl and Padmanabhan, 2000). RADAR uses the RF signal strength [also known as the received signal strength indicator (RSSI)] as an indicator of the distance between an AP and a receiver. The major advantage of this approach is that a consumer does not have to purchase any specialized equipment and can still benefit from a location-aware application. For example, existing devices, such as WiFi-enabled mobile phones, PDAs, or laptops, can be repurposed as a receiver or tag.

The initial RADAR system used a trilateration approach on the RSSI values, but problems with multipath led researchers to use a mapping or fingerprinting approach for localization, where an offline signal map is constructed before the operation of the system. The signal map consists of locations in a building and the signal strength of RF waves emanating from nearby WiFi APs. For example, at a location (x,y) there are signal values associated with that position, one for each detectable WiFi base station. The creation of this signal map involves a user walking to several different locations with a location tag in the building and recording the physical coordinates of each location together with the signal strength from each AP.

To determine the position of the WiFi-enabled device, the receiver measures the signal strength of each of the APs and then searches through the signal map to determine the signal strength values that best matches the signal strengths seen in the past. An NN approach is used to find the closest signal values and then the system estimates the location associated with the best-matching signal strengths.

Experiments with this approach have shown that RADAR has a median position error of about 3 meters and 90 percentile resolution of 6 meters. At least three APs are needed to be in range for effective localization. Some of the drawbacks of this approach are that changes in the environment (moving furniture, appliances, etc.) may change the signal propagation patterns in the space. This would require another site survey to be conducted to update the signal map. Some solutions to help reduce this problem are to use more APs and environmental modeling.

Based on the results of RADAR, many commercial WiFi-based indoor positioning systems have also emerged. Ekahau is a positioning system that offers 3 to 5 meter resolution using six enterprise WiFi APs (http://www.ekahau.com). Similarly, Cisco also offers a fingerprinting-based location tracking services with their WiFi APs.

Another fingerprinting approach has looked at leveraging existing GSM cell towers to localize GSM mobile phones indoors (Varshavsky et al., 2007). Their approach fingerprints signal values of all available GSM channels to provide higher dimensionality to increase localization accuracy. Similar to WiFi localization solutions, a fingerprinting approach is used to determine the location of the device from a known signal map. This approach has a known median error of 5 meters and a 90 percentile resolution of 15–20 meters. Although this solution provides effective localization with little to no additional hardware to be deployed in an environment, one major drawback is that the user has very little control of the infrastructure itself. The performance of the system relies on publicly accessible infrastructures (e.g., GSM cellular towers), which can change over time. Service providers can adjust the operation parameters of the cellular towers with little warning, thus requiring an update to the signal map.

7.3.7 Place Lab

Place Lab (LaMarca et al., 2005) is a software-based indoor and outdoor localization system developed by Intel™ Research. Place Lab runs on commodity devices such as notebooks, PDAs, and mobile phones, and determines their position using radio beacons, such as 802.11 APs, GSM cell phone towers, and fixed Bluetooth devices that are already deployed in the environment (www.placelab.org). An advantage of Place Lab is that clients can determine their location privately without having to reveal information to a central service provider. In general, clients running the Place Lab software determine their location by detecting multiple unique IDs from these existing radio beacons and referring to a map of these devices.

PlaceLab's WiFi localization approach is very similar to RADAR and Ekahau, but there are two important distinctions. The aim of Place Lab is to provide location tracking at a large scale, whereas RADAR and Ekahau are primarily for smaller indoor environments. These approaches require calibration by the installer of the system. The aim of Place Lab is to use less dense calibration data that is contributed by a community of users so there is no need for an individual to populate a signal map.

Much of the Place Lab data is derived from the war driving community. War driving is the process of driving around with a mobile device equipped with a GPS receiver and an 802.11, GSM, and/or Bluetooth radio to collect traces of wireless base stations. Most of the war driving data is time-stamped recordings containing GPS coordinates and the associated signal strength of any beacons heard at the location. Wigle.net and

Worldwidewardrive.org are some examples of war driving repositories that contain millions of known APs. These data are used to estimate the position of the wireless beacons using a centroid approach, which estimates the position of the device to be a weighted average of positions of the overheard beacons. Although this approach only infers the location of the beacons, it has the added benefit that millions of beacon estimates have already been determined. Thus, this allows the ability to scale a location tracking system much more quickly despite the loss in accuracy. This approach has shown a median accuracy of 20–30 meters in large cities.

A similar effort was also started in Japan by the Sony Computer Science Laboratory called PlaceEngine™ (http://www.placeengine.com/en). Place Engine provides a mechanism for a community of users to update 802.11 beacon positions and the ability to track the location of any WiFi-enabled device. Place Lab also inspired commercial products such as Skyhook (http://www.skyhookwireless.com/) and Navizon (http://www.navizon.com/).

7.3.8 PowerLine Positioning

Inspired by this strategy of leveraging existing infrastructure and recognizing that there are drawbacks to relying on public infrastructure or the deployment of many beacons, PowerLine Position (PLP) was developed to provide indoor localization that would work in nearly every building (Patel et al., 2006b; Stuntebeck et al., 2008). With the significant insight being to use the power line as the signaling infrastructure, PowerLine Positioning is the first example of a whole-house or whole-building indoor localization system that repurposes the electrical system. PLP requires the installation of two small, plug-in modules for every 1000 square meters. In a home, only two modules would be necessary (see Figure 7.10). These modules inject a mid-frequency (300–1600 kiloHertz), attenuated signal throughout the electrical system of the home. Both modules continually emit their respective signals over the power line, and location tags equipped with specially tuned tags sense these signals in a building and relay them wirelessly to a receiver in the building. Depending on the location of the tag, the detected signal levels provide a distinctive signature, or fingerprint, resulting from the density of electrical wiring present at the given location and the distance from the plug-in module. PowerLine Positioning is capable of providing subroom-level positioning for multiple regions of a building. The current PLP system has a median error of 0.75 meters and a 90 percentile accuracy of 1 meter.

One drawback of PLP is that it requires a complete site survey to be conducted before the deployment of the system (similar to WiFi- and

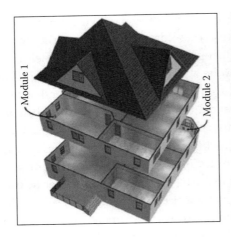

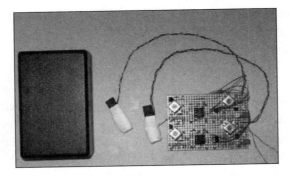

FIGURE 7.10 Top left: Placement of two signal-generating modules at extreme ends of a house. Top right: Signal generator plug-in modules. Bottom: Prototype PowerLine Positioning tag.

GSM-based fingerprinting solutions). However, the minimal infrastructure requirements and potential lower deployment costs is an important consideration for low-cost application, especially those that may be deployed in a home.

7.3.9 ActiveFloor

Many of the location tracking solutions discussed in this chapter require the attachment of a specialized tag to a device or a person carrying the tag. In addition, some of these tags require line-of-sight operation, which constrains where and how the tag is placed on a device. The tags also have to be associated with the attached device in a database. Because of these constraints, the tagging-based solution may not be appropriate for some applications. ActiveFloor is one example of a location tracking technology that is designed to locate a person in a space without an individual having to carry or wear a special tag (Addlesee et al., 1997). ActiveFloor consists of load sensors embedded within floor titles. The location trace of a person is determined through the weight distribution throughout the floor. Although ActiveFloor does not require a tag to be carried by a person, it does require significant instrumentation to a space.

ActiveFloor uses 50 cm square plywood floor tiles supported by load sensors distributed across an entire floor (see Figure 7.11). The load cells are sampled at 500 Hz, which is sufficient for most walking and running activities. The static weight of the floor and the systematic errors are calibrated out by capturing sensor values with no additional weight being applied to the surface. When an individual walks on the surface, the reaction that the load sensors produce in response to the weight and inertia of a body in contact with the floor is called the ground reaction force (GRF) (in this case, the person's foot). As a person walks on the floor, the GRF values from each load sensor can be averaged to find the location of the mass. In addition, as a person steps on the tile, the GRF is not always constant, because individuals push of with their toe and plant their heel (heel strike)

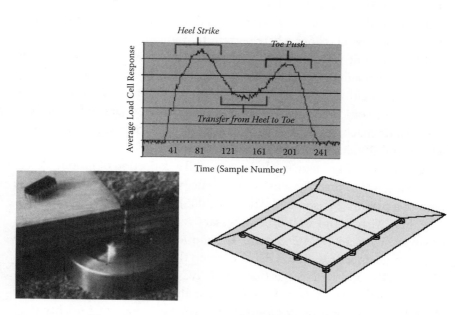

FIGURE 7.11 Top: GRF response of a single footstep. Bottom: Load sensor supporting a floor tile (http://www.cc.gatech.edu/fce/smartfloor/).

as they walk. This heel-to-toe transfer time and heel/toe GRF values can be used to calculate a footstep signature. ActiveFloor uses these features from the footstep signatures to build a hidden Markov model in order to identify the person. For further reading, a similar approach is also used by the Smart Floor project (http://www.cc.gatech.edu/fce/smartfloor/).

7.3.10 Airbus

Recent work has looked at providing passive tracking of individuals without having people carry a tag, similar in spirit to ActiveFloor. New strategies have tried to greatly reduce the amount of additional infrastructure needed for deployment. Airbus (Patel et al., 2008) is a location tracking system capable of detecting gross human movement and room transitions by sensing differential air pressure in a home (see Figure 7.12). The solution leverages central heating, ventilation, and air conditioning (HVAC) systems present in many homes. The home forms a closed air circulation circuit, where the HVAC system provides a centralized airflow source and a convenient single monitoring point for the entire circuit. Disruptions in home airflow caused by human movement through the house, especially those caused by the blockage of doorways and thresholds, result in static pressure changes in the HVAC air handler unit. The system detects and records this pressure variation using differential sensors mounted on the air filter and classifies where certain movement events are occurring, such as an adult walking through a particular doorway or the opening and closing of a door. Results have shown that the system can classify the opening and closing of specific doors with up to 80% accuracy with the HVAC in operation and 68% with the HVAC not in operation using support vector machines. These door events are used to compute a trace of where an individual is moving through the space.

An alternative strategy might be to install a collection of motion detectors in a space to directly sense the presence of a person to determine the path of a person (Wilson and Atkeson, 2005). This solution is more accurate than Airbus and would be necessary for environments that are not equipped with an HVAC system. The motion detector approach would also provide more resolution depending on the density of the installation. However, the other tradeoff is that the HVAC approach is much less obtrusive than installing motion detectors throughout a living space. Although the HVAC sensing provides location information at a lower fidelity than other tagging-based solutions, it does not require a person to carry a

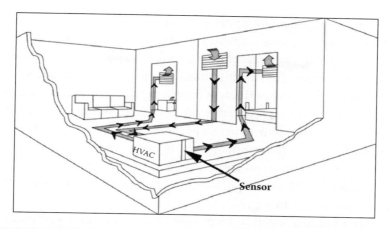
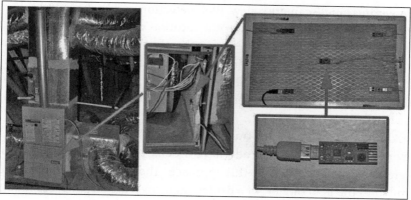

FIGURE 7.12 Top: Diagram of airflow from return and supply ducting in a home. Bottom: Instrumentation of a standard HVAC air filter with pressure sensors able to detect airflow in both directions. The air filter is then installed in the HVAC's air handler unit.

location tag and there is only a single location where the sensors reside and that has to be maintained. However, this technique does not work well for determining the identity of a person in a building. Airbus is more appropriate for applications that need to know people's presence, such as for smart heating and cooling or lighting control.

7.3.11 Tracking with Cameras

Another popular strategy to tracking the location of people is the use of cameras and computer vision techniques. The advantage of this solution is there is no requirement for a person to carry a specialized tag and it is possible to leverage existing cameras typically found in many environments (e.g., surveillance cameras, closed-circuit television, etc.), reducing the deployment burden. Existing solutions have looked at using stereo camera images for locating the position of people in a space, and the color images for inferring identities (Krumm et al., 2000). To determine the distance to any point in the environment in a stereo vision system, a stereo camera identifies where that point appears in both camera views. Traditional stereo vision algorithms (Forsyth and Ponce, 2002) rely on distinctive textures in the pair of images to determine which points from the left camera image corresponds to a particular point in the right camera image. This provides additional depth information to the color information already present in a single camera's view. Blob detection and background subtraction techniques are used to infer the location of moving objects (usually people) in the camera's view. Nonoverlapping cameras can also be used to provide location information on a planar surface (i.e., using overhead cameras) (Yang and Bobick, 2005). In addition to color histograms, more advanced face recognition approaches are used to determine the identity of a tracked person.

Although camera-based tracking provides an attractive solution for environments that may already have a deployed camera infrastructure, there are some drawbacks to this approach. The effective tracking range is limited to the field of view of a camera and covering large spaces requires coordination between adjacent views from multiple cameras. Similar to IR- and ultrasonic-based solutions, this approach does not work well in environments where there might be numerous occlusions. Tracking the position and the trajectory of a person in a fairly open space works well when using a camera, but tracking objects that might reside in cabinets and drawers would require a different technique. Finally, the stigma and privacy concerns associated with cameras may hinder the adoption of these techniques for certain applications.

7.4 CONCLUSIONS AND CHALLENGES

This chapter introduced the basic concepts of location technologies and surveyed some of the current and historical location systems. Furthermore, it described the differences between client-based and network-based positioning and identified some of the major sources of error in location systems. Finally, this chapter ends with a discussion of some of the challenges and opportunities facing developers of location systems.

There is no single location technology today that is ubiquitous, accurate, low-cost (in terms of required hardware and installation), and easy to deploy. Although a novel location technology that fits all these parameters might still become available in the future, a more realistic approach is to combine several of the existing technologies into an integrated location system. For instance, although the median error of GPS is 10 meters, the combined solution of GPS and European's global navigation satellite system called Galileo (as soon as it comes online) should yield a median accuracy of 1.5 meters.

Developers of location-aware applications are faced with the challenge of building and maintaining location-aware middleware and location-aware back-end services from limited existing solutions. These include collecting and reasoning about low-level sensor readings, storing location information on the back-end servers, and making the information available to third-party applications in a scalable and privacy-preserving manner. Although there are several existing middleware and back-end solutions available today (FireEagle; Hong and Landay, 2004), there is no well-accepted standard that is widely available for application developers.

Privacy remains as one of the main challenges for the proliferation of location services (Krumm, 2008). There is a need to hand over the control of location information disclosure to the user, without overwhelming him/her with privacy configurations, while still providing useful location services. For example, a recent study showed that users want plausible deniability in a location system (Iachello et al., 2005). Another study showed that people's preferences for disclosing location information differs based on many parameters, including the location of the user and the other person, the current user activity, and the relationship between the user and the other person (Consolvo et al., 2005). Refer to Chapter 4 for further discussion on user privacy in ubicomp.

REFERENCES

Addlesee, M., Jones, A., Livesey, F., and Samaria, F. ORL Active Floor. *IEEE Personal Communications* 4(5), 35–41, 1997.

Ascension Technology Corporation, www.ascension-tech.com.

Bahl, P., and Padmanabhan, V. RADAR: An in-building RF-based user location and tracking system. In *Proceedings of IEEE Infocom*, Los Alamitos, 2000, pp. 775–784.

Cheng, Y. C., Chawathe, Y., LaMarca, A., and Krumm, J. Accuracy characterization for metropolitan-scale Wi-Fi localization. In *Proceedings of MobiSys*, 2005, pp. 233–245.

Consolvo, S., Smith, I. E., Matthews, T., LaMarca, A., Tabert, J., and Powledge, P. Location disclosure to social relations: Why, when, & what people want to share. In *Proceedings of the Conference on Human Factors and Computing Systems: CHI '05*, Portland, OR, Apr. 2005, pp. 81–90.

Forsyth, D. A., and Ponce, J. *Computer Vision: A Modern Approach*. Prentice Hall, Upper Saddle, NJ, 2002.

Hong, J., and Landay, A. J. An architecture for privacy-sensitive ubiquitous computing. In *Proceedings of the 2nd International Conference on Mobile Systems, Applications, and Services (MobiSys 2004)*, Boston, MA, 2004, pp. 177–189.

Iachello, G., Smith, I., Consolvo, S., Abowd, G., Hughes, J., Howard, J., Potter, F., Scott, J., Sohn, T., Hightower, J., and LaMarca, A. Control, deception, and communication: Evaluating the deployment of a location-enhanced messaging service. In *Proceedings of Ubicomp* 2005, Tokyo, Japan. September 2005.

Krumm, J. A survey of computational location privacy. *Personal and Ubiquitous Computing*, 2008.

Krumm, J., Cermak, G., and Horvitz, E. RightSPOT: A novel sense of location for smart personal objects. In *Proceedings of the Fifth International Conference on Ubiquitous Computing*, Seattle, WA, 2003, pp. 36–43.

Krumm, J., Harris, S., Meyers, B., Brumitt, B., Hale, M., and Shafer, S. Multi-camera multi-person tracking for easyliving, IEEE Workshop on Visual Surveillance, July 2000.

LaMarca, A., Chawathe, Y., Consolvo, S., Hightower, J., Smith, I., Scott, I., Sohn, T., Howard, J., Hughes, J., Potter, F., Tabert, J., Powledge, R., Borriello, G., and Schilit, B. Place Lab: Device positioning using radio beacons in the wild. *In Proceedings of the International Conference on Pervasive Computing (Pervasive 2005)*, Munich, Germany, 8–13 May 2005, pp. 116–133.

LaMarca, A., and de Lara, E. *Location Systems: An Introduction to the Technology behind Location Awareness*. Morgan and Claypool Publishers, San Rafael, CA, 2008.

Patel, S. N., Kientz, J. A., Hayes, G. R., Bhat, S., and Abowd, G. D. Farther than you may think: An empirical investigation of the proximity of users to their mobile phones. In *Proceedings of Ubicomp 2006*, Orange County, CA, 2006a, pp. 123–140.

Patel, S. N., Reynolds, M. S., and Abowd, G. D. detecting human movement by differential air pressure sensing in HVAC system ductwork: An exploration in infrastructure mediated sensing. In *Proceedings of the International Conference on Pervasive Computing (Pervasive 2008)*, Springer LNCS 5013, 2008, pp. 1–18.

Patel, S. N., Truong, K. N., and Abowd, G. D. PowerLine Positioning: A practical sub-room-level indoor location system for domestic use. In *Proceedings of the International Conference on Ubiquitous Computing (Ubicomp 2006)*, Orange County, CA, 2006b, pp. 441–458.

Priyantha, N. B., Chakraborty, A., and Balakrishnan, H. The Cricket location-support system. In *Proceedings of the International Conference on Mobile Computing and Networking (Mobicom 2000)*, Boston, MA, 2000, pp. 32–43.

Rizos, C., Higgins, M. B., and Hewitson S. New global navigation satellite system developments and their impact on survey service providers and surveyors. In *Proceedings of SSC2005 Spatial Intelligence, Innovation and Praxis: The National Biennial Conference of the Spatial Sciences Institute*, September 2005.

Schilit, B., Adams, N., and Want, R. Context-aware computing applications. In *Proceedings of the Workshop on Mobile Computing Systems and Applications*, 1999.

Stuntebeck, E. P., Patel, S. N., Robertson, T., Reynolds, M. S., and Abowd, G. D. Wideband powerline positioning for indoor localization. In *Proceedings of the International Conference on Ubiquitous Computing (Ubicomp 2008)*, 2008, pp. 94–103.

Varshavsky, A., de Lara, E., LaMarca, A., Hightower, J., and Otsason, V. GSM indoor localization. *Pervasive and Mobile Computing Journal (PMC)* 3(6), 698–720, 2007.

Want, R., Hopper, A., Falcao, V., and Gibbons, J. The Active Badge location system. *ACM Transactions on Information Systems* 10, 91–102, 1992.

Ward, A., Jones, A., and Hopper, A. A new location technique for the active office. *IEEE Personal Communications* 4(5), 42–47, 1997.

Wilson, D. H., and Atkeson, C. G. Simultaneous tracking and activity recognition (STAR) using many anonymous, binary sensors. In *Proceedings of International Conference on Pervasive Computing (Pervasive 2005)*, 2005, pp. 62–79.

Yang, Z., and Bobick, A. F. Visual integration from multiple cameras. In *Proceedings of Application of Computer Vision, WACV/MOTIONS 2005*, 2005, pp. 488–493.

CHAPTER 8

Context-Aware Computing

Anind K. Dey

CONTENTS

8.1	Introduction to Context-Aware Computing	322
8.2	What Is Context?	325
8.3	What Is Context Awareness?	327
	8.3.1 Categorization of Features for Context-Aware Applications	329
8.4	Context-Aware Applications	332
	8.4.1 Historical Context-Aware Applications	332
	8.4.1.1 Tour Guides	333
	8.4.1.2 Reminders	334
	8.4.1.3 Environmental Controls	334
	8.4.2 Contemporary Context-Aware Applications	334
8.5	Designing and Implementing Context-Aware Applications	335
	8.5.1 Design Process	335
	8.5.2 Tools for Building	337
8.6	Issues to Consider when Building Context-Aware Applications	342
	8.6.1 Context Is a Proxy for Human Intent	342
	8.6.2 Context Inferencing	343
	8.6.3 Context Ambiguity	343
	8.6.4 Rules versus Machine Learning	344
	8.6.5 Privacy	344
	8.6.6 Evaluation	345
	8.6.7 End User Issues	345
8.7	Challenges in Writing Academic Papers on Context Awareness	346
8.8	Summary	348
References		348

8.1 INTRODUCTION TO CONTEXT-AWARE COMPUTING

Humans are quite successful conveying ideas to each other and reacting appropriately. This is due to many factors, including the richness of the language they share, the common understanding of how the world works, and an implicit understanding of everyday situations. When humans speak with humans, they are able to use information apparent from the current situation, or *context*, to increase the conversational bandwidth. Unfortunately, this ability to convey ideas does not transfer well when humans interact with computers. Computers do not understand our language, do not understand how the world works, and cannot sense information about the current situation, at least not as easily as most humans can. In traditional interactive or desktop computing, users have an impoverished mechanism for providing information to computers, typically using a keyboard and mouse. As a result, information must explicitly be provided to computers, producing an effect contrary to the promise of transparency in Weiser's vision of ubiquitous computing (Weiser, 1991). Users translate what they want to accomplish into specific minutiae on how to accomplish the task, and then use the keyboard and mouse to articulate these details to the computer so that it can execute their commands. This is nothing like our interaction with other humans. Consequently, computers are not currently enabled to take full advantage of the context of the human-computer dialogue. By improving the computer's access to context, users can increase the richness of communication in human-computer interaction and make it possible to produce more useful computational services.

Many research areas are attempting to address this input deficiency, but they can mainly be seen in terms of two basic approaches:

- Improving the language that humans can use to interact with computers

- Increasing the amount of situational information, or context, that is made available to computers

The first approach tries to improve interaction by allowing the human user to communicate in a much more natural manner. This type of communication is still very explicit, in that the computer only knows what the user tells it. With natural input techniques such as speech and gestures, no other information besides the explicit input is available to the computer. As we know from human–human interactions, situational information such as facial expressions, emotions, past and future events, the existence of other people in the room, and relationships to these other people are crucial to understanding what is occurring. The process of building this shared understanding between two people is called *grounding* (Clark and Brennan, 1991). Since both human participants in such an interaction share this situational information, there is no need to make it explicit. This helps to explain why a driver finds it easier to talk to a passenger than to someone on a cell phone—with the passenger, there is grounding without explicit communication. However, this need for explicitness does exist in human–computer interactions, because the computer does not share this implicit situational information or context. The goal of context-aware computing is to use context as an implicit cue to enrich the impoverished interaction from humans to computers, making it easier to interact with computers.

Researchers in context awareness want to make it easier, not harder, for users to interact with computers and the environment, by allowing users to not have to think consciously about using the computers. Weiser coined the term *calm technology* to describe an approach to ubiquitous computing, where computing moves back and forth between the center and periphery of the user's attention (Weiser and Brown, 1997). To this end, the approach to context-aware application development is to collect *implicit* contextual information through automated means, make it easily available to a computer's runtime environment, and let the application designer decide what information is relevant and how to deal with it. This is the better approach, because it removes the need for users to make all information explicit and it puts the decisions about what is relevant into the designer's hands. Furthermore, it is likely that most users will not know which information is potentially relevant and, therefore, will not know what information to provide. The application designer should have spent considerably more time analyzing the situations under which their application will be executed and can more appropriately determine what information could be relevant and how to react to it.

The need for context is even greater when we move into ubicomp environments. Mobile computing and ubiquitous computing have given users the expectation that they can access whatever information and services they want, whenever they want, and wherever they are. With computers being used in such a wide variety of situations, interesting new problems arise, and the need for context is clear: users are trying to obtain different information from the same services or systems in different situations. Context can be used to help determine what information or services to make available or to bring to the forefront for users.

Applications that use context, whether on a desktop or in a mobile or ubiquitous computing environment, are called *context-aware*. The increased availability of commercial, off-the-shelf sensing technologies is making it more viable to sense context in a variety of environments. The prevalence of powerful, networked computers makes it possible to use these technologies and distribute the context to multiple applications, in a somewhat ubiquitous fashion. Mobile computing allows users to move throughout an environment while carrying their computing power with them. Combining this with wireless communications allows users to have access to information and services not directly available on their portable computing device. The increase in mobility creates situations where the user's context, such as his or her location and the people and objects around him/her, is more dynamic. With ubiquitous computing, users move throughout an environment and interact with computer-enhanced objects within that environment. This also allows them to have access to remote information and services. With a wide range of possible user situations, we need to have a means for the services to adapt appropriately, in order to best support the human–computer and human–environment interactions. Context-aware applications are becoming more prevalent and can be found in the areas of wearable computing, mobile computing, robotics, adaptive and intelligent user interfaces, augmented reality, adaptive computing, intelligent environments, and context-sensitive interfaces. It is not surprising that in most of these areas, the user is mobile and his or her context is changing rapidly.

For example, the most common context-aware application is a mobile tour guide. A mobile tour guide is a handheld device that one receives when visiting a museum or historical site. It can present information to a user about exhibits the user is interested in, using audio, video/images, or text. Today, most mobile tour guides require users to explicitly enter the name or ID of the exhibit they want more information about. However, context-aware mobile tour guides remove this need for explicit input. They automatically sense which exhibit the user is closest to (e.g., using radio frequency ID (RFID) tags on each exhibit and an RFID reader in the tour guide) and automatically present the appropriate exhibit information.

In the remainder of this chapter, definitions of context and context awareness are provided, along with descriptions of different types of context awareness. Following these descriptions, a discussion of context-aware applications, including historical and more contemporary applications, along with design processes and tools for building context-aware applications, is presented. Then, a number of issues in building context-aware applications, including sensor fusion, privacy, and evaluation, are outlined. Finally, the chapter ends with a discussion of the pitfalls in writing academic papers about context-aware computing.

8.2 WHAT IS CONTEXT?

Realizing the need for context is only the first step toward using it effectively. Most researchers have a general idea about what context is and use that general idea to guide their use of it. However, a vague notion of context is not sufficient; in order to use context effectively, we must attain a better understanding of what context is. A better understanding of context will enable application designers to choose what context to use in their applications and provide insights into the types of data that need to be supported and the abstractions and mechanisms required to support context-aware computing.

In the work that first introduces the term *context-aware*, Schilit and Theimer (1994) refer to context as location, identities of nearby people and objects, and changes to those objects. In a similar definition, Brown et al. (1997) define context as location, identities of the people around the user, the time of day, season, temperature, etc. Ryan et al. (1998) define context as the user's location, environment, identity, and time. In a previous work, Dey (1998) enumerated context as the user's emotional state, focus of attention, location and orientation, date and time, and objects and people in the user's environment. These definitions define context by example and are difficult to apply. When you want to determine whether a type of information not listed in the definition is context, it is not clear how to use these enumerations to do so.

Merriam-Webster defines context as "the interrelated conditions in which something exists or occurs." Other definitions have simply provided synonyms for context, referring, for example, to context as the environment or situation. Some consider context to be the user's environment, whereas others consider it to be the application's environment. Brown (1996a) defined context to be the elements of the user's environment that the user's computer knows about. Franklin and Flaschbart (1998) see it as the situation of the user. Ward et al. (1997) view context as the state of the application's surroundings, and Rodden et al. (1998) define it to be the application's setting. Hull et al. (1997) included the entire environment by defining context as aspects of the current situation. As with the definitions by example, definitions that simply use synonyms for context are extremely difficult to apply in practice.

Schilit et al. (1994) claim that the important aspects of context are as follows: where you are, whom you are with, and what resources are nearby. They define context to be the constantly changing execution environment. They include the following elements of the environment:

- *Computing environment*—available processors, devices accessible for user input and display, network capacity, connectivity, and costs of computing
- *User environment*—location, collection of nearby people, and social situation
- *Physical environment*—lighting and noise level

Dey et al. (1998) define context to be the user's physical, social, emotional, or informational state. Finally, Pascoe (1998) defines context to be the subset of physical and conceptual states of interest to a particular entity. These definitions, although closer to an ideal definition, are too specific. Context is all about the whole situation relevant to an application and its set of users. It is not possible to enumerate which aspects of all situations are important, because this will change from situation to situation. For example, in some cases, the physical environment may be important, whereas in others it may be completely immaterial.

Finally, Dey and Abowd (2000a) define context as

> any information that can be used to characterize the situation of an entity. An entity is a person, place, or object that is considered relevant to the interaction between a user and an application, including the user and application themselves.

Context-aware applications look at the *who*'s, *where*'s, *when*'s, and *what*'s (i.e., what activities are occurring) of entities and use this information to determine *why* a situation is occurring. An application does not actually determine why a situation is occurring, but the designer of the application does. The designer uses incoming context to determine the *user's intent*, or why a situation is occurring, and uses this to encode some action in the application that helps to satisfy this intent. For example, in a context-aware tour guide, a user carrying a handheld computer approaches some interesting site resulting in information relevant to the site being displayed on the computer (Abowd et al., 1997). In this situation, the designer has encoded the understanding that when a user approaches a particular site (the "incoming context"), it means that the user is interested in the site (the "why") and the application should display some relevant information (the "action").

This definition of context includes not only implicit input but also explicit input. For example, the identity of a user can be implicitly sensed through face recognition or can be explicitly determined when a user is asked to type in her name using a keyboard. From the application's perspective, both are information about the user's identity and allow it to perform some added functionality. Context awareness uses a generalized model of input, including implicit and explicit input, allowing *any* application to be considered more or less context-aware insofar as it reacts to input. However, most work in context awareness is concerned with the gathering and use of *implicit* input by applications.

There are certain types of context that are, in practice, more important than others. These are *location* (where), *identity* (who), *time* (when), and *activity* (what). Location, identity, time, and activity are important context types for characterizing the situation of a particular entity. These context types not only answer the questions of who, what, when, and where, but also act as indices into other sources of contextual information. For example, given a person's identity, we can acquire many pieces of related information such as phone numbers, addresses, email addresses, a birth date, list of friends, relationships to other people in the environment, etc. With an entity's location, we can determine what other objects or people are near the entity and what activity is occurring near the entity. This attempt at a categorization of context is clearly incomplete. For example, it does not include hierarchical or containment information. An example of this for location is a point in a room. That point can be defined in terms of coordinates within the room, by the room itself, the floor of the building the room is in, the building, the city, etc. (Schilit and Theimer, 1994). It is not clear how this categorization helps to support this notion of hierarchical knowledge. How to represent and model context is still an open question that researchers are addressing.

8.3 WHAT IS CONTEXT AWARENESS?

Context-aware computing was first discussed in 1994 by Schilit and Theimer as software that "adapts according to its location of use, the collection of nearby people and objects, as well as changes to those objects over time." However, it is commonly agreed that the first research investigation of context-aware computing was the Olivetti Active Badge (Want et al., 1992) work in 1992. Since then, there have been numerous attempts to describe context-aware computing.

The first definition of context-aware applications given by Schilit and Theimer (1994) expanded the idea of context awareness from applications that are simply informed about context to applications that *adapt* themselves to context. Context-aware has become somewhat synonymous with other terms: adaptive (Brown, 1996a), reactive (Cooperstock et al., 1995), responsive (Elrod et al., 1993), situated (Hull et al., 1997), context sensitive (Rekimoto et al., 1998), and environment directed (Fickas et al., 1997). Previous definitions of context-aware computing fall into two categories: using context and adapting to context.

Let us first discuss the more general case of using context. Hull et al. (1997) and Pascoe et al. (Pascoe, 1998; Pascoe et al., 1998; Ryan et al., 1998) define context-aware computing as the ability of computing devices to detect and sense, interpret and respond to aspects of a user's local environment and the computing devices themselves. Dey et al. (Dey, 1998; Dey et al., 1998; Salber et al., 1999) have defined context awareness as the use of context to automate a software system, to modify an interface, and to provide maximum flexibility of a computational service.

The following definitions are in the more specific "adapting to context" category. Many researchers (Schilit et al., 1994; Brown et al., 1997; Dey and Abowd, 1997; Ward et al., 1997; Abowd et al., 1998; Davies et al., 1998; Kortuem et al., 1998) define context-aware applications to be applications that dynamically change or adapt their behavior based on the context of the application and the user. More specifically, Ryan (1997) defines them as applications that monitor input from environmental sensors and allow users to select from a range of physical and logical contexts according to their current interests or activities. This definition is slightly more restrictive than the previous one by identifying the method in which applications acts upon context. Brown (1998) defines context-aware applications as applications that automatically provide information and/or take actions according to the user's present context as detected by sensors. He also takes a narrow view of context-aware computing by stating that these actions can take the form of presenting information to the user, executing a program according to context, or configuring a graphical layout according to context. Fickas et al. (1997) define environment-directed (practical synonym for context-aware) applications as applications that monitor changes in the environment and adapt their operation according to predefined or user-defined guidelines. Dey and Abowd (2000a) define context awareness more generally with the following statement:

> A system is context-aware if it uses context to provide relevant information and/or services to the user, where relevancy depends on the user's task.

This definition is more general and inclusive to include all context-aware applications, including both those that adapt to context and those that display the user's context.

8.3.1 Categorization of Features for Context-Aware Applications

In a further attempt to help define the field of context-aware computing, researchers have categorized features of context-aware applications. There have been three attempts to develop such a taxonomy. The first was provided by Schilit et al. (1994) and had two orthogonal dimensions: whether the task is to obtain information or to execute a command, and whether the task is executed manually or automatically. Applications that retrieve information for the user manually based on available context are classified as *proximate selection* applications. Proximate selection is an interaction technique where a list of objects (printers) or places (offices) is presented and where items relevant to the user's context are emphasized or made easier to choose. Applications that retrieve information for the user automatically based on available context are classified as *automatic contextual reconfiguration*. It is a system-level technique that creates an automatic binding to an available resource based on current context. Applications that execute commands for the user manually based on available context are classified as *contextual command* applications. They are executable services made available due to the user's context or whose execution is modified based on the user's context. Finally, applications that execute commands for the user automatically based on available context use *context-triggered actions*. They are services that are executed automatically when the right combination of context exists, and are based on simple if-then rules.

More recently, Pascoe (1998) proposed a taxonomy of context-aware features. There is considerable overlap between the two taxonomies but some crucial differences as well. Pascoe's taxonomy was aimed at identifying the core features of context awareness, as opposed to the previous taxonomy, which identified classes of context-aware applications. In reality, the following features of context awareness map well to the classes of applications in the Schilit taxonomy. The first feature is *contextual sensing* and is the ability to detect contextual information and present it to the user, augmenting the user's sensory system. This is similar to *proximate selection*, except that in this case, the user does not necessarily need to select one of the context items for more information (i.e., the context may be the information required). The next feature is *contextual adaptation* and is the ability to execute or modify a service automatically based on the current context. This maps directly to Schilit's *context-triggered actions*. The third feature, *contextual resource discovery*, allows context-aware applications to locate and exploit resources and services that are relevant to the user's context. This maps directly to *automatic contextual reconfiguration*. The final feature, *contextual augmentation*, is the ability to associate digital data with the user's context. A user can view the data when he is in that associated context. For example, a user can create a virtual note providing details about a broken television and attach the note to the television. When another user is close to the television or attempts to use it, he will see the virtual note left previously. This feature does not exist in Schilit's taxonomy.

Pascoe and Schilit both list the ability to exploit resources relevant to the user's context, the ability to execute a command automatically based on the user's context, and the ability to display relevant information to the user. Pascoe goes further in terms of displaying relevant information to the user by including the display of context, and not just information requiring further selection (e.g., showing the user's location vs. showing a list of printers and allowing the user to choose one). Pascoe's taxonomy has a category not found in Schilit's taxonomy: *contextual augmentation*, or the ability to associate digital data with the user's context. Finally, Pascoe's taxonomy does not support the presentation of commands relevant to a user's context. This presentation is called *contextual commands* in Schilit's taxonomy.

Dey and Abowd (2000a) combine the ideas from these two taxonomies and take into account the three major differences. Similar to Pascoe's taxonomy, it is a list of the context-aware features that context-aware applications may support. There are three categories:

1. *Presentation* of information and services to a user
2. Automatic *execution* of a service
3. *Tagging* of context to information for later retrieval

Presentation is a combination of Schilit's proximate selection and contextual commands. To this, Pascoe's notion of presenting context (as a form of information) to the user has been added. Examples of the first feature is a mobile computer that dynamically updates a list of closest printers as its user moves through a building, a peripheral display of activity in a remote location, and awareness of the location or activity of a friend or family member. Automatic execution is the same as Schilit's context-triggered actions and Pascoe's contextual adaptation. Examples of the second feature is when the user prints a document and it is printed on the closest printer to the user, alarms being triggered to indicate that activity in a remote location has increased beyond a threshold, and mobile texting a friend automatically when she is within a certain distance of the friend. Tagging is the same as Pascoe's contextual augmentation. An example of the third feature is when an application records the names of the documents that the user printed, the times when they were printed and the printer used in each case. The user can than retrieve this information later to help him determine where the printouts are what he forgot to pick up.

This last taxonomy has two important distinguishing characteristics: the decision not to differentiate between information and services, and the removal of the exploitation of local resources as a feature. In most cases, it is too difficult to distinguish between a presentation of information and a presentation of services. For example, Schilit writes that a list of printers ordered by proximity to the user is an example of providing information to the user. But whether that list is a list of information or a list of services depends on how the user actually uses that information. For example, if the user just looks at the list of printers to become familiar with the names of the printers nearby, she is using the list as information. However, if the user chooses a printer from that list to print to, she is using the list as a set of services. Rather than try to assume the user's state of mind, we chose to treat information and services in a similar fashion.

This taxonomy also does not use the exploitation of local resources or resource discovery, as a context-aware feature, but instead includes this as part of the first two categories. Resource discovery is the ability to locate new services according to the user's context. This ability can be seen as no different than choosing services based on context. For example, when a user enters an office, his or her location changes and the list of nearby printers changes. The list changes by having printers added, removed, or being reordered (e.g., by proximity). Is this an instance of resource exploitation or simply a presentation of information and services? Rather than giving resource discovery its own category, it was split into two of the existing categories: presenting information and services to a user and automatically executing a service. When an application presents information to a user, it falls into the first category, and when it automatically executes a service for the user, it falls into the second category.

Definitions of context aware have provided researchers with a way to conclude whether an application is context aware or not. This has been useful in determining what types of applications to focus on. Categorizations of context-aware features provide two main benefits. The first is that it further specifies the types of applications that researchers provide support for. The second benefit is that it describes the types of features that developers should be thinking about when building context-aware applications.

8.4 CONTEXT-AWARE APPLICATIONS

With a basic understanding of context and context awareness, this section discusses historical context-aware applications that originated the field of context awareness, and more contemporary applications.

8.4.1 Historical Context-Aware Applications

As noted earlier, the Active Badge system (Want et al., 1992) is commonly viewed as the first context-aware system. In this application (Figure 8.1), users wore Active Badges, infrared transmitters that transmitted a unique identity code. As users moved throughout their building, a database was being dynamically updated with information about each user's current location, the nearest phone extension, and the likelihood of finding someone at that location (based on age of the available data). When a phone call was received for a particular user, the receptionist used the database to forward the call to the last known location of that user, rather than blindly forwarding the call to the user's office, where he may not be located. This application, along with much of the early work in context-aware computing was focused on location-aware computing, or, as they are more commonly known today, *location-based services*.

The other seminal work in context-aware computing was performed by the Ubicomp group at (then) Xerox Palo Alto Research Center (PARC), in the early 1990s. Schilit et al. coined the term *context awareness*, and built a system architecture that supported the building of context-aware

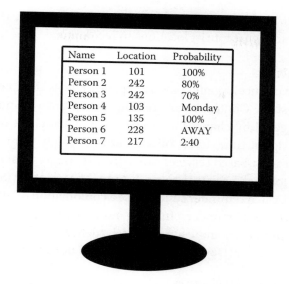

FIGURE 8.1 Rendition of the original Active Badge application showing the location and certainty of Active Badge wearers.

applications (1994, 1995), as a part of the highly influential PARCTAB work (Want et al., 1995). In the PARCTAB work, the provided systems present information to users based on proximity to services (e.g., printers and people), turn devices on or reconfigure them based on the people who are nearby, present information or services based on where the user is located, and automatically execute a service in a particular manner depending on movement or proximity of users to specific rooms or devices.

Since then, although there has been work that has applied context awareness in a huge number of domains, there are three canonical classes of applications: tour guides, reminder systems, and environmental control.

8.4.1.1 Tour Guides
As described earlier, the mobile tour guide is the most canonical context-aware application. When interacting with a mobile tour guide system, users carry a portable computing device as they travel through some area such as a museum or a city. As users go to different exhibits or tourist locations, their mobile devices present information relevant to those locations. Although early systems focused heavily on location (Bederson, 1996; Abowd et al., 1997), later systems took into account users' interests and the amount of time they spent at a tourist location or the amount of time they

had to tour in choosing what information to show (Brown, 1996b; Davies et al., 1998), and what tourist locations to recommend.

8.4.1.2 Reminders

A second canonical context-aware application is the context-aware reminder system. Context-aware reminders present reminders to individuals, triggered by changes in context. An alarm clock uses a simple contextual trigger, time, to set off an alarm, a simple form of reminder. Similarly, location-based services can deliver reminders when users are at a particular location or within some distance of each other (Schilit, 1995). More sophisticated reminder systems use a combination of different forms of context to trigger reminders. In being more sophisticated, these applications can remind users more appropriately, delivering the right reminder in the right situation (Dey and Abowd, 2000b; Ludford et al., 2006).

8.4.1.3 Environmental Controls

A third canonical context-aware application is a system to control an environment's heating and lighting, generally for the purposes of being energy efficient or saving users effort. As many people often leave lights on unnecessarily, or have to manually change heating or cooling levels to remain comfortable, many systems have been developed that can control these on behalf of users. Some are based on simple rules (Elrod et al., 1993), whereas others use more sophisticated mechanisms to learn how users use a space and sets heating and lighting accordingly (Mozer, 1998).

8.4.2 Contemporary Context-Aware Applications

Although these context-aware applications may seem quite simple today, researchers still continue to build these applications. In fact, many of these are starting to move out of the world of research and into commercial use.

Beyond these three canonical domains, researchers and developers have built applications in a wide variety of domains, including

- Interpersonal communications including instant messaging (Avrahami et al., 2008)
- Interruptibility in the office and while mobile (Fogarty et al., 2005; Iqbal et al., 2005)
- Phone calls (Schmidt et al., 2000)
- Health care (Bardram and Nørskov, 2008)

- Location-aware systems
- Agriculture (Kjær, 2008)
- Application personalization (Weiss et al., 2008)
- To peripheral displays of information (Wisneski et al., 1998)

This is by no means an exhaustive list. Rather than go into the details of each of these applications, or to provide a definitive list, both of which will be outdated soon, it is more important to point out that the field of context-aware computing is headed in the right direction. It is moving away from toy problems of little consequence to real users toward critical and fundamental issues that address aspects of everyday life and critical situations. The field has crossed a threshold in maturity, and researchers and developers are focusing their attention not on showing what is possible, but on what is compelling, and we should all be excited by this.

Another important point to note is that a greater number of researchers are building context-aware applications, but their focus is not context awareness. They are simply building a useful, compelling application that just happens to be context-aware. Although this may seem like a small distinction, context awareness will certainly have reached an appropriate level of maturity when it is commonly viewed as an application feature, rather than as the focus of an application.

8.5 DESIGNING AND IMPLEMENTING CONTEXT-AWARE APPLICATIONS

Given the focus on building compelling context-aware applications, it is important to understand how designers and developers of context-aware applications should think about building them, and what types of tools exist for building applications.

8.5.1 Design Process

The design process for building the vast majority of context-aware applications can be boiled down to a pretty simple idea: figure out what context your application needs and when you receive that context, figure out what you want to do with it. Unfortunately, it takes a little more work than this to build a context-aware application (see Table 8.1 for an example). The first step in building an application is context *specification*—determining

TABLE 8.1 Steps for Building an Application for Sharing Awareness of Remote Room Activity

	Step	Example
1	Specification	Light up different colored light emitting devices (LEDs) (red, yellow, green) depending on the amount of activity in remote location
2	Acquisition	Write software to analyze frame-to-frame changes in video camera image of remote location, and install camera
3	Delivery	Make the percentage change in activity available to interested applications, using publish-subscribe approach
4	Reception	Application wants to receive all changes in activity above a particular threshold
5	Action	Application analyzes received activity changes, and lights up the appropriate LEDs (red, low activity; yellow, moderate activity; green, high activity)

what context-aware behaviors your application will have and in which situations (or collections of context) each behavior should be executed and how. This step is often performed by studying a combination of the domain and the expected users of the application, and determining what services would be beneficial. The second step is context *acquisition*—determining what hardware and/or software sensors are required to acquire the context identified in the first step. This includes installing the software on the appropriate platform, understanding exactly what type of information the sensor provides, using an application programming interface (API) (if any) to communicate with the sensor, determining how to query the sensor, and how to be notified when changes occur, and, if applicable, store the context, combine it with other context, and perform inference on the sensor data to achieve some higher-level understanding of the situation. The third step is context *delivery*—specifying how context should be delivered from the sensors to the (possibly remote) applications that will use the context. The fourth step is context *reception*—having the application specify what context it is interested in (possibly indicating from which sensors) and receiving that context. This includes converting the context into a form usable by the application through interpretation, and then analyzing the context to determine whether this context, when combined with other available context, describes a relevant user situation for the application. Finally, the last step is *action*—analyzing all the received context to determine what context-aware behavior to execute, and then execute it. Although applications can be more complex than this (see the section on issues in developing applications), most applications can be described and built using this simple design process (or multiple instances of it for applications with multiple context-aware behaviors).

8.5.2 Tools for Building

Based on this design process, a number of toolkits or software architectures have been built to support the building of context-aware applications. Although every application developer needs to perform the specification step for his/her application, the rest of the design process may or may not be relevant depending on what the underlying toolkit offers the developer. In the best-case scenario, the specification of the relevant situations and the corresponding context-aware behaviors would be enough to cause context to flow from the appropriate sensors to applications, with relevant inferencing and interpretation occurring, and then the correct context-aware behavior being executed.

In general, there are three main alternatives used for building applications: no support, a widget- or object-based system (Figure 8.2), and a blackboard-based system (Figure 8.3). The vast majority of systems, particularly those built before 1998, were built with *no architectural support* or with highly customized support for a small class of applications. For each of these applications, the entire design process must be applied and software implemented to support the process on an individual basis. Very little, if any, code can be reused between application instances, as most of the code is customized for each application.

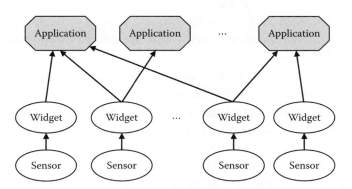

FIGURE 8.2 Widget-based system for building context-aware applications.

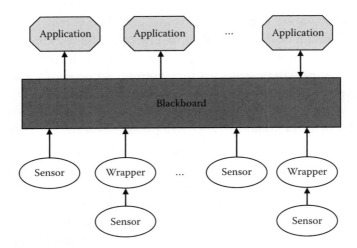

FIGURE 8.3 Blackboard-based system for building context-aware applications.

Before 1998, the exceptions were Schilit's (1995) Context-Aware Computing Architecture upon which most, if not all, the context-aware applications at Xerox PARC were based on, and the University of Kent's Stick-e Notes platform (Brown, 1996b). The Context-Aware Computing Architecture moved away from having to rewrite location sensing modules and context delivery from scratch each time, and instead allowed application developers to focus on the application details of what should be done when a user or users were in a particular location. The Stick-e Notes platform similarly provided a common infrastructure for building context-aware applications, but focused on making the authoring of applications accessible to end users (by making the process similar to that of constructing web pages). However, neither system was used much beyond the groups that developed these systems, so reuse was naturally limited. These systems showed the promise of context-aware systems and pointed the way toward reusable infrastructure.

Since 1998, a large number of context-aware infrastructures have been built, using either the widget-based approach or the blackboard-based approach. These infrastructures were developed primarily to make context-aware system components reusable, to provide a persistent executable infrastructure that could support multiple applications running at the same time and to simplify the building of context-aware applications.

The general *widget-based approach* is based on the model of building graphical user interfaces (GUIs). In the early 1980s, there was no common model for handling user input, and no support for reuse of widgets. By the mid 1980s, we began to see GUI toolkits that supported reuse, including infrastructure for handling input events and a component or widget library with widgets that could be used across multiple applications. These toolkits made it fundamentally easier to build GUIs and provided the following benefits:

- They hide specifics of physical interaction devices from the application's programmer so that those devices can change with minimal impact on applications. Whether the user points and clicks with a mouse or fingers and taps on a touchpad or uses keyboard shortcuts does not require any changes to the application.
- They manage the details of the interaction to provide applications with relevant results of user actions. Widget-specific dialogue is handled by the widget itself, and the application often only needs to implement a single callback to be notified of the result of an interaction sequence.
- They provide reusable building blocks of presentation to be defined once and reused, combined, and/or tailored for use in many applications. Widgets provide encapsulation of appearance and behavior. The programmer does not need to know the inner workings of a widget to use it.

In a similar manner, in the late 1990s and early 2000s, context-aware toolkits, such as the Context Toolkit (Dey et al., 2001) and Java Context-Aware Framework (Bardram, 2005) adopted a similar model and took a context widget approach. A context widget is a software component that provides applications with access to context information from their operating environment. In the same manner GUI widgets insulate applications from some presentation concerns, context widgets insulate applications from context acquisition concerns by wrapping sensors with a uniform interface. Context widgets provide the following benefits:

- They provide a separation of concerns by hiding the complexity of the actual sensors used from the application. Whether the presence of people is sensed using Active Badges, floor sensors, video image processing, or a combination of these should not impact the design of the application.

- They abstract context information to suit the expected needs of applications. A widget that tracks the location of a user within a building or a city notifies the application only when the user moves from one room to another, or from one street corner to another, and does not report less significant moves to the application. Widgets provide abstracted information that we expect applications to need most frequently.

- They provide easy access to context data through querying and notification mechanisms (i.e., publish-subscribe) available through a common, uniform interface for accessing context. No matter what type of context is being sensed from the environment, all widgets make it available in the same manner.

- They provide reusable and customizable building blocks of context sensing. A widget that tracks the location of a user can be used by a variety of applications, from tour guides to office awareness systems. Furthermore, context widgets can be tailored and combined in ways similar to GUI widgets. For example, a Presence widget senses the presence of people in a room. A Meeting widget may be built on top of a Presence widget and assume a meeting is beginning when two or more people are present.

In addition to context widgets, these infrastructures provided components for combining context information, making higher-level inferences from low-level context, collecting or aggregating context from multiple sources, and discovering appropriate components to use. As a collection, they not only provide components that can be composed to create individual applications, but serve as a persistent distributed infrastructure that can serve multiple applications at the same time. This approach is becoming much more mainstream, with Microsoft Windows 7's inclusion of a Sensor and Location Platform that provides an API to access location and light sensors, among others, to enable the building of context-aware applications.

The main contrasting approach to the widget approach is the blackboard-based approach. Blackboards originated from the Linda programming language and tuple space model from the early 1980s (Gerlernter, 1985). Linda supported a small number of functions for interacting with a globally accessible persistent storage system, the tuple space. Executable components can place information into the storage system, and read or remove this information. Later blackboard systems added the ability to notify components when information of interest has been added to the blackboard.

A number of infrastructures that support the building of context-aware applications use this blackboard approach including EQUIP (Greenhalgh, 2002), the Event Heap (Johanson and Fox, 2002), and the Open Agent Architecture (Cheyer and Martin, 2001). In contrast to the widget-based approach in which requests for information are often made directly to a particular component or indirectly through a discovery mechanism, the blackboard approach allows requests for information to come to and be handled by a single, centralized tuple space. In general, the blackboard-based context systems tend to be straightforward implementations of traditional blackboard systems, optimized for performance and augmented to support XML encoding and decoding of information.

Both approaches have their advantages and disadvantages. Blackboard approaches are often too inefficient particularly as the amount of data in the tuple space grows and searching for specific tuples becomes harder, can often be too tightly integrated with the applications they are supporting to make it easy to modify those applications, can often suffer from synchronization issues as the order of tuple space operations is not necessarily guaranteed, and generally do not support complex data structures. The widget-based model can often be too rigid, from a robustness and configurability standpoint, to handle the dynamic nature of context-aware applications that need to connect and disconnect from different components as a user moves around and his/her context changes, and can provide a more complex model to deal with from an application developer's point of view.

However, these two approaches are not mutually exclusive either. From an infrastructure developer's perspective, both approaches require sources of context input, either to create widgets around or to place context into a tuple space. The context widget approach provides a nice model for utilizing context sources, regardless of which approach is actually being used to support applications. On the other hand, the blackboard model provides a simpler abstraction for these context sources to actually deliver their context to context consumers. This model can be used in the widget approach, where widgets can be told to connect to a particular centralized component (e.g., a discovery system), which then determines how to route context from a widget to a consumer(s). From an application developer's perspective, the ability to both abstract away the details of the underlying sensing technology via context widgets and to not worry about making individual connections to components via blackboards or context widgets and a discovery mechanism, makes things easier. In fact, there has been some recent work in augmenting the context widget model to include an advanced discovery and connection mechanism that leverages the advantages of both the context widget and blackboard approaches (Dey and Newberger, 2009).

8.6 ISSUES TO CONSIDER WHEN BUILDING CONTEXT-AWARE APPLICATIONS

Now that we have discussed the design process and different approaches for supporting context-aware applications, we will discuss aspects that designers need to consider when building applications.

8.6.1 Context Is a Proxy for Human Intent

The holy grail of context awareness is to divine or understand human intent. Applications would use this human intent to adapt appropriately by providing information or taking some actions. However, context information is only a proxy for this intent. In museum tour guides, a user standing in front of a particular exhibit may not represent interest in the exhibit, but may simply represent that a user has her back turned to the exhibit and is having a conversation with a friend there. Additional context is required to understand human intent. This is true for almost any situation and application. No matter how much context you can acquire for an application, you will always need more to truly determine the intent of the user. When designing an application, a developer must appropriately scope the types of situations that the application should sense and understand, realizing that the application may make an incorrect adaptation when a situation arises that is outside the scope or understanding of the designed application. Applications should express a certainty in their beliefs about their sensed information and inferences, as the Active Badge application did (Figure 8.1), and, if that certainty is not above an appropriate threshold, ask for confirmation of this information before taking action, exemplified by the Microsoft agent Clippy and the Lookout system (Horvitz, 1999).

8.6.2 Context Inferencing

Context-aware systems take data as input, and then determine how to adapt or respond to these data. Context inferencing is the act of making sense of these input data from sensors and other sources, to determine or infer the user's situation. Once the user's situation has been inferred, then the application can take an appropriate action. As described in the previous subsection, the sensed input is often not enough to infer the situation appropriately, so this brings up additional issues such as how to resolve ambiguity or uncertainty in context, and the role of rules and machine learning. These issues are discussed in the following subsections.

8.6.3 Context Ambiguity

Context-aware systems have many sources of ambiguity or errors: context sensors can sense incorrectly, fail, or be unsure about what they sensed; context inferencing systems can inaccurately reach conclusions about a situation or be unsure about their inferences; and applications can take an incorrect action or be unsure about what action to take. However, the vast majority of context-aware systems pretend that this source of ambiguity does not exist, and behave as if there are no errors or ambiguity. One important decision for designers of context-aware infrastructures and applications is whether to model context as being ambiguous or not. Acting as if there is no ambiguity in a system simplifies the design and use of a system, but leads to brittle systems in the face of ambiguity. In contrast, accepting that ambiguity exists means that the modeled systems better represent the real world, but are more challenging to create and use. Ambiguity is often specified as a single number representing the likelihood or certainty of a particular context value.

One approach to dealing with context ambiguity is to combine multiple disparate sources of the same type of context to improve the accuracy or dependability of the provided context. This is commonly known as sensor fusion (Wu, 2003). For example, in activity recognition, a hidden Markov model (see Chapter 9) can be used with different sensors and the fused results can be represented as a confusion matrix over the set of possible activities (Lester et al., 2006). An alternate approach is to allow users to manually disambiguate ambiguity in context (Dey and Mankoff, 2005). Rather than rely on an automated approach, this approach leverages a user's knowledge of the situation to help resolve and remove any ambiguity in the sensed or inferred context. A user may be presented with, for example, an *N*-best list, a list of the *N* most likely interpretations of context, ranked by likelihood, and asked to select the "correct" interpretation.

8.6.4 Rules versus Machine Learning

Context-aware applications are most commonly designed from a set of if-then rules: *if* the application senses a particular situation, *then* it should perform a particular action. Rules are easy to create because all the knowledge for each rule is represented in a homogenous format, and rule-based systems are relatively easy to build because there is a large number of existing rules engines that determine when a rule has been satisfied. Rules are also relatively intuitive and are thus easy to work with. However, rule-based systems also have some disadvantages: they are prone to conflicts between rules due to hidden dependencies between rules, making it challenging to understand the impact of adding or removing a rule particularly as the number of rules grows larger—they are notoriously difficult to debug when the number of rules grows larger because it is difficult to understand or follow the flow of control between rules, and they are a source of inefficiency because a rules engine must examine all rules to determine which rule, if any, should be executed. Finally, rules are rigid, meaning that they are also brittle. Small exceptions to a rule mean that a rule will not be fired or a different rule will be fired (than the one the developer would like).

A common alternative approach is to apply machine learning. Rather than create a series of rules about how an application should adapt its behavior, instead, an application developer can collect data on the types of situations that a user will experience and the types of adaptation desired. Machine learning can then be applied to learn the *probabilistic* relationships between the situations and adaptations, rather than have these relationships be hard-coded and deterministic. This still requires that the application or supporting infrastructure provide the ability to perform context inferencing to map the sensor data to user situations. Alternatively, machine learning can be applied to learn the relationships between the sensor data and the adaptations, skipping the middle step of context inferencing. The disadvantages of applying machine learning are that these relationships may be difficult to learn, may require a large amount of data to learn, are difficult to debug, and may not be intuitive to the application developer or end user.

8.6.5 Privacy

Context-aware systems have the capability and often the need to collect tremendous amounts of information about individuals. An inherent danger in collecting this information is in releasing information to the wrong person or during the wrong situation. In particular, as context systems are distributed across a computer network or networks, there is a real concern that context information could either be disseminated inappropriately or disseminated to a component that is not completely trustworthy, and may disseminate the information further. Developers of context-aware infrastructure need to ensure that data are only shared between components that actually have a real need to share and use that information. Similarly, developers of context-aware applications need to ensure that a user's privacy is maintained and that information is not being used inappropriately, or in a manner that the user deems inappropriate. The issue of privacy is very complex, and is the subject of another chapter within this book. There are no easy answers about how to support privacy.

8.6.6 Evaluation

Context-aware applications, like most ubicomp applications, are difficult to evaluate. Because context-aware applications are context-dependent, they can rarely be tested in a laboratory environment. The contexts or situations of interest often cannot be simulated. Even in field evaluations, these situations may be rare enough that they can be challenging to test and evaluate in anything other than a qualitative manner. How to evaluate such applications has been the subject of much ongoing research (e.g., Carter et al., 2008; O'Neill et al., 2007; Oh et al., 2007; Scholtz and Consolvo, 2004).

8.6.7 End User Issues

The goal of any context-aware application is to provide users with appropriate support based on their current situation. Until recently, the development of context-aware applications has limited user involvement to observation in the formative stages and being active users after deployment. However, there are two end user issues that developers should consider when building their applications. The first is *intelligibility* of applications—how a user forms an understanding of a context-aware application's behavior (Bellotti and Edwards, 2001). As context primarily consists of implicit input, it is much more challenging for a user to understand that an application took some action based on nonexplicit input, to understand what nonexplicit input that was, and to understand what action was even taken (in some cases). Unlike typical non-ubicomp applications, context-aware applications do not always have a mechanism for providing appropriate feedback to users to indicate that some action has been taken or why. This is challenging enough for rules-based systems, but even more challenging for machine learning–based systems because they are usually not amenable to generating explanations. This is an active area of research (e.g., Assad et al., 2007; Dey and Newberger, 2009; Lim et al., 2009; Tullio et al., 2007), with a workshop series on the topic of explanation-aware computing.

A second end-user issue of concern is *control*. Developers often design their applications based on knowledge of a particular domain. However, context-aware applications also need to be personalized to their users, rather than use a "one-size-fits-all" approach. Rather than developing individual variations of applications for each individual user, developers should consider putting the control for how an application should behave in the hands of users, that is, letting users create the rules that govern a context-aware application's behavior. Particularly for long-lasting applications when there can be many changes (to the environment, to the situations that a user experiences, and/or to the preferences that a user has for application behavior), users should have the ability to modify and customize the application to their needs. For example, iCAP allows end users to visually create rules-based context-aware applications (Dey et al., 2006), and a CAPpella supports end users in building simple context-aware applications using a programming by demonstration approach (Dey et al., 2004). By addressing both intelligibility and control, developers can significantly increase the usability of context-aware applications.

8.7 CHALLENGES IN WRITING ACADEMIC PAPERS ON CONTEXT AWARENESS

To complete this chapter, a discussion of the pitfalls that researchers often fall into when writing about context-aware systems is presented. The two most common topics in context-aware research are applications and infrastructure, and the discussion of pitfalls will center around these two topics.

From an application standpoint, there are two main issues that authors continue to have when writing papers about context awareness. The first is building one of the canonical applications described above (tour guide, reminder system, environmental control) without a novel contribution. Far too many researchers build a simple prototype of one of these applications, and try to publish a paper about it. With so many papers already published about these applications, it is *very difficult* to make a novel contribution.

The second issue is related to the first issue, but also applies to application design. It is not enough to just build an application; the application must also have a novel facet to it: whether the application itself is in a novel domain that involves a unique set of challenges, has a unique set of users that brings on some unique challenge, solves a unique and significant problem for users, or has some interesting scaling aspects (number of users, number of services provided or amount of context used, or period of deployment that reveals interesting long-term usage characteristics and issues). Not only should the application be built and studied, but in the write-up of the work, the authors *must* declare their contribution that differentiates the work from the vast amount of existing literature on context-aware applications.

From a development standpoint, there are numerous context-aware infrastructures that have been built and written about. When designing yet another context infrastructure, there are two pitfalls that researchers tend to fall into. The first is in failing to make a novel contribution that forms an improvement beyond existing infrastructures, and the second is in failing to demonstrate that the contribution is actually a contribution. First, with so many toolkits and infrastructures that have been written about that make it easier to build context-aware applications, or that better support a particular issue (e.g., privacy, modeling, ambiguity) or application domain, it is challenging to make a novel contribution. The biggest pitfall is to ignore the vast literature and to reinvent the wheel, claiming a contribution where an existing system already does what the proposed system is claiming to do. Much of this can be addressed in the planning stages for a research project, simply by performing an in-depth literature review—see Bauldauf et al. (2007) and Chen and Kotz (2000) as a potential starting point. Often, a new infrastructure is built to support some application because the appropriate existing infrastructures are either not publicly available or the existing infrastructures are determined to not be appropriate. Researchers then try to publish the infrastructure that was mainly built to support research and development of a particular application, and the novelty of the infrastructure is marginal at best.

The second pitfall that infrastructure researchers fall into is failing to demonstrate the contribution of their novel infrastructure or novel aspects of their infrastructure. If your infrastructure makes it faster to build context-aware applications or easier to build context-aware applications, you must either provide a well-reasoned argument why this is the case, or better yet, perform some evaluation that concretely demonstrates that the infrastructure has this benefit. If your infrastructure addresses a novel or interesting issue such as privacy or ambiguity, you must demonstrate how your infrastructure addresses the problem in the context of an actual application. Too often, authors describe their infrastructure and either neglect to describe an application built with the infrastructure or leave too little space in which to adequately describe the application. When writing about the infrastructure, do not make the mistake of describing the entire infrastructure in great detail—that takes too much space. Instead, describe the novel aspects in detail, and provide more cursory details of the less novel aspects. Then, be sure to save space for a rigorous evaluation, demonstrating that the infrastructure provides the novel benefits that you are claiming.

8.8 SUMMARY

This chapter introduced the concept of context awareness, a core aspect of ubiquitous computing. It presented a historical perspective on context awareness, including definitions of context and taxonomies of context awareness. It surveyed a wide variety of context-aware applications, and discussed different approaches for building context-aware applications. This chapter highlighted a number of open research questions and challenges for the field of context-aware computing. Finally, this chapter presented advice to those looking to publish papers and to conduct research in context-aware computing, pointing out pitfalls that can be avoided to increase the likelihood of producing novel and publication-worthy research.

Context awareness is a central feature of ubicomp systems. Although the field of context-aware computing is already growing rapidly, as more and more sensors proliferate throughout the world, its growth will accelerate and its importance will only increase. Addressing the open questions described in this chapter is necessary for a future world of ubicomp.

REFERENCES

Abowd, G. D., Atkeson, C. G., Hong, J., Long, S., Kooper, R., and Pinkerton, M. (1997). Cyberguide: A mobile context-aware tour guide. *ACM Wireless Networks* 3(5), 421–433.

Abowd, G. D., Dey, A. K., Orr, R. J., and Brotherton, J. (1998). Context-awareness in wearable and ubiquitous computing. *Virtual Reality* 3, 200–211.

Assad, M., Carmichael, D., Kay, J., and Kummerfield, B. (2007). PersonisAD: Distributed, active, scrutable model framework for context-aware services. In *Proceedings of Pervasive 2007*, pp. 55–72.

Avrahami, D., Fussell, S. R., and Hudson, S. E. (2008). IM waiting: Timing and responsiveness in semi-synchronous communication. *Proceedings of CSCW 2008*, pp. 285–294.

Bardram, J. E. (2005). The Java Context Awareness Framework (JCAF)—A service infrastructure and programming framework for context-aware applications. *Proceedings of Pervasive 2005*, pp. 98–115.

Bardram, J. E., and Nørskov, N. (2008). A context-aware patient safety system for the operating room. *Proceedings of Ubicomp 2008*, pp. 272–281

Bauldauf, M., Dustdar, S., and Rosenberg, F. (2007). A survey on context-aware systems. *International Journal of Ad Hoc and Ubiquitous Computing* 2(4), 263–277.

Bederson, B. B. (1995). Audio augmented reality: A prototype automated tour guide. In *Proceedings of Human Factors in Computing Systems (CHI 95)*, New York, ACM Press, pp. 210–211.

Bellotti, V., and Edwards, K. (2001). Intelligibility and accountability: Human considerations in context-aware systems. *Human-Computer Interaction* 16(2), 193–212.

Brown, M. G. (1996a). Supporting user mobility. In *Proceedings of the IFIP Conference on Mobile Communications (IFIP '96)*, Canberra, Australia, pp. 69–77.

Brown, P. J. (1996b). The Stick-e Document: A framework for creating context-aware applications. In *Proceedings of the Electronic Publishing '96*, pp. 259–272.

Brown, P. J., Bovey, J. D., and Chen, X. (1997). Context-aware applications: From the laboratory to the marketplace. *IEEE Personal Communications* 4(5), 58–64.

Brown, P. J. (1998). Triggering information by context. *Personal Technologies* 2(1), 1–9.

Carter, S., Mankoff, J., Klemmer, S., and Matthews, T. (2008). Exiting the cleanroom: On ecological validity and ubiquitous computing. *HCI Journal* 23(1), 47–99.

Chen, G., and Kotz, D. (2000). A survey of context-aware mobile computing research. Dartmouth Computer Science Technical Report TR2000-381, Dartmouth College, Hanover, NH.

Cheyer, A., and Martin, D. (2001). The open agent architecture. *Journal of Autonomous Agents and Multi-Agent Systems* 4(1), 143–148.

Clark, H., and Brennan, S. (1991). Grounding in communication. In: Resnick, L., Levine, L., and Teasley, S. (Eds.), *Perspectives on Socially Shared Cognition*. APA Books, Washington, DC, pp. 127–149.

Cooperstock, J. R., Tanikoshi, K., Beirne, G., Narine, T., and Buxton, W. (1995). Evolution of a reactive environment. In *Proceedings of the 1995 ACM Conference on Human Factors in Computing Systems (CHI '95)*, pp. 170–177.

Davies, N., Mitchell, K., Cheverst, K., and Blair, G. (1998). Developing a context-sensitive tour guide. 1st Workshop on Human Computer Interaction for Mobile Devices.

Dey, A. K. (1998). Context-aware computing: The CyberDesk project. In *Proceedings of the AAAI 1998 Spring Symposium on Intelligent Environments* (AAAI Technical Report SS-98-02), pp. 51–54.

Dey, A. K., and Abowd, G. D. (1997). CyberDesk: The use of perception in context-aware computing. In *Proceedings of the 1997 Workshop on Perceptual User Interfaces (PUI '97)*, pp. 26–27.

Dey, A. K., and Abowd, G. D. (2000a). Towards a better understanding of context and context-awareness. Workshop on the What, Who, Where, When and How of Context-Awareness, affiliated with the 2000 ACM Conference on Human Factors in Computer Systems (CHI 2000).

Dey, A. K., and Abowd, G. D. (2000b). CybreMinder: A context-aware system for supporting reminders. *Proceedings of the 2nd International Symposium on Handheld and Ubiquitous Computing (HUC)*, pp. 172–186.

Dey, A. K., and Mankoff, J. (2005). Designing mediation for context-aware applications. *ACM Transactions on Computer-Human Interaction* 12(1), 53–80.

Dey, A. K., and Newberger, A. (in press). Support for context intelligibility and control. In *Proceedings of CHI 2009*.

Dey, A. K., Abowd, G. D., and Salber, D. (2001). A conceptual framework and a toolkit for supporting the rapid prototyping of context-aware applications. *Human-Computer Interaction* 16(2), 97–166.

Dey, A. K., Abowd, G. D., and Wood, A. (1998). CyberDesk: A framework for providing self-integrating context-aware services. *Knowledge Based Systems* 11(1), 3–13.

Dey, A. K., Hamid, R., Beckmann, C., Li, I., and Hsu, D. (2004). aCAPpella: Programming by demonstration of context-aware applications. In *Proceedings of CHI 2004*, pp. 33–40.

Dey, A. K., Sohn, T., Streng, S., and Kodama, J. (2006). iCAP: Interactive context-aware prototyping. In *Proceedings of Fourth International Conference on Pervasive Computing*, pp. 254–271.

Elrod, S., Hall, G., Costanza, R., Dixon, M., and des Rivieres, J. (1993). Responsive office environments. *Communications of the ACM* 36(7), 84–85.

Fickas, S., Kortuem, G., and Segall, Z. (1997). Software organization for dynamic and adaptable wearable systems. In *Proceedings of the 1st International Symposium on Wearable Computers (ISWC'97)*, pp. 56–63.

Fogarty, J., Hudson, S. E., Atkeson, C. G., Avrahami, D., Forlizzi, J., Kiesler, S., Lee, J. C., and Yang, J. (2005). Predicting human interruptibility with sensors. *ACM Transactions on Computer-Human Interaction* 12(1), 119–146.

Franklin, D., and Flaschbart, J. (1998). All gadget and no representation makes Jack a dull environment. In *Proceedings of the AAAI 1998 Spring Symposium on Intelligent Environments* (AAAI Technical Report SS-98-02), pp. 155–160.

Gerlernter, D. (1985). Generative communication in Linda. *ACM Transactions on Programming Languages and Systems* 7(1), 80–112.

Greenhalgh, C. (2002). EQUIP: An extensible platform for distributed collaboration. Workshop on Advanced Collaborative Environments 2002.

Horvitz, E. (1999). Principles of mixed-initiative user interfaces. In *Proceedings of CHI 1999*, pp. 159–166.

Hull, R., Neaves, P., and Bedford-Roberts, J. (1997). Towards situated computing. In *Proceedings of the 1st International Symposium on Wearable Computers (ISWC '97)*, pp. 146–153.

Iqbal, S. T., Adamczyk, P. D., Zheng, X. S., and Bailey, B. P. (2005). Towards an index of opportunity: Understanding changes in mental workload during task execution. In *Proceedings of the SIGCHI Conference on Human Factors in Computing Systems*, pp. 311–320.

Johanson, B., and Fox, A. (2002). The event heap: A coordination infrastructure for interactive workspaces. In *Proceedings of the 4th IEEE Workshop on Mobile Computing Systems and Applications*, pp. 83–93.

Kjær, K. E. (2008). Designing middleware for context awareness in agriculture. In *Proceedings of the 5th Middleware Conference Doctoral Symposium*, pp. 19–24.

Kortuem, G., Segall, Z., and Bauer, M. (1998). Context-aware, adaptive wearable computers as remote interfaces to 'intelligent' environments. In *Proceedings of the 2nd International Symposium on Wearable Computers (ISWC '98)*, pp. 58–65.

Lester, J., Choudhury, T., and Borriello, G. (2006). A practical approach to recognizing physical activities. In *Proceedings of Pervasive 2006*, pp. 1–16.

Lim, B., Dey, A. K., and Avrahami, D. (2009). Why and why not explanation improve the intelligibility of context-aware intelligent systems. In *Proceedings of CHI 2009*, in press.

Ludford, P., Frankowski, D., Reily, K., Wilms, K., and Terveen, L. G. (2006). Because I carry my cell phone anyway: Improving location-based reminder systems. In *Proceedings of ACM Conf. on Human Factors in Computing Systems (CHI) 2006*, pp. 889–898.

Mozer, M. C. (1998). The neural network house: An environment that adapts to its inhabitants. In Coen, M. (Ed.), *Proceedings of the American Association for Artificial Intelligence Spring Symposium on Intelligent Environments*, pp. 110–114.

O'Neill, E., Lewis, D., McGlinn, K., and Dobson, S. (2007). Rapid user-centered evaluation for context-aware systems. In *Proceedings of Interactive Systems: Design, Specification, and Verification (DS-VIS 2006)*, pp. 220–233.

Oh, Y., Schmidt, A., and Woo, W. (2007). Designing, developing, and evaluating context-aware systems. In *Proceedings of Multimedia and Ubiquitous Engineering (MUE 2007)*, pp. 1158–1163.

Pascoe, J. (1998). Adding generic contextual capabilities to wearable computers. In *Proceedings of the 2nd IEEE International Symposium on Wearable Computers (ISWC '98)*, pp. 92–99.

Pascoe, J., Ryan, N. S., and Morse, D. R. (1998). Human-computer-giraffe interaction—HCI in the field. Workshop on Human Computer Interaction with Mobile Devices.

Rekimoto, J., Ayatsuka, Y., and Hayashi, K. (1998). Augmentable reality: Situated communication through physical and digital spaces. In *Proceedings of the 2nd IEEE International Symposium on Wearable Computers (ISWC '98)*, pp. 68–75.

Rodden, T., Cheverst, K., Davies, N., and Dix, A. (1998). Exploiting context in HCI design for mobile systems. Workshop on Human Computer Interaction with Mobile Devices.

Ryan, N. (1997). MCFE metadata elements, version 0.2. Working document. University of Kent at Canterbury. Kent, UK.

Ryan, N., Pascoe, J., and Morse, D. (1998). Enhanced reality fieldwork: The context-aware archaeological assistant. In: van Leusen, M., and Exxon, S. (Eds.), *Computer Applications and Quantitative Methods in Archaeology*. Oxford.

Salber, D., Dey, A. K., and Abowd, G. D. (1999). The Context Toolkit: Aiding the development of context-enabled applications. In *Proceedings of the 1999 ACM Conference on Human Factors in Computer Systems (CHI '99)*, pp. 434–441.

Schilit, B., and Theimer, M. (1994). Disseminating active map information to mobile hosts. *IEEE Network* 8(5), 22–32.

Schilit, B. N., Adams, N. I., and Want, R. (1994). Context-aware computing applications. In *Proceedings of the 1st International Workshop on Mobile Computing Systems and Applications*, pp. 85–90.

Schilit, B. N. (1995). System architecture for context-aware mobile computing. PhD dissertation, Columbia University, New York.

Schmidt, A., Takaluoma, A., and Mäntyjärvi, J. (2000). Context-aware telephony over WAP. *Personal and Ubiquitous Computing* 4(4), 225–229.

Scholtz, J., and Consolvo, S. (2004). Toward a framework for evaluating ubiquitous computing applications. *IEEE Pervasive Computing Magazine* 3(2), 82–88.

Tullio, J., Dey, A. K., Fogarty, J., and Chalecki, J. (2007). How it works: A field study of non-technical users interacting with an intelligent system. In *Proceedings of CHI 2007*, pp. 31–40.

Want, R., Hopper, A., Falcao, V., and Gibbons, J. (1992). The Active Badge location system. *ACM Transactions on Information Systems* 10(1), 91–102.

Want, R., Schilit, B. N., Adams, N. I., Gold, R., Petersen, K., Goldberg, D., Ellis, J. R., and Weiser, M. (1995). The PARCTAB ubiquitous computing experiment. Technical Report CSL-95-1. XEROX Palo Alto Research Center, Palo Alto, CA.

Ward, A., Jones, A., and Hopper, A. (1997). A new location technique for the active office. *IEEE Personal Communications* 4(5), 42–47.

Weiser, M. (1991). The computer for the 21st century. *Scientific American* 265(3), 94–104.

Weiser, M., and Brown, J. S. (1997). The coming age of calm technology. In: Denning, P. J., and Metcalfe, R. M. (Eds.), *Beyond Calculation: The Next Fifty Years of Computing*. Springer-Verlag, New York, NY.

Weiss, D., Duchon, M., Fuchs, F., and Linnhof-Popien, C. (2008). Context-aware personalization for mobile multimedia services. In *Proceedings of the International Conference on Mobile Computing and Multimedia (MoMM 2008)*, pp. 267–271.

Wisneski, C., Ishii, H., Dahley, A., Gorbet, M., Brave, S., Ullmer, B., and Yarin, P. (1998). Ambient displays: Turning architectural space into an interface between people and digital information. In *Proceedings of CoBuild 1998*, pp. 22–32.

Wu, H. (2003). Sensor fusion for context-aware computing using Dempster-Shafer theory. PhD dissertation, Carnegie Mellon University, Pittsburgh, PA.

CHAPTER 9

Processing Sequential Sensor Data

John Krumm

CONTENTS

9.1	Introduction	354
9.2	Tracking Example	356
9.3	Mean and Median Filters	358
9.4	Kalman Filter	359
	9.4.1 Linear, Noisy Measurements	360
	9.4.2 Linear, Noisy Dynamics	361
	9.4.3 All Parameters	362
	9.4.4 Kalman Filter	363
	9.4.5 Discussion	364
9.5	Particle Filter	365
	9.5.1 Problem Formulation	366
	9.5.2 Particle Filter	368
	9.5.3 Discussion	369
9.6	Hidden Markov Model	370
	9.6.1 Problem Formulation	371
	9.6.2 Hidden Markov Model	373
	9.6.3 Discussion	374
9.7	Presenting Performance	376
	9.7.1 Presenting Continuous Performance Results	376
	9.7.2 Presenting Discrete Performance Results	378
9.8	Conclusion	380
References		380

9.1 INTRODUCTION

Ubiquitous computing (ubicomp) applications are normally envisioned to be sensitive to context, where context can include a person's location, activity, goals, resources, state of mind, and nearby people and things. Context is often inferred with sensors that periodically measure some aspect of the user's state. For instance, a global positioning system (GPS) sensor can repeatedly measure a person's location at some interval in time. This is an example of sequential sensor data in that it is a sequence of sensor reading of the same entity spread out over time. Unfortunately, sensors are never perfect in terms of noise or accuracy. For example, a GPS sensor gives noisy latitude/longitude measurements, and sometimes the measurements are wildly inaccurate (outliers). In addition, sensors often do not measure the necessary state variables directly. Although a GPS sensor can be used to infer a person's velocity and even mode of transportation (Patterson et al., 2003), it cannot directly measure these states. This chapter is aimed at introducing fundamental techniques for processing sequential sensor data to reduce noise and infer context beyond what the sensor actually measures. The techniques discussed are not necessarily on the cutting edge of signal processing, but they are well-accepted approaches that have proven to be fundamentally useful in ubicomp research. Because of their wide acceptance and usefulness, you should feel comfortable using them in your own ubicomp work. Specifically, this chapter discusses mean and median filters, the Kalman filter, the particle filter, and the hidden Markov model (HMM). Each of these techniques processes sequential sensor data, but they all have different assumptions and representations, which are highlighted to help the reader make an intelligent choice.

This chapter concentrates on processing sequential sensor measurements, because it is usually necessary to make repeated measurements to keep up with possibly changing context in ubicomp applications. For instance, a person's location usually changes with time, so location must be measured repeatedly. Sequential measurements mean that processing techniques can take advantage of both a sense of the past and a sense of the future, which will be explained next.

A sense of the past is useful because context does not change completely randomly, but instead shows some coherence over time. Although an isolated sensor measurement might lead to an uncertain conclusion about a person's context, repeated measurements give more certainty in spite of noisy measurements, partly because context normally does not change very quickly compared to how often measurements can be taken. Thus, it often makes sense to average together the last few measurements, which is a simple means of exploiting continuity, and one which this chapter examines in Section 9.3. In general, a new measurement triggers a reestimation of the context state variables that is sensitive not only to the new measurement, but also to previous measurements and/or estimates. This historical perspective means the techniques can exploit expectations about the continuity of the state variables. The mean filter fulfills expectations that the output varies relatively slowly and that it should be sensitive to current and past measurements. All the techniques in this chapter pay attention to past measurements or past state estimates in some way.

A sense of the future is usually realized with a dynamic model of the measured process, something that the mean and median filters do not have. For instance, if a person is walking in a certain direction, it is most likely he or she will continue in that direction, because turns are somewhat rare. Like a sense of the past, a sense of the future can help smooth out noise in the measurements. A dynamic model can also keep track of more than just what is measured. For example, the Kalman filter, as explained in Section 9.4, tracks both location and speed based only on location measurements. Paying attention to higher level, albeit unmeasured, state variables means that the processing technique can exploit reasonable assumptions about the behavior of the whole system rather than just the directly measured parts of it. In the Kalman filter, a speed estimate can be used to mitigate the effect of a wildly distant location measurement, because the user might not have been able to move to the distant point given the current speed. As a bonus, these higher level variables become available for use in a context-sensitive system. For instance, even if a system is measuring location only, speed estimates help the processing and can be used as a context output. The activity inference project takes this to the extreme: Lester et al. (2006) were able to distinguish among various activities (e.g., sitting, standing, walking) by measuring acceleration, sound, and barometric pressure. Some of the processing techniques in this chapter explicitly estimate these extra state variables to improve their performance, and the estimates are a useful by-product for inferring context.

A running example in this chapter helps highlight the assumptions, processing, and output of each technique. The next section introduces the example, followed by discussions of each technique in subsequent sections.

9.2 TRACKING EXAMPLE

The problem presented in this chapter focuses on tracking a moving person in the (x,y) plane based on noisy (x,y) measurements taken at an interval of 1 second. The simulated actual path and simulated noisy measurements are shown in Figure 9.1, where the unit of measurement is 1 meter. The path starts at the center of the spiral. Pretend that the person is carrying a location sensor, possibly a GPS, if he is outside. There are 1000 measurements spread evenly in time from beginning to end. The measurements are noisy versions of the actual path points. As is common in modeling noisy measurements, the measurements are taken as the actual values plus added Gaussian noise with zero mean. The Gaussian probability distribution is the familiar bell-shaped curve. "Noise" is not noise in the audible sense, but represents random errors in the measurements. Engineers and researchers use Gaussian noise, as opposed to other probability distributions, partly because it is often an adequate model and partly because it is theoretically convenient. The zero mean assumption implies that the sensor is unbiased (no constant offset error), and this is easily realizable

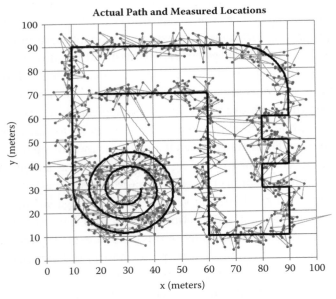

FIGURE 9.1 The actual path, in black, starts at the center of the spiral. The noisy, measured points are gray. One goal of processing the measured points is to estimate the actual path. These data form the basis of the running example in this chapter.

by subtracting any constant offset before processing the measurements. In this tracking example, the standard deviation of the noise is $\sigma = 3$ meters. There are also ten outliers placed randomly along the path, which were taken as the regular noisy measurements plus Gaussian noise with a standard deviation of 15 meters.

This example is helpful because it is easy to visualize and understand. Even though it is clearly in the domain of location measurement, it demonstrates some more general characteristics of the type of inference problems that appear in ubicomp. It consists of sequential measurements that are corrupted by noise, and there are dynamic models of expected behavior that can be used to combat the noise. In this particular example, there are assumptions about the smoothness of the path that are described in subsequent sections. These assumptions can be expressed in terms of other state variables that may be of interest, such as the speed of the person in this case. This example does not illustrate the processing of discrete state variables such as the person's mode of transportation or the number of people traveling together. However, Section 9.6 explains how to convert the problem into discrete variables and how to process them.

In mathematical terms, the actual locations are given by vectors $x_i = (x_i, y_i)^T$, where x_i is a column vector giving the ith location in the sequence, and i goes from 1 to the number of measurements. Note that the boldface x_i is a vector, and the non-boldface x_i is one of its scalar components. The x_i are unknown and the goal of the processing is to estimate them from the noisy measurements and to possibly infer other state variables such as velocity. The noisy measurements are represented by z_i, which are the same as the measurement vectors, but with independent, zero-mean Gaussian noise added to each component:

$$z_i = x_i + v_i \qquad v_i \sim N\left(0, \begin{bmatrix} \sigma^2 & 0 \\ 0 & \sigma^2 \end{bmatrix}\right) \qquad (9.1)$$

Here, v_i is a random Gaussian noise vector with zero mean and a diagonal covariance matrix. Since the covariance matrix is diagonal, this is the same as adding independent Gaussian noise to each element of x_i. Note that Equation (9.1) is not part of any algorithm for inferring x_i, but just an assumption on the relationship between the measurements z_i and actual x_i values. It also describes the simulated measurement data used in this chapter's running example.

This example is intentionally simple so it is easy to visualize and thus some of the simpler techniques will work on it. This chapter will show how each technique performs on this example, and simple techniques, such as the mean and median filter (discussed next), appear to work better than some of the more sophisticated methods. However, each of the techniques presented has tunable parameters that affect their performance, and the presented results do not necessarily represent the optimal state of tune. This means that each of the techniques has the potential to give more accuracy. Moreover, some of the more sophisticated techniques have the ability to give additional information, such as speed estimates, that simpler techniques do not give. Thus, judging the techniques merely by their accuracies presented in this chapter would be naïve.

One danger of a specific, running example is that it appears to limit the applicability of the methods to just that example. However, these techniques have been used for a wide range of sequential sensor data and a wide range of inferred state variables.

9.3 MEAN AND MEDIAN FILTERS

One of the simplest and most effective means of filtering out noise is to average together multiple samples. For the example problem, the mean filter estimates x_i by the average of z_i and the most recent $n-1$ measurements, that is,

$$\hat{x}_i = \frac{1}{n} \sum_{j=i-n+1}^{i} z_j \qquad (9.2)$$

Here $\hat{x}_i$ represents the estimated value of x_i. Equation (9.2) represents a sliding window of width n, where n is the number of values to use to compute the mean. This filter is simple to implement and works well to reduce noise, as shown in Figure 9.2, where $n = 10$. The primary disadvantage is that it introduces lag in the estimate, because the average is taken over mostly measurements that come before z_i. One way to reduce the lag is to use a weighted average whose weights decrease in value for increasingly older measurements. Note that the mean filter in Equation (9.2) is a "causal filter," because it does not look ahead in time to future measurements to estimate x_i. This is true of all techniques discussed in this chapter.

In addition to lag, another potential problem with the mean filter is its sensitivity to outliers. In fact, just a single measured point, placed far enough away, can move the mean to any location. The median is a more robust version of the mean, and it still works if up to half the data is outliers.

$$\hat{x}_i = \text{median}\{z_{i-n+1}, z_{i-n+2}, \ldots, z_{i-1}, z_i\} \qquad (9.3)$$

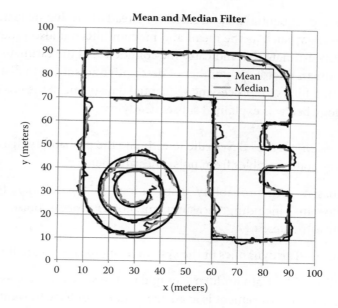

FIGURE 9.2 The actual path is in black. The paths estimated by the mean and median filters are in dark gray and light gray, respectively. The median filter produces fewer large excursions from the true path, because it is more robust to outliers.

Figure 9.2 shows the estimated path based on the median. There are places where the mean drifts relatively far from the actual path due to an outlier, whereas the median stays closer.

The mean and median filters are simple, yet effective techniques for processing sequential sensor data. However, they do suffer from lag, and they do not intrinsically estimate any higher level variables such as speed. One could estimate speed with a numerical derivative, but this is very sensitive to the noise in the original measurements. The remaining techniques discussed in the chapter work to reduce lag with a dynamic model, and the Kalman and particle filters (Sections 9.4 and 9.5) can estimate higher level state variables in a principled way.

9.4 KALMAN FILTER

The Kalman filter is a big step up in sophistication from the mean and median filters discussed above. It explicitly accounts for sensor noise (as long as the noise is additive Gaussian), and it explicitly models the system's dynamics. The Kalman filter introduces probability to the problem of processing sequential measurements. Instead of producing just an estimated result $\hat{x}_i$ such as the mean and median filters, it also produces an uncertainty estimate that indicates the confidence of the estimate.

The Kalman filter has a number of descriptive adjectives to distinguish it from other filters for processing sequential data (some of the less obvious terminology below is explained subsequently):

- Bayesian—It uses Bayes rule to estimate the probability distribution of the state variables from noisy measurements z_i.
- Gaussian—All probability distributions in the Kalman filter are Gaussians, including measurement noise, process noise, and the state estimates.
- Linear—The dynamics of the system being measured are linear.
- Online—The Kalman filter can update the state estimate as soon as the latest measurement is available. It does not have to wait until all the data are available, like a batch filter would.

One of the most enduring references on Kalman filtering is the book by Gelb (1974). The mathematical notation in this section is the same as in Gelb's book.

9.4.1 Linear, Noisy Measurements

The mean and median filters can only estimate states that are directly measured. For instance, if the measurements are (x,y) coordinates as in this chapter's example, the mean and median filters cannot be used to directly, in a principled way, estimate velocity. The Kalman filter makes a distinction between what quantities are in the measurement vector z_i and the unknown state vector x_i. In the example for the Kalman filter, these two vectors are

$$z_i = \begin{pmatrix} z_i^{(x)} \\ z_i^{(y)} \end{pmatrix} \quad x_i = \begin{pmatrix} x_i \\ y_i \\ s_i^{(x)} \\ s_i^{(y)} \end{pmatrix} \tag{9.4}$$

where $z_i^{(x)}$ and $z_i^{(y)}$ are the ith measurements of the x and y coordinates, x_i and y_i are the unknown actual coordinates, and $s_i^{(x)}$ and $s_i^{(y)}$ represent the unknown velocity vector's x and y components. The state vector contains variables that are not directly measured—velocity, in this case.

To express the full relationship between the actual state and measurements, the Kalman filter adds a measurement matrix H and zero-mean Gaussian noise:

$$z_i = H_i x_i + v_i \qquad v_i \sim N(0, R_i) \qquad (9.5)$$

Here, the measurement matrix H_i translates between the state vector and the measurement vector. Because the measurements are related to the actual state by a matrix, the measurements are said to be linear. In the example, H_i simply deletes the unmeasured velocity and passes through the (x,y) coordinates:

$$H_i = \begin{bmatrix} 1 & 0 & 0 & 0 \\ 0 & 1 & 0 & 0 \end{bmatrix} \qquad (9.6)$$

The noise vector v_i has the same dimensions as the measurement vector z_i, and is distributed as zero-mean Gaussian noise with a covariance vector R_i. In the example, the noise covariance is independent of i. Because this example is a simulation, the Kalman filter has the advantage of knowing the exact noise covariance, from Equation (9.1):

$$R_i = \begin{bmatrix} \sigma^2 & 0 \\ 0 & \sigma^2 \end{bmatrix} \qquad (9.7)$$

9.4.2 Linear, Noisy Dynamics

The mean and median filters have no model of how the state variables change over time. The Kalman filter does have such a model. It says that the state at time i is a linear function of the state at time $i-1$ plus zero-mean, Gaussian noise:

$$x_i = \phi_{i-1} x_{i-1} + w_{i-1} \qquad w_i \sim N(0, Q_i) \qquad (9.8)$$

The system matrix φ_{i-1} gives the linear relationship between the state at time $i-1$ and i. For the example, the system matrix is

$$\phi_{i-1} = \begin{bmatrix} 1 & 0 & \Delta t_i & 0 \\ 0 & 1 & 0 & \Delta t_i \\ 0 & 0 & 1 & 0 \\ 0 & 0 & 0 & 1 \end{bmatrix} \qquad (9.9)$$

Here, Δt_i is the time elapsed between measurements $i - 1$ and i. With this matrix and the state vector of the example, the equation $x_i = \varphi_{i-1} x_{i-1}$ says that $x_i = x_{i-1} + \Delta t_i s_i^{(x)}$ and similarly for y_i. This is just standard physics for a particle moving in a straight line at constant velocity. The system equation also says that the velocity stays constant over time. Of course, this is not true for the example, and is usually not true for any system that controls its own trajectory. In general, the system noise w_i helps account for the fact that φ_{i-1} is not an exact model of the system. w_i is zero-mean, Gaussian noise with covariance Q_i. For the example, Q_i is

$$Q_i = \begin{bmatrix} 0 & 0 & 0 & 0 \\ 0 & 0 & 0 & 0 \\ 0 & 0 & \sigma_s^2 & 0 \\ 0 & 0 & 0 & \sigma_s^2 \end{bmatrix} \quad (9.10)$$

This implies that the physical model connecting position and velocity is correct (which it is). But it also implies that velocity is subject to some noise between updates, which helps compensate for the fact that otherwise velocity is assumed to be constant. Setting the actual values of Q_i is not straightforward. For the example, σ_s^2 was chosen to represent the variance in the actual velocity from measurement to measurement. Q_i is modeling the fact that velocity is not constant in the example. If it were, the trajectory would have to be straight. With Q_i given as Equation (9.10), Equation (9.8) states that the velocity changes randomly between time steps, with the changes distributed as a zero mean Gaussian.

9.4.3 All Parameters

Implementing the Kalman filter requires the creation of the measurement model, system model, and initial conditions. Specifically, it requires

H_i = measurement matrix giving measurement z_i from state x_i (Equation 9.5)

R_i = measurement noise covariance matrix (Equation 9.5)

φ_{i-1} = system matrix giving state x_i from x_{i-1} (Equation 9.8)

Q_i = system noise covariance matrix (Equation 9.8)

$\hat{x}_0$ = initial state estimate

P_0 = initial estimate of state error covariance

The initial state estimate can usually be estimated from the initial few measurements. For this chapter's example, the initial position came from z_0, and the initial velocity was taken as zero. A reasonable estimate of P_0 for this example is

$$P_0 = \begin{bmatrix} \sigma^2 & 0 & 0 & 0 \\ 0 & \sigma^2 & 0 & 0 \\ 0 & 0 & \sigma_s^2 & 0 \\ 0 & 0 & 0 & \sigma_s^2 \end{bmatrix} \qquad (9.11)$$

9.4.4 Kalman Filter

The Kalman filter proceeds in two steps for each new measurement z_i. The result of the two steps is the mean and covariance of the estimated state, $\hat{x}_i^{(+)}$. The first step is to extrapolate the state and the state error covariance from the previous estimates. These are pure extrapolations with no regard for the measurement, and they depend only on the system model. The $(-)$ superscript indicates an extrapolated value.

$$\hat{x}_i^{(-)} = \phi_{i-1} \hat{x}_{i-1}^{(+)} \qquad (9.12)$$

$$P_i^{(-)} = \phi_{i-1} P_{i-1}^{(+)} \phi_{i-1}^T + Q_{i-1} \qquad (9.13)$$

The second step is to update the extrapolations with the new measurement, giving the mean and covariance of the new state estimate, $\hat{x}_i^{(+)}$ and $P_i^{(+)}$:

$$K_i = P_i^{(-)} H_i^T \left(H_i P_i^{(-)} H_i^T + R_i \right)^{-1} \qquad (9.14)$$

$$\hat{x}_i^{(+)} = \hat{x}_i^{(-)} + K_i \left(z_i - H_i \hat{x}_i^{(-)} \right) \qquad (9.15)$$

$$P_i^{(+)} = (I - K_i H_i) P_i^{(-)} \qquad (9.16)$$

where I is the identity matrix and K_i is the Kalman gain matrix.

Applying these equations to this chapter's example gives the result shown in Figure 9.3, where the plotted value is $\hat{x}_i^{(+)}$. The estimated path in dark gray does not follow turns very well, partly because the system model assumes the path is a straight line, with only the noise process to account for turns.

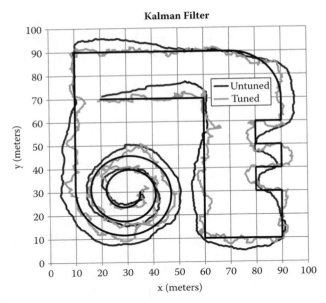

FIGURE 9.3 The Kalman filter result, in dark gray, tends to overshoot turns because its dynamic model assumes a single, straight line path. The lighter gray line is a version of the Kalman filter tuned to be more sensitive to the data. It follows turns better, but is also more sensitive to noise.

9.4.5 Discussion

The major difference between the Kalman filter and the mean and median filters is that the Kalman filter has a dynamic model of the system to keep up with changes over time. Besides fixing the lag problem, the dynamic model counterbalances the measurement model to give a tunable tradeoff between a designer's belief in the predictive dynamics versus sensor readings. This tuning is reflected in the choice of the elements of the measurement and system covariance matrices. Although the measurement covariance is relatively simple to estimate based on sensor characteristics, the system covariance is more difficult, and it represents an opportunity for tuning the tradeoff between dynamics and measurement. Figure 9.3 shows two results of the Kalman filter: the less accurate one in dark gray reflects a principled choice of σ_s, which specifies the noise in the predictive velocity model. The inferred path tends to have trouble tracking curves and corners, because the model is biased too much toward straight line paths. The more accurate inferred path in light gray comes from an optimal choice of σ_s based on an exhaustive sequence of test values of σ_s. The price of this more responsive model is a wigglier path, because this filter is more sensitive to the measurements. Of course, using a textbook example makes such a search easy. The problem of tuning is more difficult for real applications.

The dynamic model can include parameters beyond the ones that are measured. In the example, the state vector for the Kalman filter added velocity as a state parameter, even though the measurements had location only. An obvious extension to the example would be to include acceleration. The ability to track nonmeasured parameters is a benefit, because the nonmeasured parameters may be useful for context inference.

The Kalman filter can also make sensor fusion straightforward. For the example in this section, an accelerometer could add valuable data about location and velocity. This would involve augmenting the state vector, system model, and measurement model with acceleration.

Another advantage of the Kalman filter is that it gives an estimate of its own uncertainty in the form of the covariance matrix $P_i^{(+)}$. A knowledge of uncertainty is useful for ubicomp systems. For instance, a Kalman filter tracker might indicate that a person is equally likely in the kitchen or the living room, which could affect whether certain automatic actions are triggered.

One of the main limitations of the Kalman filter is the linearity of the dynamic model. Some processes are inherently linear, such as the radar traces of ballistic missiles to which Kalman filters were applied a long time ago. System dynamics are often not linear. For instance, a ball bouncing off a wall could not be represented by the state system matrix φ_{i-1} in the Kalman filter. Likewise, the combinations of distances and angles in pose estimation are not linear. The extended Kalman filter can sometimes solve nonlinear problems by linearizing the system around the estimated state. Until ubicomp advances to the ballistic missile stage, applications of the basic Kalman filter in the field will be relatively rare.

The Kalman filter is also unsuitable for representing discrete state variables such as a person's mode of transportation, knowledge of the world, or goals. Fortunately, for nonlinear problems with a mix of continuous and discrete state variables, the particle filter is a practical, although computationally more expensive, solution. This is the topic of the next section.

9.5 PARTICLE FILTER

The particle filter is a more general version of the Kalman filter, with less restrictive assumptions and, because of that, more computational demand. Unlike the Kalman filter, the particle filter does not require a linear model for the process in question, and it does not assume Gaussian noise. One of the main advantages of the particle filter for ubicomp applications is that it can easily represent an arbitrary mix of continuous and discrete variables along with a rich model of how these variables interact and affect sensors. In terms of the list of properties given for the Kalman filter, the particle filter is Bayesian, non-Gaussian, nonlinear, and online.

One good example of a particle filter for a ubicomp application is the work of Patterson et al. (2003). They created a rich model of a person's location, velocity, transportation mode, GPS error, and the presence of a parking lot or bus stop. The only sensor used was GPS for location and velocity. Using a particle filter, they could infer all the other variables. Note that their state space contained both continuous variables (location, velocity, and GPS error) and discrete variables (transportation mode, presence of parking lot or bus stop).

The particle filter has its name because it represents a multitude of possible state vectors, which can be thought of as particles. Each particle can be considered a hypothesis about the true state, and its plausibility is a function of the current measurement. After each measurement, the most believable particles survive, and they are subject to random change in state according to a probabilistic dynamic model. An easy-to-understand introduction to particle filtering is the chapter by Doucet et al. (2001), and this chapter uses their notation. Hightower and Borriello (2005) give a useful case study of particle filters for location tracking in ubicomp.

As in the Kalman filter section (Section 9.4.4), the subsections below explain how to implement a particle filter using this chapter's tracking example.

9.5.1 Problem Formulation

The particle filter is based on a sequence of unknown state vectors, x_i, and measurement vectors, z_i, which is the same as the Kalman filter, both in general and in this chapter's example. As a reminder, for the example, the state vector represents location and velocity, and the measurement vector represents a noisy version of location. The particle filter parallels the Kalman filter with a probabilistic model for measurements and dynamics, although both are more general than the Kalman filter.

The probability distribution $p(z_i|x_i)$, which you must provide, models the noisy measurements. This can be any probability distribution, which is more general than the Kalman filter formulation, which was $z_i = H_i x_i + v_i$. The measurement probability distribution gives a probabilistic relationship between the actual state x_i given the measurement z_i. This models the realistic scenario where the measurement can be some complicated function of the state, with some uncertainty. To stay consistent, the particle filter example will use the same measurement model as the Kalman filter, which states

$$p(z_i|x_i) = N\big((x_i, y_i)^T, R_i\big) \tag{9.17}$$

This says that the measurement is normally distributed around the actual location (x_i, y_i) with the same covariance matrix used for the Kalman measurement model in Equation (9.7). However, $p(z_i|x_i)$ could be much more interesting and expressive. For instance, with knowledge of a map, $p(z_i|x_i)$ could express the fact that measurements in some regions are less accurate than measurements in other regions (e.g., the problem of GPS losing satellite signals in buildings, tunnels, and urban canyons).

Another probability distribution you must provide is $p(x_i|x_{i-1})$, which models the dynamics. It gives the distribution of the current state x_i given the previous state x_{i-1}. This is also more general than the Kalman filter, whose dynamics model is $x_i = \varphi_{i-1} x_{i-1} + w_{i-1}$. It is not necessary to write down $p(x_i|x_{i-1})$, but the particle filter requires sampling from it. That is, given an x_{i-1}, the particle filter algorithm requires the generation of a random x_i, subject to $p(x_i|x_{i-1})$. For this chapter's example, the assumption is that the location changes deterministically as a function of the velocity, and that the velocity is randomly perturbed with Gaussian noise:

$$\begin{aligned}
x_{i+1} &= x_i + s_i^{(x)} \Delta t_i \\
y_{i+1} &= y_i + s_i^{(y)} \Delta t_i \\
s_{i+1}^{(x)} &= s_i^{(x)} + w_i^{(x)} \quad w_i^{(x)} \sim N(0, \sigma_s^2) \\
s_{i+1}^{(y)} &= s_i^{(y)} + w_i^{(y)} \quad w_i^{(y)} \sim N(0, \sigma_s^2)
\end{aligned} \tag{9.18}$$

This is the same as the assumed dynamic model for the Kalman filter in Equation (9.8), but expanded for each component of the state vector to emphasize that the update does not need to be linear. It also does not need to be Gaussian. For consistency with the example, however, this dynamic model is both linear and Gaussian. The generality of $p(x_i|x_{i-1})$ would allow for more domain knowledge in the model. For instance, with knowledge of a map, $p(x_i|x_{i-1})$ could express the fact that it is more likely to accelerate going down hills and decelerate going up hills.

The other component of the particle filter formulation is a prior distribution $p(\boldsymbol{x}_0)$ on the state vector before any dynamics can be applied. For the example, let the prior be a normal distribution around the first measurement:

$$p(\boldsymbol{x}_0) = N(\boldsymbol{z}_0, R_i) \qquad (9.19)$$

Although the variables in the example are all continuous, note that these distributions could involve a mix of continuous and discrete variables, and they could interact in interesting ways. For instance, consider appending a discrete home activity variable to the state vector $\boldsymbol{x}_i$, where the activities can be {sleeping, eating, studying}. Assuming that the subject engages in only one of these activities at a time, the measurement model of the particle filter $p(\boldsymbol{z}_i|\boldsymbol{x}_i)$ could express the fact that certain activities are more likely to occur in or near certain rooms of the house.

9.5.2 Particle Filter

The particle filter works with a population of N particles representing the state vector: $\boldsymbol{x}_i^{(j)}$, $j = 1, \ldots, N$. Unlike the other methods in this chapter, the particle filter actually involves generating random numbers, which are necessary to instantiate the values of the particles. This sampling means writing a routine that generates sample random numbers adhering to the relevant probability distributions.

Although there are several variations of the particle filter, one of the most popular is called the bootstrap filter, introduced by Gordon (1994). It starts by initializing N samples of $\boldsymbol{x}_0^{(j)}$ generated from the prior distribution $p(\boldsymbol{x}_0)$. In the example, a plot of these particles would cluster around the first measurement $\boldsymbol{z}_0$, because the prior term in Equation (9.19) is a normal distribution with mean $\boldsymbol{z}_0$.

With the initialization complete, for $i > 1$, the first step is "importance sampling," which uses the dynamic model to generate random $\tilde{\boldsymbol{x}}_i^{(j)}$ from $p(\boldsymbol{x}_i|\boldsymbol{x}_{i-1})$. This propagates the particles forward according to the assumed dynamic model, but without any guidance from the new measurement $\boldsymbol{z}_i$.

The next step computes "importance weights" for each particle according to the measurement model:

$$\tilde{w}_i^{(j)} = p\left(\boldsymbol{z}_i \middle| \tilde{\boldsymbol{x}}_i^{(j)}\right) \qquad (9.20)$$

A larger importance weight indicates a particle that is better supported by the measurement. The importance weights are then normalized so they sum to one.

Finally, the last step in the loop is the "selection step," which samples a new set of particles $x_i^{(j)}$ from the $\tilde{x}_i^{(j)}$ based on the normalized importance weights. This involves picking N of $\tilde{x}_i^{(j)}$ at random, where the probability of picking $\tilde{x}_i^{(j)}$ is proportional to its weight $\tilde{w}_i^{(j)}$. This step tends to eliminate unlikely particles, and it is not unusual to pick the same $\tilde{x}_i^{(j)}$ more than once if its weight is relatively large. The selection step is the last step in the loop. The algorithm returns to the importance sampling step for the next value of i to process the next measurement.

To compute an estimate of x_i at any point, which was the original goal, compute the weighted mean of the particles:

$$\hat{x}_i = \sum_{j=1}^{N} \tilde{w}_i^{(j)} \tilde{x}_i^{(j)} \qquad (9.21)$$

This works for continuous variables. For discrete variables, a viable approach is weighted voting. Similar equations work for computing other expected values, such as variance. This is useful for estimating the uncertainty of the estimate.

Applied to this chapter's example problem, the particle filter gives the result shown in Figure 9.4.

9.5.3 Discussion

The main problem with the particle filter is computation time. In general, using more particles (larger N) helps the algorithm work better, because it can more completely represent and explore the space of possible state vectors. But often, adding enough particles to make a noticeable improvement in the results also causes a significant increase in computation time. Fox (2003) gives a method for choosing N based on bounding the approximation error.

Because the particle filter allows such a rich state representation, it is tempting to add state variables. In this chapter's example, it would be interesting to add state variables governing the mode of transportation (e.g., walking vs. driving), intended route, and intended destination. However, increasing the dimensionality of the state vector usually

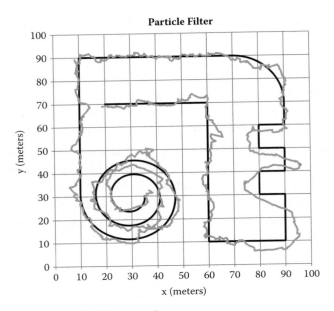

FIGURE 9.4 The particle filter result.

requires adding more particles to account for the larger state space, leading to increased computation time. One interesting remedy for this is the Rao-Blackwellized particle filter (Murphy and Russell, 2001). It uses a more conventional filter, such as Kalman, for the state variables where it is appropriate, and it uses a particle filter for the other state variables. For instance, Kalman could cover location and velocity, and a particle filter could track the higher level states.

9.6 HIDDEN MARKOV MODEL

The HMM works only for discrete valued state space variables. This is in contrast to the Kalman filter, which works only for continuous valued variables, and the particle filter, which works for a mix of discrete and continuous variables. The HMM has been primarily the tool of speech understanding researchers, where the problem is to find the set of words that best accounts for a speech signal. In fact, the classic tutorial paper on HMMs by Rabiner (1989) was aimed at speech understanding, although the paper is useful to learn about the HMM for any application.

As in the previous sections, this section uses the chapter's tracking example to illustrate how to use an HMM.

9.6.1 Problem Formulation

As with the previous filtering methods, the measurements are represented by z_i, where the subscript represents the time the measurement was taken. Instead of a continuous state x_i, the HMM works with discrete states $X_i^{(j)}$. Here, the subscript again refers to time. The superscript indexes through the M possible states. In an activity recognition system, $X_i^{(j)}$ could represent different modes of transportation at time i, that is, $X_i^{(j)} \in$ {bus, foot, car}, as in Patterson et al. (2003). Here, $j = 1$ means "bus," $j = 2$ means "foot," and $j = 3$ means "car," and there are $M = 3$ possible states.

In this chapter's example, the natural way to represent the state is with a continuous coordinate $(x_i, y_i)^T$. For the HMM, however, with its requirement for discrete state, the example splits the coordinate plane into small cells, as shown in Figure 9.8. Each cell in the example is a square whose sides are 1 meter long. Since the space extends to 100 meters along both axes, there are $M = 10,000$ cells. The goal of HMM is to estimate which of these cells contains the tracked object after each continuous measurement z_i.

Both the Kalman filter and particle filter have a measurement model. In the Kalman filter, the measurement model, given by Equation (9.5), says that the measurement is a linear function of the state plus additive Gaussian noise. In the particle filter, the measurement model is a general, conditional probability $p(z_i|x_i)$. The measurement model for the HMM is similar to the one for the particle filter, except that it gives the discrete probability of each state, given the measurement: $P(X_i^{(j)}|z_i)$. P (uppercase) indicates a discrete probability distribution. In the HMM terminology, the measurement model is used to compute the "observation probabilities." Given a measurement z_i, there is one observation probability value for each possible state $X_i^{(j)}$. These observation probabilities must sum to 1; that is,

$$\sum_{j=1}^{M} P\left(X_i^{(j)}|z_i\right) = 1 \qquad (9.22)$$

In the example, the measurements have Gaussian noise, as in Equation (9.1). This means that the observation probabilities at time i are concentrated in the cells around the one containing z_i. In fact, the amount of probability in each cell is properly computed by integrating the 2-D Gaussian over the boundaries of each cell. An adequate approximation, used in the example here, is to compute the value of the Gaussian at the center of each cell and then normalize the probabilities so they sum to 1. Note that since the Gaussian goes on forever (infinite support), all cells have some nonzero probability.

Both the Kalman filter and particle filter have a dynamic model. For the Kalman filter, the dynamic model says that the new state is a linear function of the old state, plus additive Gaussian noise, as given by Equation (9.8). The particle filter is more general, specifying the dynamic model as a conditional probability distribution $p(x_i|x_{i-1})$. The dynamic model for the HMM is expressed in terms of transition probabilities, $P(X_{i+1}^{(k)}|X_i^{(j)}) = a_{jk}$. This gives the probability of transitioning from state j to state k between successive measurements. In the {bus, foot, car} example, the transition probabilities could reflect the fact that it is improbable to transition directly from a bus to a car, because there is usually some foot travel in between. The transition probabilities going from one state to all the others must sum to 1:

$$\sum_{k=1}^{M} a_{jk} = 1 \qquad (9.23)$$

In another example, the LOCADIO WiFi location system, Krumm and Horvitz (2004) attempted to infer whether a person was moving based on WiFi signal strengths measured from the laptop they were carrying. High variance in the measured signal strengths was found to indicate movement. However, the raw inferences from signal strength variance indicated that people would transition between moving and still much more often than in reality, as shown in Figure 9.9. An HMM was used to reduce the number of transitions. Here, the transition probabilities are given by the diagram in Figure 9.5, where the self-transition probabilities are quite large, which help prevent spurious transitions between the two states.

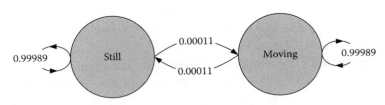

FIGURE 9.5 These transition probabilities were used in an HMM to smooth the transitions in Figure 9.9.

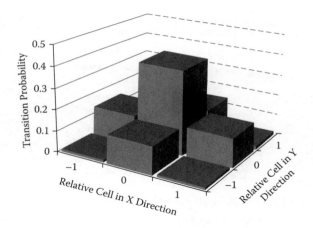

FIGURE 9.6 The HMM transition probabilities for each cell look like this, with the most probable transition being back to the same cell. The only other nonzero transition probabilities are to the immediate neighbors.

For this chapter's tracking example, there are $M = 10{,}000$ states, one for each cell. The transition probabilities reflect the fact that it is more likely to transition to a nearby cell than a distant cell. In fact, an examination of the actual path shows that the probability of staying in the same cell from time i to $i+1$ is about 0.40. The only other nonzero transition probabilities are into the cells immediately surrounding the current cell, as shown in Figure 9.6. The example uses the values in Figure 9.6 to fill in the transition probabilities a_{jk}.

The last element of the HMM is a set of initial state probabilities, $P(X_0^{(j)})$. With no knowledge of the initial state, it is reasonable to specify a uniform distribution over all states, that is, $P(X_0^{(j)}) = 1/M$. For the tracking example, the initial state probabilities were spread in a Gaussian pattern around the first measurement.

9.6.2 Hidden Markov Model

An example HMM is summarized in Figure 9.7. The time index starts at zero on the left and increases to the right, with a total of $N = 3$ measurements. The rows represent $M = 8$ possible states. At the beginning, there are M initial state probabilities. The subsequent columns represent observation probabilities that are sensitive to the measurements z_i. Connecting the states in each column are transition probabilities. The goal of the HMM is to find an optimal path starting with the initial state probabilities,

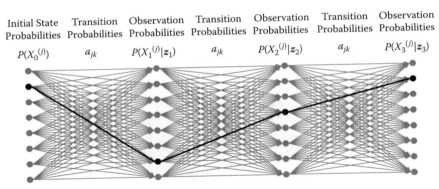

FIGURE 9.7 HMM considers all possible states at each time step. This HMM has $M = 8$ states and $N = 3$ measurements. The Viterbi algorithm is an efficient way to find the most probable path through the observation probabilities and transition probabilities. The dark lines represent one possible path.

through the transition and observation probabilities. The mostly likely path is the one that maximizes the product of these probabilities.

One way to find the optimal path is an exhaustive search through the M^N possible paths. Although this is could be feasible for the HMM in Figure 9.7, the tracking example has $M = 10{,}000$ and $N = 1000$ for a total of 1×10^{4000} different possible paths. Instead, finding the optimal path traditionally makes use of the Viterbi algorithm, explained for the HMM by Rabiner (1989). The Viterbi algorithm uses dynamic programming to efficiently find the most probable path.

Results of the HMM on this chapter's example are shown in Figure 9.8.

9.6.3 Discussion

The HMM is sometimes a very helpful add-on to a system that infers discrete states as a means of reducing the frequency of transitions between states. For example, one part of the LOCADIO system attempted to infer if a user was walking or still, based on the variance of measured WiFi signal strengths. The raw measurements of variance indicated very frequent transitions between the two states. The system used a simple two-state HMM with transition probabilities that approximated the realistic probabilities of making such a transition, leading to a much smoother output, as shown in Figure 9.9.

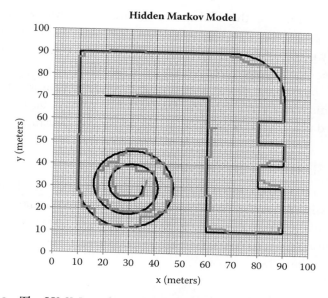

FIGURE 9.8 The HMM works with discrete state variables. For this example, each discrete state is one of 10,000 1 × 1 meter cells. The gray line shows the result of the HMM.

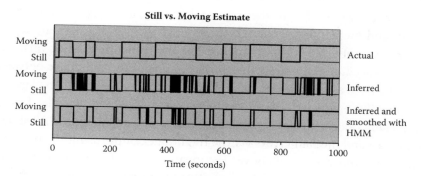

FIGURE 9.9 An HMM can be a useful add-on to a process that infers the discrete state of an object. Here, an HMM was used to smooth the inferred transitions between the states of "still" and "moving" inferred from WiFi signal strengths for a person carrying a laptop inside a building.

One limitation of the HMM is state duration. Although the self-transition probability a_{jj} is adjustable, its value leads to a certain probability density governing how long the subject will stay in state j, as explained by Rabiner (1989). There are techniques to get around this problem, also explained by Rabiner (1989), but they are not as elegant as the original HMM. This was not a problem in the tracking example, because the subject moved at a constant speed along the path (never pausing).

9.7 PRESENTING PERFORMANCE

There are a few standard ways to present the performance of algorithms for processing sequential sensor data. Having de facto standards is important, because it allows a fair comparison of different research results. This section discusses performance measures for both continuous and discrete state measurements.

9.7.1 Presenting Continuous Performance Results

Continuous state variables usually exist in a so-called metric space where there is a defined distance (metric) between the values. For instance, it is easy to define the distance between pairs of temperatures, pressures, light levels, speeds, etc. In the tracking example, the distance is simply Euclidean distance. In general, the error between the actual state vector x_i and the estimated state vector $\hat{x}_i$ is the scalar distance between the vectors: $e_i = \|\hat{x}_i - x_i\|$. For these scalar distances, it is easy to compute the mean and median error over all the estimates, both of which are legitimate means of assessing performance. For one-dimensional state variables such as speed, note that e_i is the absolute value of the difference between the estimated and the actual value. This is desirable, because then negative and positive errors will not cancel each other out, which could lead to an artificially optimistic aggregate performance estimate. Figure 9.10 shows the mean and median errors for the tracking example using the processing algorithms discussed above. The mean error is the expected error in the probabilistic sense, and it is sensitive to outliers. For instance, if one or a few errors are extremely large, these will pull the mean to a noticeably larger value, possibly misrepresenting the fact that the algorithm is usually close to the correct result. The median is much less sensitive to outliers.

A more detailed description of the error is based on the cumulative error distribution. This shows, for different values of the error e^*, how

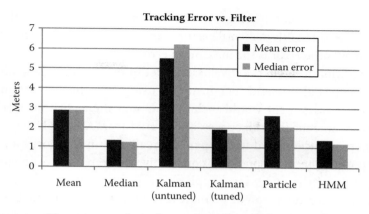

FIGURE 9.10 The mean and median error of six different processing algorithms applied to the noisy data in Figure 9.1.

often the resulting error is less than or equal to e^*. Figure 9.11 shows the cumulative error distributions for the example algorithms in this chapter. A steeply rising curve indicates better performance. The data used for this type of plot are also used to find percentile errors. For example, the 90th percentile error is the error value where the cumulative error curve crosses

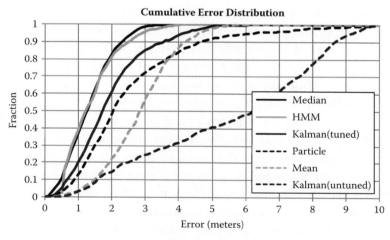

FIGURE 9.11 The cumulative error distribution gives a more detailed view of how the errors are spread. It is also useful for reading the percentile errors. For instance, the 90th percentile error for HMM is about 2.5 meters, meaning that the error is less than or equal to 2.5 meters 90% of the time.

0.9, and it means that the errors will be less than or equal to this value 90% of the time. The 50th percentile error is the same as the median, and the 100th percentile error is the maximum error.

In addition to a quantitative assessment of performance, a qualitative assessment is also useful. For instance, using this chapter's example, a qualitative assessment would include how well the algorithm works on straight segments, smooth curves, and sharp turns. Although the qualitative performance is hard to compare from algorithm to algorithm, it can be important, because sometimes qualitative differences have an effect on the quality of the upper level application. For instance, in the tracking example, sharp turns might be indicative of a subject's erratic behavior, and therefore an algorithm that tracks such turns could be better than one that tracks more accurately overall for certain applications.

Note that the results presented in this section are merely illustrations of how to present performance. They should not be used to declare that one processing algorithm is better than another. Each algorithm can be tuned for better performance, and the results presented here are not necessarily indicative of each algorithm's potential performance.

9.7.2 Presenting Discrete Performance Results

Discrete results normally come from an attempt to classify a situation into one of a plurality of categories. A good example from ubicomp is given by Lester et al. (2006), who use a sensor package to classify a subject's activity into one of sitting, standing, walking, walking up stairs, walking down stairs, riding elevator down, riding elevator up, and brushing teeth. Another common ubicomp application is to detect which "activity of daily living" a subject is engaged in. The Lester et al. (2006) paper gives its results using the time-tested and informative confusion matrix, adapted for this chapter in Table 9.1. The confusion matrix shows, for each discrete category, how often the actual category was classified into each of the whole set of categories, including the correct one. Table 9.1 shows that their classification worked well in most cases, with generally large numbers on the diagonal, meaning most activities were not confused with other activities. The best result is for "going up stairs," where the classification was correct 95% of the time. The weakest result was for standing, which was recognized correctly 55% of the time. Standing was confused with sitting 24% of the time, and with brushing teeth 10% of the time.

TABLE 9.1 Confusion Matrix[a] Showing How Often Each Activity Was Classified into Each Possible Activity

		Inferred Activities						
	Sitting	Standing	Walking	Upstairs	Downstairs	Elevator down	Elevator up	Brushing teeth
Sitting	75%	24%	1%	0%	0%	0%	0%	0%
Standing	29%	55%	6%	1%	0%	4%	3%	2%
Walking	4%	7%	79%	3%	4%	1%	1%	1%
Up stairs	0%	1%	4%	95%	0%	0%	1%	0%
Down stairs	0%	1%	7%	0%	89%	2%	0%	0%
Elevator down	0%	2%	1%	0%	8%	87%	1%	0%
Elevator up	0%	2%	2%	6%	0%	3%	87%	0%
Brushing teeth	2%	10%	3%	0%	0%	0%	0%	85%

Actual Activities

If the classification worked perfectly, all the off-diagonals would be 0%, and the diagonals would be 100%.
[a] Data adapted from Lester et al., 2006. With kind permission of Springer Science + Business Media.

9.8 CONCLUSION

This chapter examined some algorithms for processing sequential sensor measurements. The algorithms were the mean and median filters, the Kalman filter, the particle filter, and the HMM. The process being measured is assumed to show some continuity in time, and the algorithms presented take advantage of this by looking at past measurements to help make a state estimate. For all but the mean and median filters, the algorithms use some type of dynamic model to improve accuracy. Researchers in signal processing and machine learning continue to develop new methods to process sequential sensor data, but the well-established methods in this chapter have proven useful for many ubicomp tasks. They serve as at least a good starting point for new sequential inference tasks in ubicomp. Part of the fun of using them is determining how your problem fits with the assumptions and limitations of each technique.

REFERENCES

Doucet, A., Freitas, N. D., and Gordon, N., An introduction to sequential Monte Carlo methods, in *Sequential Monte Carlo Methods in Practice*, Doucet, A., Freitas, N. D., and Gordon, N., Eds., Springer, New York, 2001.

Fox, D., Adapting the sample size in particle filters through KLD-sampling, *International Journal of Robotics Research* 22(12): 985–1003, 2003.

Gelb, A., *Applied Optimal Estimation*, Analytical Sciences Corporation, MIT Press, Cambridge, MA, 1974.

Gordon, N. J., *Bayesian Methods for Tracking*, Imperial College, University of London, London, 1994.

Hightower, J., and Borriello, G., Particle filters for location estimation in ubiquitous computing: A case study, in *Ubiquitous Computing*, Springer, Nottingham, UK, 2004, pp. 88–106.

Krumm, J., and Horvitz, E., Locadio: Inferring motion and location from Wi-Fi signal strengths, First Annual International Conference on Mobile and Ubiquitous Systems: Networking and Services (Mobiquitous 2004), Cambridge, MA, USA, 2004, pp. 4–13.

Lester, J., Choudhury, T., and Borriello, G., A practical approach to recognizing physical activities, in *Pervasive Computing*, LNCS 3968, Springer-Verlag, Dublin, Ireland, 2006, pp. 1–16.

Murphy, K., and Russell, S., Rao-Blackwellised particle filtering for dynamic Bayesian networks, in *Sequential Monte Carlo Methods in Practice*, Doucet, A., Freitas, N. D., and Gordon, N., Eds., Springer-Verlag, New York, 2001, pp. 499–515.

Patterson, D., et al., Inferring high-level behavior from low-level sensors, in *Ubiquitous Computing, Ubicomp 2003*, Seattle, WA, USA, 2003, pp. 73–89.

Rabiner, L. R., A tutorial on hidden Markov models and selected applications in speech recognition, *Proceedings of IEEE* 77(2): 257–286, 1989.

Index

A

Absolute location, 287
ACM Mobicom, 29
ACM Transactions on Office Information Systems (TOIS), 28–30
Active Badge, 20–22, 145–147, 290, 328, 332
Active BAT, 22–23, 305–306
ActiveFloor, 312–314
Activity-based computing (ABC), 47–48
ActivityStudio, 84
Address-based location, 289
Ad hoc sensor networks, 43, 44, 75
Advanced Encryption Standard (AES), 137
Advanced User Resource Annotation (AURA), 169, 175, 184, 186
AES, 137
Affinity diagramming, 192–193
Agent systems, 49
Air traffic control studies, 211
Airbus tagless solution, 314–315
Airline passenger check-in system, 11–12
Altman, Irwin, 119–120
Amazon Kindle, 6, 32
Ambient displays, 168, 172, 265–266
Ambient Orb, 268–269
Ambient trolley, 269
Ambient Umbrella, 268–269
Ambient user interface (AUI), 264–269
Ambient wood, 72–73
Analysis of variance (ANOVA), 190
Anonymization, 154–155
Anticollision protocol, 138–140
Apple iPhone design, 249
Apple Newton, 2
Application implementation and evaluation, 77–78
Arduino, 87
Artificial intelligence (AI), 54, 122
AR user interface, 252, 260
ARToolkit, 86
Assisted GPS, 290
Asymmetric public key cryptography, 50
Atmospheric signal delay, 299, 303
ATOM, 31–32
Attacker model, 153
Audi Multi Media Interface (MMI), 260
Audio input, 276–277
AudioPad, 262
Augmented reality, 17
AuraOrbs, 269
Authentication, 50–51, 128, 129, 137
Autoethnography, 218
Automatic contextual reconfiguration, 329, 330
Automotive applications, 260, 276
Autonomic computing, 48–49
Aware Home, 19

B

Bacon, Francis, 204
Bandwidth reuse, 4
Barcode reading application, 169
Batteries, 42–43
Berkeley smart dust mote, 260
Between-subjects design, 170
Binary tree walking protocol, 139
Biometric sensors, 40, 270
Blackboard-based context-aware design approach, 340–342
BlueStar, 147

382 ■ Index

Bluetooth
    discovery protocol, 43
    headset applications, 277
    identifiers, 50, 138
    Privacy Guardian, 143
    range, 291
    SMCube, 262
BMW, 276
Bodily privacy, 104
Bonferroni technique, 191
Bonjour, 43
Bootstrap filter, 368
Border crossings, 105–106
Brandeis, Louis, 101–102, 114, 122
"Bridging Physical and Virtual Worlds" project, 14
Broderson, Robert, 15
Brush, A. J., 165, 169
BusMobile, 269

C

Calm technology, 5, 54, 239, 264, 323
Cambridge Fast Ring, 21
Camera-based location tracking, 315–316
"Can You See Me Now" game, 56–57
CAPpella, 345
CardSpace, 129
CareNet, 168, 172, 179, 184, 186, 269
Cautious approach, 57–58
Cellular phones
    ethnographic studies, 215–216
    iPhone design, 249
    PC compatibility, 32
    proximity study, 165–166, 172, 178, 184, 192, 193
    smart phones, 2, 3, 31
Chaum, David, 128–129
Clandestine scanning, 136
Classroom 2000, 18–19
Client-based location systems, 289–290
Client-server systems, 44
Clinical applications, 39–40, 77, 86, 271
Clippy, 342
Clock synchronization error, 299
CodeBlue, 87
Cohen's kappa, 192

Collective perception, 24
Compensating study participants, 185–186
Computer-supported collaborative work (CSCW), 204, 210–214
Computer vision tools, 86
Computing paradigms, 38–39
Confab, 131–134
Conference venues, 28–30
Confusion matrix, 378
Consent form, 182
Context, 325–327, 354
    ambiguity, 343
    human–computer interaction and, 322–323
    inferencing, 343
    location, identity, time, and activity, 327
    proxy for human intent, 342
Context-aware applications, 47, 321–325
    blackboard-based design approach, 340–342
    categorization of features for, 329–332
    contemporary applications, 334–335
    design process, 335–337
    end user issues, 345
    environmental controls, 334
    evaluating, 345
    historical applications, 328, 332–334
    Java framework (JCAF), 77, 85
    location-based services, 332, *See also* Location technologies
    power management, 166
    privacy issues, 344–345
    reminders, 333
    testing user reactions, 52
    tools for building, 337–342
    tour guides, 25–26, 46–47, 59, 169, 324, 333
    user interfaces, 252
    widget-based design approach, 338–342
    Xerox PARC research and, 10–11
Context-Aware Computing Architecture, 338
Context-aware reminders, 334
Context Aware Toolkit (CAT), 85

Context awareness, 10–11, 324, 327–329, 354
  input model, 327
  privacy and, 127
  rules vs. machine learning, 344
  using and adapting to context, 328–329
  writing academic papers on, 346–347
Context Toolkit, 85, 339
Context-triggered actions, 329, 330
Contextual adaptation, 330
Contextual augmentation, 330
Contextual command applications, 329, 330
Contextual resource discovery, 330
Contextual sensing, 330
Context widgets, 339–340
Contingency management, 49
Continuous performance measures, 376–378
Continuous Speech Recognition Consortium, 276
Control condition, 171
Cooltown, 13–15, 67
CORBA middleware, 66
CRAWDAD, 88
Cricket, 23, 147, 306–307
Critical technical practice, 231
Culturally Embedded Computing Group, 231
Cumulative error distribution, 376–377
Cyber foraging, 42
Cyberguide, 25

D

Datafountain, 267
Data leakage, 137
Data mining technology, 126
Data processing, sequential sensor data, *See* Sequential sensor data processing
Dataset resources, 87–88
Day Reconstruction Method, 178
Daylight Display, 269
Dead reckoning, 297–298
Deafness support applications, 169, 173
Debugging ubicomp systems, 63

Dependent variables, 170
Descriptive statistics, 187–189
Designers Augmented Reality Toolkit, 86
Designing ubicomp systems, *See* Ubiquitous computing systems, design and development
Design-oriented ethnography, 210–214
  data analysis, 223–228
  fieldwork, 219–223
  planning, 214–219
  practical implications, 228–232
Design patterns, 249
Deterministic anticollision protocols, 139
DexterNet, 86
DiamondTouch, 258, 261–262
Diaries, 177–178
Differential GPS, 303
Diffuse illumination, 263
Digital Desk, 261
Digital rights management systems, 130
Digital scrapbook, 275
Direct manipulation, 251
Disabilities, 69, 169, 173
Discrete performance measures, 376–378
Display Cube, 274
Distance estimation techniques, 292–294
Documenting ubicomp systems, 81–83
Dragon NaturallySpeaking, 274
Driver's Privacy Protection Act, 116–117
d.tools, 87
Dynabook, 5

E

EasyLiving, 26, 27
Eavesdropping, 136, 138
Ekahau, 309, 310
Elder support applications, 20, 168, 172, 179, 184, 186
Electronic books, 32
Embodiment, 256
Encryption, 137–138
Energy issues, 42–43
Environmental controls, 334
Environment-associated location errors, 299
EO pad, 3

EQUATOR Component Toolkit (ECT), 85
Eras of modern computing, 2
E-reader, 6
Ethical issues
    claims from study findings, 198–199
    study participant treatment, 181–182
Ethnography, 180, 203–206
    analytic sensibility, 209–210, 225, 228
    authenticity issues, 213–214
    autoethnography, 218
    data collection and analysis, 223–228
    design implications, 228–230
    design-oriented, 210–214
    Evans-Pritchard's Nuer studies, 206–210
    fieldwork, 219–223
    future directions, 230–231
    generalization, 217–218
    interpretation orientation, 208, 209
    objectivity and reflexivity concerns, 220–223
    participant access and recruitment, 218–219
    participant observation approach, 208, 209
    planning, 214–219
    practical design implications, 228–232
    sampling, 216–218
Ethnomethodology, 212–213, 230–231
    policy of indifference, 221–223
EuroPARC, 133
European global navigation satellite system, 317
European Union (EU) Data Protection Directive, 117
Evaluating ubicomp systems, 74–79
    context-aware applications, 345
    ethnography, 203–232, See also Ethnography
    field studies, 161–199, See also Field studies
    heuristic evaluation, 196
    location technology factors, 301
    proof-of-concept, 75–76
    simulation models, 75–76
    usability testing, 249
Evans-Pritchard, Edward, 206–210
EventHeap, 63
Exception handling, 49
Exemplar, 84
Exertion interfaces, 252
Experience sampling methodology, 176–177
Explicit input, 327

F

Face recognition, 316
Fair Information Principles, 115–116
Fault tolerance of ubicomp systems, 61–62
Felten, E., 153
Field studies, 161–164
    context-aware applications, 344
    control condition, 171
    current behavior, 165–166
    data analysis, 186–194
    design, 169–173, 194–195
    ethical issues, 181, 198–199
    ethnography, 203–232, See also Ethnography
    experimental design, 170–171
    generalization, 217–218
    negative results, 197
    novelty effects, 180
    ongoing feedback, 196
    pilot, 195
    planning, 214–219
    proof of concept, 166–168, 180, See also Proof-of-concept
    prototype use, 168–169
    qualitative (unstructured data) analysis, 191–194, 223–228
    quantitative (statistical) data analysis, 187–191
    research questions, 164, 194
    research team, 197
    researcher objectivity and reflexivity, 220–223
    running the study, 197–198
    sampling, 216–218
    scenic fieldwork, 214
    scripts, 195
    study design document, 194–195
    study length, 180–181

technology tips, 196–197
types, 164–169
using existing resources, 196
Field studies, data collection, 173–174, 198
    diaries, 177–178
    ethnographic fieldwork, 224–225
    experience sampling methodology, 176–177
    interviews, 178–180, 224
    logging, 174
    quantitative and qualitative data, 173
    surveys, 175–176
    unstructured observation, 180
Field studies, participants, 181–186
    access and recruitment, 218–219
    activities, 169–173
    comfort and safety, 197–198
    compensation, 185
    ethical treatment, 181
    number, 184, 218
    profiles, 182–183
FingerMouse, 273
Fingerprinting, 295–297, 308–310, 372
FireFox, 273
Fitts' law, 258–259
Flowmenu, 272
Foot scale, 6
Ford, Henry, 248
Fourier transform infrared (FTIR)-based displays, 263
Freedom Network, 129
Front-projected surface user interfaces, 261–262
Future of ubiquitous computing, 31–33

G

GAIA OS services, 66–67
Galileo, 317
Gaussian noise, 356
Gaussian probability distribution, 356
Geertz, Clifford, 209
Generalizability of ethnography results, 217–218
Genius design, 249
Geocoder, 289
Geographic coordinates, 288–289

Georgia Institute of Technology, 2, 18–20, 25
Gestural user interface, 252
Gesture input, 271–274
Global positioning system (GPS), 47, 302–303
    accuracy, 302, 317
    Assisted GPS, 290
    "Can You See Me Now" game, 56–57
    causes of errors, 299, 303
    client-based location system, 290
    driving trajectory field study, 168
    location privacy protection, 147
    market for, 286
    network-assisted location system, 290
    privacy and, 126
    RADAR indoor location application, 26
    receivers, 303
    satellite system, 302–303
    sequential sensor data, 354
Goffman, Erving, 109
GPS satellites, 302–303, *See also* Global positioning system
Graceful degradation, 49–51
Graffiti, 261, 273
Graphical user interface (GUI), 237, 242, *See also* Ubicomp user interface
    direct manipulation capability, 251
    widget-based approach, 338–340
    WIMP model, 242, 251–252
Grounded theory, 193
Grounding, 323
GSM towers, 27, 310
GUIDE project, 25–26, 46–47, 59

H

Handheld and Ubiquitous Computing (HUC) conference, 30
Hardware resources, 87
Hash chains, 142
Hash locks, 141–142
Heads-up display, 17
Hearing disability support applications, 169, 173
Heterogeneous execution environments, 44–45

Hidden Markov models (HMMs), 273, 276, 314, 343, 370–376
Hippocratic databases, 131, 132
Home technology study, 165, 182–183, 186, 193
Horvitz, E., 168
Hospital applications, 39–40, 77, 270
Hotmobile conference, 30
House_n PlaceLab, 87
HP Labs, 2, 13–15, 67
HSAPD, 15
Human–computer interaction (HCI), 204, See also Ubicomp user interface
   context, 322–323, 342, See also Context-aware applications
   design-oriented ethnographic approach, 210–214, 228–232, See also Ethnography
   field studies, 163, See also Field studies
   Fitts' law, 258–259
   inventing the future, 244
   privacy and, 127–128
   ubicomp systems design considerations, 55–60
   Weiser's ubicomp vision, 4
Human–Computer Interaction for Mobile Devices (MobileHCI) conference, 30
Human PacMan, 252
HVAC system-based location monitor, 314–315
Hyperbolic lateration, 294
Hypothesis, 170, 215

I

IBM Research, 2, 11–12
IBM ViaVoice, 275
iCAP, 345
Identity management tools, 129
IEEE Pervasive Computing and Communication conference, 30
iHospital system, 270
Implicit input, 327
Inch scale, 6
Independent samples t-test, 190
Independent variables, 170

Inferential statistics, 189–191
InfoCanvas, 268
InfoPad, 15–16
Information appliance, 5
Information Art, 269
Information decoration, 267
Information percolator, 267
Information privacy, 103, See also Privacy
Informed consent, 182
InfoSpaces, 131
Infrared-based location technologies, 6–8, 304–305, 332, See also Active Badge
Infrared pen, 9
Inkpen, K., 165
Institutional review board (IRB), 182
Intel Research, 27–28, 87
Interaction design, 151–152, 246–251
   design patterns, 249
   genius design, 249
   related fields and specialties, 247
   systems design, 248–249
   user-centered design, 247–248
Interface usability metrics, 278
International Symposium on Wearable Computing (ISWC), 30
Interpersonal privacy, 119–121
Interrater reliability, 192
Interval scale data, 188
Interview-based data collection, 178–180, 224
Invisible computing, 5, 48–49
   literal vs. effective, 254–255
   privacy and, 124–125
Invisible Train Games, 252, 260
Ionospheric delay, 299
iPhone design, 249
iROS, 63, 85–86
iStuff Mobile, 84–85
ITRON, 12–13

J

J9 Virtual Machine, 12
Japanese technologies, 12–13
Java Context-Awareness Framework (JCAF), 77, 85, 339
Jeremjenko Dangling String, 266

Jini, 43
jMax, 85
Journals and publications, 30
Julius, 276

K

K-anonymity, 129, 149
K-NN algorithm, 297
Kalman filter, 355, 359–365
Karlsruhe University, 23–24, 30
Kaye, Alan, 5
Kindle, 6, 32
Krumm, J., 168

L

Laboratory user studies, 163, 170, 196
LAN-based location technology, 20–23
Lancaster University, 25–26
Latitude, 288–289
Leading questions, 175, 198
Learning-based context awareness, 344
Legal issues, 69
 privacy, 101–102, 114–118
Lemming Location-Enhanced Instant
  Messenger, 134
Length units, 6
Level of measurement, 187–189
Likert scale, 175, 188
Linda, 340–341
Liveboard, 6, 9, 19
Live Wire, 266
Living Laboratories, 18–20
LOCADIO, 372, 374
Local ad hoc communication, 4
Location and Context Aware (LoCA), 30
Location determination approaches, 290, 327
 dead reckoning, 297–298
 distance estimation, 292–294
 fingerprinting, 295–297
 hyperbolic lateration, 294
 proximity, 291
 reference points, 290, 291
 triangulation, 295
 trilateration, 291–294
Location errors, 298–300

Location information, 287
Location privacy protection, 145–149
Location representation, 288–289
Location technologies, 20–23, 46–47, 285–287, *See also* Global positioning system (GPS)
 ActiveFloor tagless solution, 312–314
 audio-based navigation, 276–277
 camera-based tracking, 315–316
 "Can You See Me Now" game, 56–57
 challenges, 316–317
 client-based, 289–290
 device proxy for user, 287
 driving trajectory field study, 168
 evaluation factors, 301
 fingerprinting, 309–310
 GSM towers, 27, 310
 hospital application, 39–40, 270
 HVAC system approach, 314–315
 indoor GPS (RADAR), 26
 infrared-based Active Badge system, 332
 IR-based Active Badge system, 20–22, 145–147, 290, 304–305
 Lemming Location-Enhanced Instant Messenger, 134
 middleware and back-end services, 317
 network-based, 290
 phone call routing applications, 286
 phone proximity study, 165–166, 172, 178, 184, 192, 193
 plausible deniability, 134, 317
 PowerLine positioning, 311–312
 privacy concerns, 134, 317, *See also* Privacy
 public domain datasets, 87–88
 RSSI-based technologies, 26
 speed estimates and, 355
 systems, 300–302
 tour guide applications, 25–26, 46–47, 59, 169
 ultrasound/RF-based (Active BAT and Cricket), 22–23, 305–307
 ultrawideband (UWB) Ubisense system, 307–308
 "Uncle Roy All Around You" game, 59
 war driving, 310–311

Whereabouts Clock, 226–227
WiFi access point mapping, 27
WiFi signal fingerprinting, 296, 308–309, 372, 374
Logging-based data collection, 174
Longitude, 288–289
Lookout system, 342

M

Machine learning-based context awareness, 344
Macromedia Director, 86
Magic Touch, 271, 272
Maintenance and support of ubicomp systems, 69, 72
Marx, Gary T., 104–105
Massachusetts Institute of Technology (MIT), 2, 16–17, 23, 87
Max/MSP, 85
Mean, 188
Mean filters, 358–359, 376
MediaCup, 23
Median, 188
Median filters, 358–359, 376
Medical applications, 39–40, 77, 86, 270
Medical databases, 129
MeetingMachine, 86
Memento, 275, 276–277
Memory Amplifiers, 102
Memory prosthetic aids, 26
Menu selection, 251
Meridians, 288
MERL DiamondTouch, 258, 261–262
Metaphor, 256
Microsoft Research, 26–27
Microsoft Surface, 258, 273
Mix-Net, 128
Mix zones, 147
MobiCom, 30
Mobile ad hoc sensor networks (MANETs), 43, 75
Mobile Computing, Applications, and Services (Mobi CASe), 30
MobileHCI conference, 30
Mobile sensors, 270
MobiQuitous, 30
MobiSys, 29–30

Modeling ubicomp system, 75–76
Motes, 27, 87
Motion capture suit, 271
Mouse gestures, 273
Moven motion capture suit, 271
Multicast DNS (mDNS), 43
Multi Media Interface (MMI), 260
Multimodal user interface, 252
Multipath signal distortions, 299, 303
MyExperience, 85, 176, 196
MyLifeBits, 26

N

NASA Task Load Index, 176
Natural language interface, 251
Navizon, 311
Nearest neighbor (NN) algorithm, 297
Network-based location system, 290
Network routing schemes, 44
Network Simulator (NS-2), 75
Network transient connectivity, 60–61
Neural networks, 273
Newton, Isaac, 83
Nominal variables, 187–189
Norman, Donald, 246
Novelty effects, 180
Nuer, 206–210
Null hypothesis, 190

O

OECD Fair Information Principles, 115–116
"Off-the-shelf" ubicomp components, 64–67, 87
Olivetti Research, 2, 20–23, 304, 328, *See also* Active Badge
ONTRACK, 276–277
Opacity tools, 128–129
Open-ended questions, 175
Opera, 273
Optimistic approach, 57–58
Orb, Ambient, 268–269
Ordinal variables, 188
Outliers, 188, 358, 376
Oyster, 271

P

p value, 190
Pad, 6, 8–9, 32, 261
Palm Pilot, 3
Pandora project, 21
Partially connected ubicomp system, 60
Participant observation, 208, *See also* Ethnography
Participant profile, 183
Participatory design, 68
Particle filter, 365–370
Password protection, RFID tags, 141–142
Path perturbation, 147
Patterns, design, 249
PawS, 132–133
PC compatible cell phone applications, 32
PCOM, 86
PDAs, *See* Personal digital assistants
Peer caches, 62
Peer-to-peer video services, 15
Pen-based user interface, 9, 252, 273
Perception API (PAPI), 24
Percepts, 24
Peripheral vision, 265
Personal border crossings, 105–106
Personal digital assistants (PDAs), 2, 31, 260–261
Personal Server, 27
Pervasive computing, 11–12
"Pervasive" conference, 29
Pessimistic approach, 57–58
Phenomenology, 213
Phicons, 255
Phone call routing, 286
Phone Proximity study, 165–166, 172, 178, 184, 192, 193
Photocopier use study, 225–226
Photography, privacy and, 102
Physiological sensors, 40, 270–271
Pilot study, 195
Pitt, William, 101
Place Lab, 27, 87, 310–311
PlaceEngine, 311
Plan 9 operating system, 86
Plan B, 86
Platform for Policy Preferences (P3P), 131
Plausible deniability, 134
Plutarch, 79–80
Policy of indifference, 221–223
Policy tools, 130–131
Posey, 255
Power consumption, 42–43
    context-aware power management, 166
    power-aware cord, 266–267
Power foraging, 42
PowerLine Position, 311–312
Privacy, 49–51, 95–97
    client-based location system, 290
    commodity approach, 112
    context-aware application issues, 344–345
    context awareness and, 127
    defining, 100–107
    Fair Information Principles, 115–116
    human–computer interaction and, 127–128
    impact assessments, 154
    interpersonal, 119–121
    invisible computing and, 124–125
    legal context, 101–102, 114–118
    motivating, 108
    novel ubicomp challenges, 122–128
    personal border crossings, 105–106
    photography and, 102
    public attitudes, 108–113
    RFID technology and, 117, 135–145
    risk models, 153–154
    security and, 97–98
    self-determination rights, 112–113
    Solove's taxonomy, 106–107
    study participant data, 182
    threat model, 153
    ubicomp system deployment concerns, 69
    ubicomp system design guidelines, 151–155
    wearable technology and, 104
Privacy beacons, 132–133
Privacy enhancing technology, 128
    anonymization methods, 154–155
    identity management tools, 129
    location information protection, 145–149
    opacity tools, 128–129
    policy tools, 130–131

protecting smart spaces, 131–134
RFID technology and, 134–145, *See also* RFID tag privacy issues
transparency tools, 129–130
Privacy Guardian, 143
Privacy tags, 131–132
Processing (language), 85
Project Aura, 276
Proof-of-concept, 75–76, 166–168, 180
Proprietary toolkits, 84–86
Proprietary ubicomp components, 64–67
Prosser, William L., 114
Prosthetic memory aid, 26
Prototype ubicomp systems, 53–54, 75
  prototyping tools, 84–86
  user experience field study, 168–169
Proximate selection, 329, 330
Proximity Browser, 10–11
Proximity sensing, 291
Pseudorandom number generator (PRNG), 138
Public domain datasets, 87–88
Public domain toolkits, 86–87
Public key cryptography, 50, 137
Publishing ubicomp research, 80–81
PureData, 85

Q

Qualitative data analysis, 191–194
  affinity diagramming, 192–193
  ethnography data, 223–228
  grounded theory, 193
  simple coding, 191–192
Qualitative data collection, 173
Quantitative data collection, 173
Questionnaire for User Interaction Satisfaction, 176

R

RADAR, 26, 308–310
Radialpoint, 129
Radio-based location technology, 22–23
Radio frequency identification technology, *See* RFID tag privacy issues
Radiofrequency (RF) fingerprinting, 295–297

Rao-Blackwellized particle filter, 370
Rapid prototypes, 54
RAVE, 133
ReacTable, 263
Reality mining, 155
Real-Time Kinematic GPS, 303
Real world knowledge, 55
Rear-projected surface user interfaces, 262–264
Received signal strength indication (RSSI), 26, 309
Recovery of state, 62
Reference points (location), 290, 291
  coordinate errors, 299
  GSM towers, 27, 310
Reflexivity, 221
Relative location, 287
Releasing ubicomp systems, 79
Reminder applications, 333
Remote monitoring, 72
Repeated-measures design, 170
Replication of state, 62
Research question, 164, 194
Research team, 197
Resource-aware computing, 42
Resource constraints, 42–43
  RFID tags, 137
REXplorer, 169, 172, 193
RFID sensor systems, 270–271
RFID tag privacy issues, 117, 134–136
  access control/deactivation, 140–145
  anticollision protocols, 138–140
  encryption solutions, 137–138
  novel security challenges, 137
  password-less access control, 144–145
  passwords and hash locks, 141–142
  privacy issues, 117
  proxy devices, 142–144
  types of threats, 136–137
Rocketbook, 5
Routing applications, 286
Rule-based context-aware systems, 344

S

Sakamura, Ken, 12
Scenic fieldwork, 214
Scribe4Me, 169, 173

Seamful design, 56
Security issues, 49–51, 97–98, *See also* Privacy
    device and user authentication, 50–51
    identity management tools, 129
    RFID tags, 117, 135–145, *See also* RFID tag privacy issues
    theft and tampering, 50
    threat model, 153
    trust, 49–50
Self-determination over privacy, 112–113
Self-positioning systems, 147
Self-projected surface user interfaces, 263
Semantic rubicon, 55
Semistructured interview, 178
Sense Cam, 26
Sensor fusion, 343, 365
Sensors, 39, 270–271
    ad hoc networks, 43, 44, 75
    biometric or physiological, 40, 270–271
    context-aware application design, 336
    GPS, 354
    mobile, 270
    off-the-shelf components, 87
    power supply issues, 43
    public domain toolkits, 86–87
    RFID systems, 270–271
    sequential data processing, 353–380, *See also* Sequential sensor data processing
    software, 270
    ubicomp systems design considerations, 56
    wearable components, 86, 87
SensVest, 270
Sequential sensor data processing, 353–380
    confusion matrix, 378
    continuous performance measures, 376–378
    discrete performance measures, 378
    hidden Markov model, 370–376
    Kalman filter, 355, 359–365
    mean and median filters, 358–359, 376
    particle filter, 365–370
    past and future sense, 354–355
    tracking example, 356–358

Service discovery, 43
Shake Well Before Use, 274
Shamir, Adi, 144
Shared secrets theory, 144
SharePic, 261–262
Sharp Zaurus, 3
Signal strength attenuation, 293–294, 308–310, 372, 374
Signal time-of-flight estimation, 292–293
Significance tests, 189–191
Sign language, 271
Simulation, 75–76, 168, 172
Siren, 260
Skimming attacks, 136
Skyhook, 311
SmartBoard, 262
Smart Dust, 102
Smart dust mote, 260
Smart floor, 19, 312–314
Smart homes (Aware Home project), 19–20
Smart-Its, 23–24, 274
Smart kitchen of the future, 98–99
Smart Labels, 102
Smart Micrel Cube (SMCube), 262
Smart phones, 2, 3, 31
    pervasive computing solutions, 12
    surface user interface technology, 260
Smart rooms, 44–45
    GAIA OS services, 66–67
    iROS, 63, 85–86
    privacy enhancing tools, 131–134
Snowballing method, 219
Softbook, 6
Software sensors, 270
Solove, D. J., 106–107
Sony SmartSkin, 262
Spatial bandwidth reuse, 4
Speech-enabled augmented reality (SEAR), 276
Speech input, 274–277
Speed estimates, 355
Sphinx, 275, 276
SPINE, 86
Standard deviation, 188
Stanford Interactive Workspaces, 63, 85–86
Stargate, 28

State recovery or replication, 62
Statistical data analysis, 187–191
    descriptive statistics, 187–189
    inferential statistics, 189–191
    level of measurement, 187–189
    significance tests, 189–191
Statistical databases, 129
Stick-e Notes, 338
Sticky policy, 131–132
Study design document, 194–195
Suchman, Lucy, 210–211, 225–226
Surface user interface (SUI), 257–265
    Fitts' law, 258–259
    gestural input, 273
    input/output coupling issues, 264
    larger interfaces, 261–264
    PDAs, 260–261
    tactile feedback, 264
    touchscreen technology, 259–260
Surveillance, 123–124
Survey-based data collection, 175–176
Swissair, 11–12
Symbolic location, 287
SyncTap, 274

T

T-Engines, 12–13
Tab, 6–11, 261, 333
Tablet PC, 258, 261
Tactile feedback, 264
Tampering, 50
TANGerINE, 262
Tangible and Embedded Interaction (TEI), 30
Tangible user interface (TUI), 253–257
TeamAwear, 166–167, 172, 175, 179
TeamPad, 25
Technical documentation, 81–83
Territorial privacy, 104
Theft, 50
Thick description, 209
Third party ubicomp solutions, 64–67, 84–86
Threat model, 153
TiltType, 274
Time-of-flight estimation, 292–293
Tivoli, 9

Tokyo University, 2, 12–13
Tolerance for ignorance, 56
Topiary, 85
Tort law, 102, 114
Total internal reaction (TIR), 264
Touch-based user interface, 252
Touch-screen technology, 259–260, *See also* Surface user interface
    tactile feedback, 264
TouchLight, 263
Tour guide applications, 25–26, 46–47, 59, 169, 324, 333
Transient network connectivity, 60–61
Triangulation, 295
Trilateration, 291–294
TRON, 12
Tropospheric delay, 299
Trust, 49–50
Tuple space, 63, 340
Twiddler, 16–17

U

Ubicomp computing model, 4–5, 238–239
    calm technology, 5, 54, 239, 264, 323
    invisible computing, 5, 48–49, 254–266
"Ubicomp" conference, 29
Ubicomp dissemination venues, 28–30
Ubicomp publishing and conference venues, 28–30
Ubicomp user interface (UUI), 237–241, 251–253
    alternate classes, 252
    ambient user interface (AUI), 264–269
    gesture input, 271–274
    GUI WIMP model, 242, 251–252
    implicit and explicit inputs, 243
    input/output coupling issues, 264
    input technologies, 269–277
    inventing the future, 243–244
    sensor input, 270–271
    speech and audio input, 274–277
    surface user interface (SUI), 257–265
    tangible user interface (TUI), 253–257
    usability metrics, 278

usability testing, 249
wearable input devices, 16–17
Ubicomp user interface (UUI), design considerations, 55–60, 244–246
   challenges, 240–242, *See also* Privacy
   GUI WIMP model, 242–243
   interaction design, 151–152, 246–251, *See also* Interaction design
   rules, 245–246
   widget-based approach, 338–340
Ubiquitous computing systems, 37–38, 238
   application implementation and evaluation, 77–78
   deployment, 67–74
   documenting, 81–83
   evaluating, 74–79
   evaluating, field studies, 161–199, *See also* Field studies
   legal and regulatory considerations, 69
   maintenance and support considerations, 69, 72
   off-the-shelf components, 64–67
   proof-of-concept, 76–77
   prototypes, 75
   public domain datasets, 87–88
   reasons for building, 51–52
   releasing and maintaining, 79
   research communication, 80–81
   research concerns, 40, 52
   simulation models, 75–76
   user interface, *See* Ubicomp user interface
   volatile environments, 73–74
Ubiquitous computing systems, design and development, 41, 51
   debugging, 63
   design patterns, 249
   ethnographic implications, 228–230
   fault tolerance, 61–63
   handling transient connections, 60–61
   hardware components, 84, 87
   ignorance tolerance, 56
   interaction design, 151–152, 246–250, *See also* Interaction design
   libraries and toolkits, 84, 86–87
   partial connectivity, 60
   participatory design, 68

   privacy protection guidelines, 151–155
   prototypes, 53–54
   prototyping tools, 84–86
   real world knowledge, 54–55
   research targets, 52
   seamful design, 56
   sensors, 56
   setting objectives, 52–54
   user interface, *See* Ubicomp user interface (UUI), design considerations
   user understanding and perceived needs, 56–60
   value sensitive design, 152
Ubiquitous computing systems, topics and challenges, 41–42
   autonomic computing, 48–49
   fluctuating usage, 45–48
   heterogeneous environments, 44–45
   invisible computing, 48–49
   resource constraints, 42–43
   security, 49–51, *See also* Privacy; Security issues
   volatile environments, 43–44
UbiSense, 307–308
Ultrasonic-based location technology, 22–23, 305–306
Ultrawideband location-tracking application, 307–308
Umbrella, Ambient, 268
"Uncle Roy All Around You," 59
Unistrokes, 261
University of California (UC) Berkeley, 2, 15–16
University of Karlsruhe, 23–24, 30
University of Kent, 338
University of Tokyo, 2
Unobservability tools, 128
Unstructured observation, 180
UPnP, 43
Urp, 254, 261
Usability testing, 249, *See also* Evaluating ubicomp systems; Field studies
User authentication, 50–51
User-centered design (USD), 247–248
User field studies, *See* Field studies
User interfaces, *See* Ubicomp user interface

V

Variables
    dependent and independent, 170
    levels of measurement, 187–189
Verne, Jules, 76
ViaVoice, 275
Video-based gestural input, 273
Video data collection, 224–225
Video Privacy Protection Act, 117
Video prototypes, 54
Video services, HSAPD, 15
Virtual Earth, 289
Visual Calendar, 269
Volatile environments, 43–44, 73–74
VoxForge, 276

W

War driving, 310–311
Warren, Samuel, 101–102, 114, 122
Watermarking systems, 129–130
Wearable Action Recognition Database, 88
Wearable computing, 16–17, 26
    ISWC symposium, 30
    physiological monitoring systems, 270
    privacy issues, 104
    TeamAwear system, 166–167, 172, 175, 179
Wearable Health Monitoring Systems (WHMS), 87
Wearable sensors, 86, 87
Web presence system (Cooltown), 14
Web Sphere, 12
Weiser, Mark, 3–6, 54, 76, 96, 238, 243, 254, 261, 266–267, 323
Weka, 86
Westin, Alan, 103, 109
Whereabouts Clock, 226–227
WHMS, 88

Widget-based approach, 338–342
WiFi access point mapping-based location technology, 27
WiFi range, 291
WiFi signal fingerprinting, 296, 308–309, 372, 374
Wiimote, 274
WiMAX, 15
Windows, Icons, Menus, and Pointer (WIMP), 242, 251–252
Wireless communication power issues, 43
Wiring, 85
Within-subjects design, 170–171, 197
Wizard of Oz techniques, 52, 54, 168, 172
Workshop on Mobile Computing Systems and Applications (WMCSA), 29–30

X

Xerox EuroPARC, 133
Xerox Palo Alto Research Center (PARC), 2, 3–11, 332, 338
    Active Badge location technology, 20–22, 145–147
    "Bridging Physical and Virtual Worlds" project, 14
    Liveboard, 6, 9
    proof-of-concept concept, 75
    Tab and Pad applications, 6–9

Y

Yard scale, 6
Yellow Pages, 289

Z

Zero Knowledge systems, 129